Statistics for Social and Behavioral Sciences

More information about this series at http://www.springernature.com/series/3463

Barry C. Arnold • José María Sarabia

Majorization and the Lorenz Order with Applications in Applied Mathematics and Economics

Barry C. Arnold
Department of Statistics
University of California
Riverside, CA, USA

José María Sarabia
Department of Economics
University of Cantabria
Santander, Spain

ISSN 2199-7357 ISSN 2199-7365 (electronic)
Statistics for Social and Behavioral Sciences
ISBN 978-3-030-06720-5 ISBN 978-3-319-93773-1 (eBook)
https://doi.org/10.1007/978-3-319-93773-1

This Springer imprint is published by the registered company Springer International Publishing AG part of Springer Nature.
The registered company address is: Gewerbestrasse 11, 6330 Cham, Switzerland

To Darrel, Lisa and Kaelyn (BCA)

To Cori, José María and Belén (JMS)

Preface to the Second Edition

In the 31 years since the first edition was published, there has been considerable growth in the volume of research dealing with inequality and the Lorenz order. This is especially true, quite naturally in the Econometrics literature. Robin Hood retains a significant presence, and majorization still provides strong motivation for consideration of Lorenz curves in discussions of inequality. There has been significant growth in the number of summary measures of inequality that are now considered, and much more attention is now paid to discussion of flexible parametric families of Lorenz curves. In the present edition, such issues are addressed in some depth in Chaps. 5, 6 and 10. In the current Chap. 7 will be found a more detailed discussion of multivariate Lorenz ordering than was included in Chap. 5 of the first edition. The important concept of a Lorenz zonoid is prominent in the discussion. Alternative Lorenz surfaces are also considered. The first edition ended with a chapter which included a potpourri of situations in which, sometimes surprisingly, majorization and/or Lorenz ordering played an, often important, role in the analysis. Over the years, the number of such examples has naturally increased. There were ten examples of such applications in the final chapter of the first edition. There are now 24, spread over two chapters. These examples reinforce the belief that majorization and Lorenz ordering continue to find new areas of application often only faintly related to income inequality but where variability comparisons are of importance. Going back to our hero, Robin Hood, pictured on page 7, wherever inequality (or variability) is to be found, Robin Hood will appear.

We are grateful to many colleagues for helpful discussion of material in this book, and we are grateful to the editorial staff of Springer for their patience and encouragement while the book was developing.

Riverside, CA, USA
Santander, Cantabria, Spain
March 21, 2018

Barry C. Arnold
José María Sarabia

Preface to the First Edition

My interest in majorization was first spurred by Ingram Olkin's proclivity for finding Schur convex functions lurking in the problem section of every issue of the *American Mathematical Monthly*. Later my interest in income inequality led me again to try and "really" understand Hardy, Littlewood and Polya's contributions to the majorization literature. I have found the income distribution context to be quite convenient for discussion of inequality orderings. The present set of notes is designed for a one quarter course introducing majorization and the Lorenz order. The inequality principles of Dalton, especially the transfer or Robin Hood principle, are given appropriate prominence.

Initial versions of this material were used in graduate statistics classes taught at the Colegio de Postgraduados, Montecillos, Mexico, and the University of California, Riverside. I am grateful to students in these classes for their constructive critical commentaries. My wife Carole made noble efforts to harness my free-form writing and punctuation. Occasionally I was unmoved by her requests for clarification. Time will probably prove her right in these instances also. Peggy Franklin did an outstanding job of typing the manuscript, and patiently endured requests for innumerable modifications.

Riverside, CA, USA
July 1986

Barry C. Arnold

Contents

List of Figures

List of Tables

Chapter 1
Introduction

The theory of majorization is perhaps most remarkable for its simplicity. How can such a simple concept be useful in so many diverse fields? The plethora of synonyms or quasi-synonyms for variability (diversity, inequality, spread, etc.) suggest that we are dealing with a basic conception which is multifaceted in manifestation and not susceptible to a brief definition which will command universal acceptance. Yet there is an aspect of inequality which comes close to the elusive universal acceptance. The names associated with this identifiable component of inequality are several. Effectively, several authors happened upon the same concept in different contexts. Any list will probably do injustice to some group of early researchers.

1.1 Early Work About Majorization

In the early accessible English language literature on the subject the names of Muirhead, Lorenz, Dalton and Hardy, Littlewood and Polya stand out. Majorization or Lorenz ordering is the name we attach to the partial order implicitly or explicitly described by these authors. The arena in which inequality measurement was discussed was broad. Lorenz and Dalton did their work in the context of income inequality. It is convenient to use income inequality as our standard example, but the reader is enjoined to recall that any of many other fields of application might serve as well. In fact, concurrent and possible earlier discovery of relevant concepts may well have occurred in other fields.

Hardy et al.'s (1959) book remains a fertile source of mathematical results relating to inequality. The passage of time and perhaps infelicitous choice of notation make that book even less accessible to the student. An extensive survey of majorization is now available in Marshall et al. (2011). It is much more accessible, but it is overwhelming in scope. Theorems are proved in all extant versions. The present book is designed to give a briefer introduction to the material. Hopefully

B. C. Arnold, J. M. Sarabia, *Majorization and the Lorenz Order with Applications in Applied Mathematics and Economics*, Statistics for Social and Behavioral Sciences,
https://doi.org/10.1007/978-3-319-93773-1_1

the reader will be stimulated to pursue some topics further in Marshall, Olkin and Arnold's book or in earlier sources. Citations to the literature will be given to help establish the temporal sequence of the development of key ideas. It would, however, be presumptuous to try to improve upon the detailed bibliography supplied by Marshall, Olkin and Arnold.

The concepts related to majorization and the Lorenz order to be discussed in this book are, in a sense, purely mathematical. Indeed, Hardy et al. (1959) treated them just that way. Nevertheless, we, however, believe that fleshing the ideas out by setting them in some real world context is helpful for motivation and suggestive of useful extensions and generalizations. So, in our introductory sections we will usually speak of inequality in the income context. In later chapters we will vacillate. Sometimes results will be presented from a purely mathematical point of view. Other times a (sometimes superficial) economic dressing is added. It is hoped that this will encourage the reader to think of analogous applications in his/her field of interest, and that it will not give the erroneous impression that only in economics is inequality, diversity, variability or what have you, of interest.

Before Lorenz's (1905) important paper, there had been several suggestions regarding how inequality might best be measured. Lorenz felt that all of the summary measures then under consideration constituted, in effect, too much condensation of the data. Each provided a snapshot of some aspect of inequality. But as with the blind men and the elephant, the beast itself remained imperfectly mirrored by the unidimensional views. As a more full bodied view, he proposed a curve which has come down to us as the Lorenz curve. Actually it provides almost no condensation of the data. It determines the distribution up to scale. Pushing our blind man and the elephant theme to (or beyond) its limits, it provides a hologram of the beast with only the actual scale unknown. There are many functionals of distributions which essentially divide the class of all distributions into equivalence classes under change of scale. What is special about the Lorenz curve is that a natural partial order for inequality is derivable from it. This partial order based on nested Lorenz curves was suggested by Lorenz in his paper and has proved to be the most widely accepted partial ordering relating to inequality. It is, of course, the main subject of this book. Let us look back at its genesis in the publication of the American Statistical Association of 1905.

Lorenz points out negative features of most of the simple summary measures of inequality with a small accolade for Bowley's measure of dispersion. Bowley's measure is touted as the best numerical measure as yet suggested, but it is clear that Lorenz considers it the best of a rotten bunch. He is more concerned with more informative graphical measures. A popular graphical technique was (and remains) available. Vito Pareto proposed plotting log income against the logarithm of the survival function of the distribution. There is good reason to believe the slope of such a chart (called a Pareto chart) will only provide a good measure of inequality when the actual distribution is of the Pareto form [i.e., $P(X > x) = (x/\sigma)^{-\alpha}$, $x > \sigma$]. In fact, the chart will only be linear for such distributions. However, taking logarithms has a way of masking some of the deviations from the Pareto distribution. In addition, much income data does seem to fit the Pareto model reasonably well.

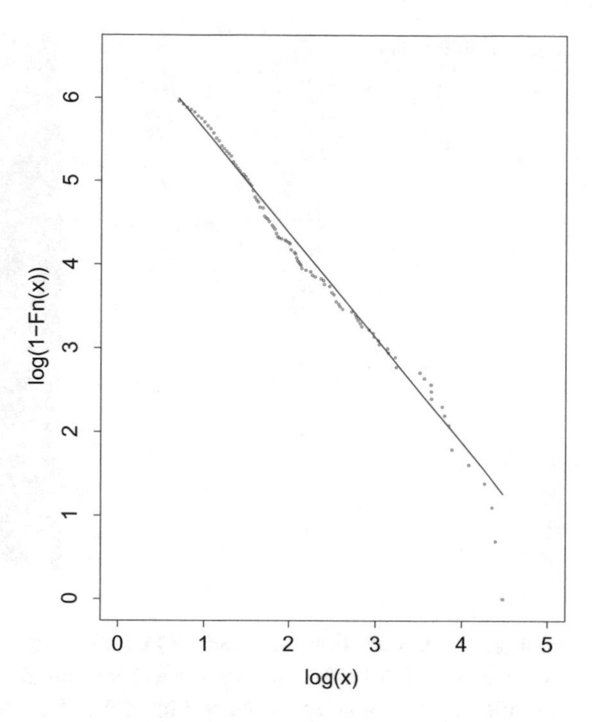

Fig. 1.1 Pareto chart using the Forbes data of the 400 richest people in the world

Thus, it was generally possible to compare income distributions in terms of the slopes of their approximately linear Pareto charts. Figure 1.1 shows a typical Pareto chart, of the points $(\log x, \log(1 - F_n(x)))$, this one using the Forbes data on the 400 richest people in the world.

Figure 1.2 shows the Italian, sociologist, economist, and political scientist Vilfredo Pareto.[1]

Lorenz rightly sensed that "logarithmic curves are more or less treacherous." He proposed a graphical technique which did not involve such a transformation. To each number between 0 and 100, thought of as a percentage, say t, he proposed to associate the percentage of the total income which accrued to the poorest t percent of the population. He used percentage of income as the abscissa of the point on the curve. Subsequently, it has become customary to reflect Lorenz's curve about the 45° line, i.e., use percentage of income as the ordinate. In modern notation, the Lorenz curve is denoted by $L(u)$, whereas the curve originally plotted by Lorenz would be denoted by $L^{-1}(u)$. Lorenz observed that such curves will be typically bow shaped. He proposed the following "rule of interpretation." "As the bow is bent, so concentration" (i.e., inequality) "increases." Presumably this rule of interpretation is self evident, for Lorenz gives no hint of justification. Was this insight or serendipity? Muirhead's (1903) work was already published, but the

[1] https://commons.wikimedia.org/wiki/File:Vilfredo_Pareto.gif.

Fig. 1.2 Vilfredo Pareto

economic interpretation provided by Dalton's transfer principles was still 15 years away. It seems doubtful that Lorenz had any precise mathematical formulation for his rule of interpretation. It seemed logical to him, and it has survived well simply because it really does capture important aspects of inequality. But Dalton's work was needed to clarify this.

Hugh Dalton, later to become chancellor of the exchequer in England for the Labor government elected in 1946, was a practical economist. Perhaps because of his practicality, he demanded precision in definitions. He did not achieve a precise definition of inequality in his pioneering work of 1920, but he did point out some key ideas regarding desirable properties of inequality measures. He tried to isolate operations on income distributions which would "clearly" increase inequality and then strove to identify measures of inequality which would be monotone under such transformations. One might quibble that we are merely replacing one difficult problem by another possibly more difficult one, i.e., how can we agree on what transformations will clearly increase inequality? It would appear in retrospect that only one of Dalton's inequality principles commands wide, almost universal, acceptance. It is known as the Pigou–Dalton transfer principle, but in this book we will use the more evocative name of the Robin Hood axiom. When Robin and his merry hoods performed an operation in the woods they took from the rich and gave to the poor. The Robin Hood principle asserts that this decreases inequality (subject only to the obvious constraint that you don't take too much from the rich and turn them into poor). Most would agree that a Robin Hood operation decreases inequality. If we are to judge by names, since a Robin Hood heist is known as a progressive transfer, it appears that most approve of such Robin Hood actions. Presumably the sheriff of Nottingham was not a right thinker in this regard. It turns out that the Robin Hood axiom is intimately related to the order proposed by

Lorenz. A chance for one last Sherwood metaphor is provided by Lorenz's comment about inequality increasing as the bow is bent. Unfortunately Robin's bow bending efforts decrease rather than increase inequality, but we can forgive Lorenz his lack of prescience here.

Dalton's second principle was that multiplication of incomes by a constant greater than one should decrease inequality. Most popular inequality measures do not satisfy this criterion. In fact, arguments regarding reexpression of incomes in new currency units suggest that a desirable property of inequality measures would be scale invariance. We will accept this revised second principle. Dalton's third principle, which may be simply phrased as "giving everyone a dollar will decrease inequality," is, as we shall see, a consequence of Principle 1 and revised Principle 2 (scale invariance). His fourth principle is that the measure of inequality should be "invariant under cloning." Specifically, consider a population of n individuals and a related population of kn individuals which consists of k identical copies (with regard to income) of each of the n individuals in the original population. The population of size kn can be called the cloned population. Principle four requires that the measure of inequality yield the same value for the cloned population as it does for the original population. Mathematically this requires that the measure of inequality for the population should be a function of the sample distribution function of the population. Most common measures of inequality satisfy this last principle.

Now, how are we to get from Dalton and Lorenz or, if you will, Robin Hood and Lorenz's bow to the titular topic of this book, namely, majorization? And where does Muirhead fit in? It is not easy to guess dates. But one might speculate that the ideas contained in Hardy et al.'s (1929) paper had been formulated several years earlier, perhaps just when Dalton was enunciating his principles. Be that as it may, their brief 8-page paper in the *Messenger of Mathematics* contains the fetus of majorization theory fully formed and, indeed, more. How Topsy has grown; from part of Hardy, Littlewood and Polya's 8 pages to all of Marshall, Olkin and Arnold's 795 pages. Perhaps the main reason for the slow recognition of the nexus of the concepts of Lorenz ordering and majorization is to be found in the mathematician's proclivity to arrange numbers in decreasing order as opposed to the statistician's tendency to use increasing order! So that both may feel at home, we will phrase our definition both ways.

1.2 The Definition of Majorization

What then is majorization? We could continue to speak of incomes in a finite population, but it is a convenient time to temporarily drop the economic trappings in favor of clear mathematical statements.

Majorization, in this book, is a partial order defined on the positive orthant of n dimensional Euclidean space, to be denoted by $\mathbb{R}_n^+$. For a vector $\underline{x} \in \mathbb{R}_n^+$ we denote its (increasing) order statistic by $(x_{1:n}, x_{2:n}, \ldots, x_{n:n})$, i.e., the x_i's written in increasing order. Its decreasing order statistic will be denoted by

$x_{(1:n)}, x_{(2:n)}, \ldots, x_{(n:n)}$. Thus $x_{1:n}$ is the smallest of the x's, while $x_{(1:n)}$ is the largest. Of course, $x_{i:n} = x_{(n-i+1:n)}$. Along with most mathematicians, HLP (an acronym for Hardy, Littlewood and Polya that we will henceforth adopt) used the decreasing order statistics in their original definition.

Definition 1.2.1 Let $\underline{x}, \underline{y} \in \mathbb{R}_n^+$. We will say that $\underline{x}$ majorizes $\underline{y}$ and write $\underline{x} \geq_M \underline{y}$ if

$$\sum_{i=1}^{k} x_{(i:n)} \geq \sum_{i=1}^{k} y_{(i:n)}, \quad k = 1, 2, \ldots, n-1 \tag{1.1}$$

and $\sum_{i=1}^{n} x_{(i:n)} = \sum_{i=1}^{n} y_{(i:n)}$ [equivalently if $\sum_{i=1}^{k} x_{i:n} \leq \sum_{i=1}^{k} y_{i:n}$, $k = 1, 2, \ldots, n-1$ and $\sum_{i=1}^{n} x_i = \sum_{i=1}^{n} y_i$].

Remark Majorization can instead be defined as a partial order on the elements of n-dimensional Euclidean space. When reading other publications dealing with majorization, it is important to verify whether the partial order is defined on $\mathbb{R}_n^+$ or $\mathbb{R}_n$.

Now, how can we write a book about that? First of all, let us see what it says in the income context. Suppose we have two populations of n individuals each, with equal total incomes assumed without loss of generality to be 1. If we plot the points $(k/n, \sum_{i=1}^{k} x_{i:n})$ and $(k/n, \sum_{i=1}^{k} y_{i:n})$, we see that $\underline{x} \geq_M \underline{y}$ if and only if the Lorenz curve of $\underline{y}$ is wholly nested within that of $\underline{x}$. Thus, $\underline{x} \geq_M \underline{y}$ if and only if $\underline{x}$ is more unequal than $\underline{y}$ in the ordering proposed by Lorenz (by bow bending). In $\mathbb{R}_n^+$ we may thus speak of majorization and the Lorenz ordering essentially interchangeably.

Where does Robin Hood fit into this majorization business? When Robin Hood does his work, he performs a progressive transfer. The new income distribution $\underline{y}$ is related to the old distribution $\underline{x}$ by

$$\underline{y} = P\underline{x}$$

where P is a very simple doubly stochastic matrix. Using Muirhead's (1903) paper HLP (1959) showed that $\underline{x} \geq_M \underline{y}$ ($\underline{x} \geq_L \underline{y}$, if we wish to speak of the Lorenz order), if and only if $\underline{y}$ can be obtained from $\underline{x}$ by Robin Hood in a finite number of operations. So if we want an inequality measure to honor Dalton's first principle, then it must be a function on $\mathbb{R}_n^+$ which preserves the majorization partial order. Such functions are known as Schur convex functions.

Schur convex functions antedate the concept of majorization by a decade. Schur (1923) spoke of an averaging of $\underline{x}$ to be any $\underline{y}$ which was of the form $\underline{y} = P\underline{x}$ where P is doubly stochastic. He identified the class of functions which preserved the partial ordering defined by such averaging. HLP (1959) showed that Schur's averaging partial order was equivalent to majorization. Actually Muirhead (1903) had considered what was to become known as majorization in a much earlier paper, although he considered it as a partial order on vectors of non-negative integers.

As HLP (1959) noted in their book, there is a good reason to expect that any theorem which provided inequalities for vectors in $\mathbb{R}_n^+$ using summation may well have a generalization for non-negative functions involving integration. The analog of the order statistic for a vector in $\mathbb{R}_n^+$ is provided by the increasing rearrangement of a non-negative function. They did not provide the necessary extension of Schur's averaging, but even that is possible using the concept of a balayage. One can, in fact, parlay the exercise into even more abstract settings, but we will satisfy ourselves here with the mere mention of such possibilities.

Other chapters in this book will address how transformations affect Lorenz curves (in an economic context this can be viewed as discussion of the effects of various taxation policies), the development of suitable inequality indices and investigation of parametric families of Lorenz curves.

One important extension which does not take us into esoteric territory, but nevertheless into a thicket of problems, involves multivariate majorization, to be discussed in Chap. 7. The basic ambiguity is a consequence of the fact that there really is no compelling ordering of vectors to play the role that the order statistics played in the development of (univariate) majorization. In addition, inevitably, stochastic versions of majorization had to evolve. Some discussion of these concepts will be provided in Chap. 8. Chapter 9 explores, albeit superficially, the relationship between majorization and several related partial orderings including stochastic dominance. Inequality analysis in popular families of income distributions is the subject of Chap. 10. The concluding chapters catalog a variety of applications of majorization.

Fig. 1.3 Robin Hood statue outside of Nottingham Castle, Nottingham, UK

As an entertainment, one might attempt to enumerate how many surrogates of the majorization partial ordering are described in this book. Important ones are introduced on pages 211 and 251. But, there are others sprinkled around.

Figure 1.3 shows a Robin Hood statue, located outside of Nottingham Castle in Nottingham, UK.[2]

[2]https://commons.wikimedia.org/wiki/File:Robin_Hood_Memorial.jpg.

Chapter 2
Majorization in $\mathbb{R}_n^+$

As mentioned in Chap. 1, the name majorization seems to have appeared first in HLP (1959), the idea had appeared earlier (HLP, 1929) although unchristened. Muirhead who dealt with $\mathbb{Z}_n^+$ (i.e., vectors of nonnegative integers) already had identified the partial order defined in (1.1) (i.e. majorization). But he, when he needed to refer to it, merely called it "ordering." Perhaps it took the insight of HLP to recognize that little of Muirhead 's work need necessarily be restricted to integers, but the key ideas including Dalton's transfer principle were already present in Muirhead's paper. If there was anything lacking in Muirhead's development, it was motivation for the novel results he obtained. He did exhibit the arithmetic-geometric mean inequality as an example of his general results, but proofs of that inequality are legion. If that was the only use of his "inequalities of symmetric algebraic functions of n letters," then they might well remain buried in the Edinburgh proceedings. HLP effectively rescued Muirhead's work from such potential obscurity. In the present book theorems will be stated in generality comparable to that achieved by HLP and will be ascribed to those authors. Muirhead's priority will not be repeatedly asserted. HLP restricted attention to $\mathbb{R}_n^+$, but the restriction to the positive orthant can and will be often dispensed with. First let us establish the relationship between majorization as defined by HLP (i.e., (1.1)) and averaging as defined by Schur. Recall that $\underline{x}$ is an average of $\underline{y}$ in the Schur sense if $\underline{x} = P\underline{y}$ for some doubly stochastic matrix P.

2.1 Basic Result

There are certain very simple linear transformations, $\underline{x} = A\underline{y}$, for which it is obvious that $\underline{x} \leq_M \underline{y}$. The very simplest case is when A is a permutation matrix. But, there are other quite straightforward cases. Going back to our Robin Hood scenario of Chap. 1, we would like to show that the linear transform associated with a Robin

B. C. Arnold, J. M. Sarabia, *Majorization and the Lorenz Order with Applications in Applied Mathematics and Economics*, Statistics for Social and Behavioral Sciences, https://doi.org/10.1007/978-3-319-93773-1_2

Hood operation leads to a new vector $\underline{x}$ which is majorized by the old vector $\underline{y}$. In order to obtain results valid in $\mathbb{R}_n$ rather than just $\mathbb{R}_n^+$, it is convenient to admit the possibility of negative incomes in our financial scenario. To begin with we consider a special kind of Robin Hood operation in which money is taken from a relatively rich individual i and given to the individual whose wealth is immediately below that of individual i in the ranking of wealth. Call this an elementary Robin Hood operation.

Because of the fact that permutation preserves majorization, we may, without loss of generality, assume that our vector $\underline{x}$ has coordinates arranged in increasing order. Now, what we want to show is that every Robin Hood operation is equivalent to a series of elementary Robin Hood operations and permutations and, then that every multiplication by a doubly stochastic matrix is equivalent to a finite series of Robin Hood operations (and thus to a longer series of elementary Robin Hood operations and permutations). Then if the elementary Robin Hood operations induce majorization, so does multiplication by a doubly stochastic matrix.

First let us ascertain that an elementary Robin Hood operation induces majorization. The old and new incomes (possibly negative) are related by

$$\underline{x} = A\underline{y} \tag{2.1}$$

where, for some i and some $\lambda \in [0, \frac{1}{2}]$,

$$\begin{aligned} a_{i,i} &= 1-\lambda, \quad & a_{i,i+1} &= \lambda, \\ a_{i+1,i} &= \lambda, & a_{i+1,i+1} &= 1-\lambda \end{aligned} \tag{2.2}$$

and

$$a_{jk} = \delta_{jk} \quad \text{otherwise,}$$

where δ_{jk} denotes the usual Kronecker delta. Note that we require $\lambda \in [0, \frac{1}{2}]$ in order to have Robin Hood's operation not disturb the ordering of the vector. It is evident from the basic Definition 1.2.1 that with A defined by (2.2), and $\underline{y}$ and $\underline{x}$ related by (2.1) we have $\underline{x} \leq_M \underline{y}$. Now a more general Robin Hood operation involves taking the income of a relatively poor individual and a relatively rich individual, perhaps individuals i and j in the increasing ranking, and redividing their combined wealth among them, subject only to the constraint that individual i not become more rich than individual j.

The effect of such an operation is to raise up the Lorenz curve (the plot of points $(k/n, \sum_{i=1}^{k} x_{i:n})$. The effect of an elementary Robin Hood operation is to raise the Lorenz curve at one point. It is a simple matter to verify that one may move from a given Lorenz curve to a higher one by successively raising the curve at individual points subject to the constraint that at no time is the convexity of the Lorenz curve destroyed. A simple example will exhibit the manner in which this may be achieved.

Consider a population of eight individuals whose ordered wealths are

$$2,\ 3,\ 5,\ 11,\ 13,\ 18,\ 23,\ 25. \qquad (A)$$

Suppose that Robin Hood take 8 from the individual with wealth 23 and gives it to the individual with wealth 5. The resulting set of ordered wealths is

$$2,\ 3,\ 11,\ 13,\ 13,\ 15,\ 18,\ 25. \qquad (B)$$

It is not difficult to reach (B) from (A) using only elementary Robin Hood operations. For example, one could successively transform the set of ordered wealths as follows.

2, 3, 5, 11, 13, 18, 23, 25
2, 3, 5, 11, 13, 20.5, 20.5, 25
2, 3, 5, 11, 15.5, 18, 20.5, 25
2, 3, 5, 13, 13.5, 18, 20.5, 25
2, 3, 9, 9, 13.5, 18, 20.5, 25
2, 3, 9, 9, 15.5, 16, 20.5, 25
2, 3, 9, 12, 12.5, 16, 20.5, 25
2, 3, 9, 12, 13, 15.5, 20.5, 25
2, 3, 9, 12, 13, 18, 18, 25
2, 3, 10, 11, 13, 18, 18, 25
2, 3, 10, 12, 12, 18, 18, 25
2, 3, 10, 12, 15, 15, 18, 25
2, 3, 11, 11, 15, 15, 18, 25
2, 3, 11, 13, 13, 15, 18, 25

Each line above was obtained from the preceding line by an elementary Robin Hood operation. The reader might ponder on the problem of determining the minimal number of elementary Robin Hood operations required to duplicate a given general Robin Hood operation. The sequence described above is not claimed to be parsimonious with regard to the number of steps used. Duplication of a general Robin Hood operation may require a countable number of elementary Robin Hood operations (see Exercise 1).

But now we may make a remarkable observation. A vector $\underline{x}$ majorizes $\underline{y}$ by definition, if the Lorenz curve of $\underline{y}$ is obtained from that of $\underline{x}$ by raising it at one or several points. It is evident that this too can be accomplished one point at a time (subject to the restriction that convexity is not violated). Let $\mathcal{E}$ be the class of all $n \times n$ doubly stochastic matrices corresponding to elementary Robin Hood operations, and let $\mathcal{P}$ be the class of all $n \times n$ permutation matrices. It is then clear that if $\underline{x} \geq_M \underline{y}$, then $\underline{y}$ can be obtained from $\underline{x}$ by a string of elementary Robin Hood operations and/or relabelings. Thus $\underline{y} = Q\underline{x}$ where Q is a product of a countable number of matrices chosen from the class of doubly stochastic matrices $\mathcal{E} \cup \mathcal{P}$. Q is then necessarily itself doubly stochastic. Thus we have proved that if $\underline{x} \geq_M \underline{y}$, then $\underline{y} = P\underline{x}$ for some doubly stochastic matrix P. The converse statement was proved in HLP essentially as follows.

Suppose that $\underline{y} = P\underline{x}$ for some doubly stochastic matrix P where without loss of generality $\underline{x}$ and $\underline{y}$ are in ascending order. Then for any $m \leq n$, we may define

$$k_j = \sum_{i=1}^{m} p_{ij} \leq 1, \quad j = 1, 2, \ldots, n,$$

and we have

$$\begin{aligned}
\sum_{i=1}^{m} y_i &= \sum_{j=1}^{n} k_j x_j \\
&\geq \sum_{j=1}^{m-1} k_j x_j + x_m \left(m - \sum_{j=1}^{m-1} k_j \right) \\
&\quad \left(\text{since } \sum_{j=1}^{n} k_j = m \right) \\
&\geq \sum_{j=1}^{m-1} k_j x_j + x_m + \sum_{j=1}^{m-1} (1 - k_j) x_j \\
&\quad (\text{since } k_j \leq 1) \\
&= \sum_{j=1}^{m} x_j .
\end{aligned}$$

Thus $\underline{x} \geq_M \underline{y}$. We may consequently state

Theorem 2.1.1 (HLP) $\underline{x} \geq_M \underline{y}$ *if and only if* $\underline{y} = P\underline{x}$ *for some doubly stochastic matrix* P.

The matrix P referred to in Theorem 2.1.1 is not necessarily unique. For a given $\underline{x}$, $\underline{y}$ the class of matrices P for which $\underline{y} = P\underline{x}$ is convex, but it can be a singleton set or in other cases may be nontrivial and have several extreme points.

An alternative characterization of majorization is then available if we utilize Birkhoff's (1946) observation that the class of $n \times n$ doubly stochastic matrices coincides with the convex hull of the set of $n \times n$ permutation matrices. We thus may state that $\underline{x} \geq_M \underline{y}$ if and only if $\underline{y} = \sum_{\ell=1}^{k} \gamma_\ell P_\ell \underline{x}$ for some set of permutation matrices $P_1, P_2, \ldots, P_k$ and some set of γ_ℓ's satisfying $\gamma_\ell \geq 0$, $\ell = 1, 2, \ldots, k$ and $\sum_{\ell=1}^{k} \gamma_\ell = 1$. Farahat and Mirsky (1960) showed that k can be chosen to be $n^2 - 2n + 2$ and that, in general, no smaller number will suffice. An alternative statement of this fact is that for a given vector $\underline{x}$ the class of vectors $\underline{y}$ majorized by $\underline{x}$ is the convex hull of the set of points obtained by rearranging the coordinates of $\underline{x}$.

2.2 Schur Convex Functions and Majorization

Functions which preserve the majorization partial order form a large class. They were first studied by Schur (1923) and are, in his honor, called Schur convex functions. We may join with Marshall and Olkin in lamenting the use of the term convex rather than monotone, increasing, order preserving or, more modishly, isotonic, but we will persist in using the name Schur convex. Many famous inequalities are readily proved by focussing on a particular Schur convex function. If one thumbs through the problem sections of the American Mathematical Monthly one is struck by the frequency with which inequalities are proved as a consequence of the Schur convexity of some judiciously chosen function. For example, Chap. 8 of Marshall et al. (2011) catalogs a plethora of such results in a geometric setting (many dealing with features of triangles). Not every analytic inequality is a consequence of the Schur convexity of some function, but enough are to make familiarity with majorization/Schur-convexity a necessary part of the required background of a respectable mathematical analyst. After that build-up, we had better get quickly to the matter of defining the required concepts.

Definition 2.2.1 Let $A \subset \mathbb{R}_n^+$. A function $g : A \to \mathbb{R}$ is said to be *Schur convex* on A if $g(\underline{x}) \leq g(\underline{y})$ for every pair $\underline{x}, \underline{y} \in A$ for which $\underline{x} \leq_M \underline{y}$

The function g alluded to in Definition 2.2.1 is said to be strictly Schur convex if it is Schur convex and if $\underline{x} \leq_M \underline{y}$, and $\underline{x}$ not a permutation of $\underline{y}$ together imply $g(\underline{x}) < g(\underline{y})$.

As an example, consider the function

$$g(x) = \sum_{i=1}^{n} x_i^2 \tag{2.3}$$

defined on $\mathbb{R}_n$. We claim that this function is Schur convex. In verifying this fact we will use a couple of tricks of the trade. The task of determining Schur-convexity is simplified immediately by the observation that a Schur convex function must be a symmetric function of its arguments. More specifically, if g is Schur convex on A and if $\underline{x}$ and $\Pi\underline{x}$ are in A for some permutation matrix Π, then $g(\underline{x}) = g(\Pi\underline{x})$ (since, clearly, $\underline{x} \geq_M \Pi\underline{x}$ and $\Pi\underline{x} \geq_M \underline{x}$). Another simplification hinges upon the fact that if $\underline{x} \geq_M \underline{y}$, then $\underline{y}$ can be obtained from $\underline{x}$ by a finite number of Robin Hood operations. Such operations only involve two coordinates at a time. So we only need to check whether $g(\underline{x}) \leq g(\underline{y})$ whenever $\underline{x} \leq_M \underline{y}$ and $\underline{x}$ and $\underline{y}$ differ only in two coordinates which by symmetry may without loss of generality be taken to be the first two.

Thus, in order to verify Schur convexity of the function (2.3), we need to verify that if

$$(\alpha_1, \alpha_2, \alpha_3, \ldots, \alpha_n)^\top \leq_M (\beta_1, \beta_2, \alpha_3, \ldots, \alpha_n)^\top \tag{2.4}$$

(where necessarily $\alpha_1 + \alpha_2 = \beta_1 + \beta_2$ and without loss of generality $\alpha_1 \le \alpha_2$ and $\beta_1 \le \beta_2$) then $\alpha_1^2 + \alpha_2^2 + \cdots + \alpha_n^2 \le \beta_1^2 + \beta_2^2 + \alpha_3^2 + \cdots + \alpha_n^2$, i.e.

$$\alpha_1^2 + \alpha_2^2 \le \beta_1^2 + \beta_2^2.$$

Without loss of generality $\alpha_1 + \alpha_2 = \beta_1 + \beta_2 = 1$, and we may write α for α_1 and β for β_1. We then wish to verify that if

$$\frac{1}{2} \ge \alpha \ge \beta$$

(as implied by (2.4)), then

$$\alpha^2 + (1-\alpha)^2 \le \beta^2 + (1-\beta)^2.$$

This is, however, obvious since the function $z^2 + (1-z)^2$ has a non-positive derivative on $(-\infty, \frac{1}{2}]$.

Let us denote by O_n the subset of $\mathbb{R}_n$ in which the coordinates are in increasing order. Thus

$$O_n = \{\underline{x} : x_1 \le x_2 \le \cdots \le x_n\}. \tag{2.5}$$

If we wish to check for Schur convexity of a function g defined on a symmetric set A, we need only check for symmetry on A and Schur convexity on $O_n \cap A$. Since, in many applications the domains of putative Schur convex functions are indeed symmetric, it is often sufficient to restrict attention to O_n or subsets thereof.

To characterize Schur convexity on subsets of O_n we return to the Robin Hood scenario. Thinking of the coordinates of O_n as the ordered wealths of n individuals in Sherwood Forest (allowing negative wealth so as not to restrict attention to $\mathbb{R}_n^+$), we may recall that (in O_n) $\underline{x} \le_M \underline{y}$ if and only if $\underline{x}$ can be obtained from $\underline{y}$ by a countable number of elementary Robin Hood operations. Such operations involve a transfer of funds from one individual to the next richest individual without disturbing the ordering of individuals. To be Schur convex on O_n then, a continuous function g need only be appropriately monotone with respect to such elementary Robin Hood operations (cf. Lemma A.2 in Marshall et al. (2011, p. 81)). Monotonicity is particularly easy to verify in differentiable cases by inspection of the sign of the appropriate partial derivative. Schur's (1923) famous sufficient condition for Schur convexity involves such partial derivatives. Although Schur's condition has proved remarkably useful, it cannot be used to deal with less regular Schur convex functions. The existence of unfriendly Schur convex functions is alluded to in Exercise 3.

Summarizing the results described above we have

Theorem 2.2.1 *Suppose $g : O_n \to \mathbb{R}$ is continuous. g is Schur convex if and only if for every $\underline{x} \in O_n$ and every $i = 1, 2, \ldots, n-1$, the function*

$$g(x + \epsilon\, \underline{v}_i)$$

is a non-increasing function of ϵ for all ϵ such that $x + \epsilon\, \underline{v}_i \in O_n$. The vector $\underline{v}_i$ is a vector with a 1 in the ith coordinate and a -1 in the $(i+1)$st coordinate, all other coordinates being 0. [If the functions $g(x + \epsilon \underline{v}_i)$ are strictly decreasing, then we characterize strict Schur convexity.]

Theorem 2.2.2 *Suppose $g : O_n \to \mathbb{R}$ is continuous with continuous partial derivatives (on the interior of O_n). g is Schur convex on O_n if and only if for every $\underline{x}$ in the interior of O_n and every $j = 1, 2, \ldots, n-1$,*

$$\frac{\partial}{\partial x_j} g(\underline{x}) \le \frac{\partial}{\partial x_{j+1}} g(\underline{x})$$

[i.e., if $\nabla g(\underline{x}) \in O_n$, $\forall x \in int(O_n)$].

Proof Follows by legitimately differentiating the $g(x + \epsilon\, \underline{v}_i)$'s. ■

Theorem 2.2.3 (Schur's Condition) *Suppose $g : I^n \to \mathbb{R}$ is continuously differentiable where I is an open interval. g is Schur convex if and only if g is symmetric and for every $i \neq j$*

$$(x_i - x_j)\left[\frac{\partial}{\partial x_i} g(\underline{x}) - \frac{\partial}{\partial x_j} g(\underline{x})\right] \ge 0, \quad \forall \underline{x} \in I^n. \tag{2.6}$$

Proof The result follows from Theorem 2.2.2 and the assumed symmetry of $g(\underline{x})$. In fact, using the assumed symmetry of $g(\underline{x})$, one only need verify (2.6) for the particular case $(i, j) = (1, 2)$. ■

If we return to our example $g(\underline{x}) = \sum_{i=1}^{n} x_i^2$ defined on $\mathbb{R}_n$, symmetry is evident and we see that Schur's condition (2.6) takes the form

$$(x_1 - x_2)(2x_1 - 2x_2) \ge 0.$$

Actually this particular Schur convex function provides an example of an important class of Schur convex functions, the separable convex class.

Definition 2.2.2 A function $g : I^n \to \mathbb{R}$ (where I is an interval) is said to be separable convex if g is of the form

$$g(\underline{x}) = \sum_{i=1}^{n} h(x_i) \tag{2.7}$$

where h is a convex function on I. Recall that h is convex on I if $h(\alpha x + (1-\alpha)y) < \alpha h(x) + (1-\alpha)h(y)$ for every $x, y \in I$ and every $\alpha \in [0, 1]$.

Schur (1923) and HLP observed that any separable convex function is Schur convex. If h is differentiable (as "most" convex functions are), then Schur's condition (2.6) is readily verified. More generally it can be verified by considering the effect of an elementary Robin Hood operation which involves an averaging of two coordinates (this is Exercise 4).

An interesting example of a separable convex function is provided by the function

$$g(\underline{x}) = -\sum_{i=1}^{n} \log x_i \tag{2.8}$$

defined on $\mathbb{R}_n^+$. The function $-\log x$ is readily shown to be convex. A consequence of the Schur convexity of (2.8) is the celebrated arithmetic-geometric mean inequality. For any $\underline{x} \in \mathbb{R}_n^+$ it is evident that

$$(x_1, x_2, \ldots, x_n)^\top \geq_M (\bar{x}, \bar{x}, \ldots, \bar{x})^\top \tag{2.9}$$

where $\bar{x} = \frac{1}{n}\sum_{i=1}^{n} x_i$ is the arithmetic mean of the x_i's (Robin Hood can clearly eventually make everyone's wealth equal). If we evaluate the Schur convex function defined in (2.8) at each of the points in $\mathbb{R}_n^+$ referred to in (2.9), we eventually conclude that

$$\left(\prod_{i=1}^{n} x_i\right)^{1/n} \leq \frac{1}{n}\sum_{i=1}^{n} x_i \tag{2.10}$$

i.e., the geometric mean cannot exceed the arithmetic mean. Myriad proofs of (2.10) exist, a remarkable number of which can be presented in a majorization framework. As remarked at the beginning of the chapter, Muirhead (1903) provided an early example of such a proof.

Not every Schur convex function is separable convex. This is not a transparent result. A famous example of a non-separable Schur convex function is the Gini index. This is defined on $\mathbb{R}_n^+$ as follows:

$$g(\underline{x}) = \left[\sum_{i=1}^{n} (2i - n - 1)x_{i:n}\right] \Bigg/ \left(\sum_{i=1}^{n} x_i\right). \tag{2.11}$$

Another such example is provided by

$$g(\underline{x}) = \left(\sum_{i=1}^{n} x_i\right)\left(\sum_{i=1}^{n} x_i^2\right). \tag{2.12}$$

The elementary symmetric functions

$$\sum_{i=1}^{n} x_i, \quad \sum_{i \neq j} x_i x_j, \quad \sum_{i \neq j \neq k} x_i x_j x_k, \quad etc. \tag{2.13}$$

are examples of Schur concave functions (g is Schur concave if $-g$ is Schur convex). Except for the first one, ($\sum_{i=1}^{n} x_i$), they are clearly not separable concave.

The class of Schur convex functions is closed under a variety of operations. For example, consider m Schur convex functions $g_1, \ldots, g_m$ each defined on some set

$A \subset \mathbb{R}_n$. It is easily established that the following functions are necessarily Schur convex on A.

$$h_1 = \sum_{i=1}^{m} c_i g_i \quad \text{where} \quad c_i \geq 0,$$

$$h_2 = \prod_{i=1}^{m} c_i g_i \quad \text{where} \quad c_i \geq 0 \quad \text{and} \quad g_i \geq 0,$$

$$h_3 = \min_i \{c_i g_i\} \quad \text{where} \quad c_i \geq 0,$$

$$h_4 = \max_i \{c_i g_i\} \quad \text{where} \quad c_i \geq 0. \tag{2.14}$$

It is also evident that the class of Schur convex functions is closed under pointwise convergence. In all the above cases Schur convexity is verifiable directly from Definition 2.2.1.

Another useful way to construct Schur convex functions (or way to recognize Schur convexity) involves marginal transformations before applying a known Schur convex function. Specifically suppose that g is a Schur convex function defined on I^n (where I is an interval in $\mathbb{R}$) and that h is a convex function defined on $\mathbb{R}$. Define

$$\phi(\underline{x}) = g\left(h(x_1), \ldots, h(x_n)\right). \tag{2.15}$$

A sufficient condition that ϕ be Schur convex is that g be nondecreasing in addition to being Schur convex. We could use this observation to verify yet again the Schur convexity of the function $\sum_{i=1}^{n} x_i^2$ (since evidently $h(x) = x^2$ is convex and $\sum_{i=1}^{n} x_i$ is Schur convex and increasing).

In statistical applications several important Schur convex functions are constructed via integral transforms. We will satisfy ourselves with a representative example (due to Marshall and Olkin (1974)).

Lemma 2.2.1 *Let g be an integrable Schur convex function on $\mathbb{R}_n$ and suppose that $A \subset \mathbb{R}_n$ satisfies:*

$$\textit{If } \underline{x} \in A, \textit{ and } \underline{y} \leq_M \underline{x} \textit{ then } y \in A. \tag{2.16}$$

It follows that

$$\phi(\underline{x}) = \int_{A+\underline{x}} g(\underline{y})\, d\underline{y},$$

is a Schur convex function of $\underline{x}$ on $\mathbb{R}_n$.

Since measurable Schur concave functions are approximable by simple Schur concave functions (linear combinations of indicator functions), the above Lemma can be parlayed (as Marshall and Olkin did) into a proof of the following.

Theorem 2.2.4 *If g is Schur convex on $\mathbb{R}_n$ and f is Schur concave on $\mathbb{R}_n$, then*

$$\phi(\underline{x}) = \int_{\mathbb{R}_n} f(\underline{x} - \underline{y})g(\underline{y})\,d\underline{y}, \tag{2.17}$$

is Schur convex on $\mathbb{R}_n$ provided that the integral in (2.17) exists for all $\underline{x} \in \mathbb{R}_n$.

The key observation in proving Theorem 2.2.4 by way of Lemma 2.2.1 is that (2.16) is equivalent to the statement that the indicator function of A is Schur concave.

The following direct proof of Theorem 2.2.4 is attributed by Marshall and Olkin to Proschan and Cheng. Without loss of generality, assume $n = 2$ (since we only need to consider elementary Robin Hood operations). Additionally we need to only consider small heists. So for ϵ small, consider $(x_1 + \epsilon, x_2 - \epsilon) \leq_M (x_1, x_2)$ where $x_1 < x_2$. We have

$$\begin{aligned}
&\phi(x_1, x_2) - \phi(x_1 + \epsilon, x_2 - \epsilon)\\
&= \int_{\mathbb{R}^2} f(x_1 - y_1, x_2 - y_2)g(y_1, y_2)\,dy_1dy_2\\
&\quad - \int_{\mathbb{R}^2} f(x_1 + \epsilon - y_1, x_2 - \epsilon - y_2)g(y_1, y_2)\,dy_1dy_2\\
&= \int_{\mathbb{R}^2} f(u_1, u_2 + \epsilon)g(x_1 - u_1, x_2 - \epsilon - u_2)\,du_1du_2\\
&\quad - \int_{\mathbb{R}^2} f(u_1 + \epsilon, u_2)g(x_1 - u_1, x_2 - \epsilon - u_2)\,du_1du_2.
\end{aligned}$$

Since f is symmetric we can write

$$\begin{aligned}
&\int_{u_1 \leq u_2} [f(u_1, u_2 + \epsilon) - f(u_1 + \epsilon, u_2)]g(x_1 - u_1, x_2 - \epsilon - u_2)\,du_1du_2\\
&= \int_{u_1 \leq u_2} [f(u_2 + \epsilon, u_1) - f(u_2, u_1 + \epsilon)]g(x_1 - u_1, x_2 - \epsilon - u_2)\,du_1du_2\\
&= \int_{u_1 \geq u_2} [f(u_1 + \epsilon, u_2) - f(u_1, u_2 + \epsilon)]g(x_1 - u_2, x_2 - \epsilon - u_1)\,du_1du_2
\end{aligned}$$

(relabelling u_1 and u_2). Consequently,

$$\begin{aligned}
&\phi(x_1, x_2) - \phi(x_1 + \epsilon, x_2 - \epsilon)\\
&= \int_{u_1 \geq u_2} [f(u_1, u_2 + \epsilon) - f(u_1 + \epsilon, u_2)]g(x_1 - u_1, x_2 - \epsilon - u_2)\\
&\quad - g(x_1 - u_2, x_2 - \epsilon - u_1)\,du_1du_2
\end{aligned}$$

However for $u_1 \geq u_2$ we have

$$(u_1, u_2 + \epsilon) \leq_M (u_1 + \epsilon, u_2)$$

and

$$(x_1 - u_2, x_2 - \epsilon - u_1) \le_M (x_1 - u_1, x_2 - \epsilon - u_2).$$

Consequently, by the Schur concavity of f and the Schur convexity of g, the integrand is non-negative and the Schur convexity of ϕ is verified.

Other useful examples of majorization involving integral transformations are to be found in the work of Hollander et al. (1977) in the context of functions "decreasing in transposition."

The definition of majorization involves ordering the components of vectors in $\mathbb{R}_n$. If n is not large, this is not an arduous task. If n is large, then it becomes desirable to determine sufficient conditions for majorization which do not involve ordering. Of course a sufficient condition that $\underline{x} \le_M \underline{y}$ is that $g(\underline{x}) \le g(\underline{y})$ for every Schur convex function g. This is just the contrapositive of the definition of Schur convexity. This does not involve ordering, but it does involve checking a vast number of functions. The cardinality of the set of Schur convex functions may be larger than your first guess in light of the question at the end of Exercise 3. Surely we do not have to check whether $g(\underline{x}) \le g(\underline{y})$ for every Schur convex function g. The pathological ones alluded to in Exercise 3 certainly don't have to be checked (why?). In 1929 HLP verified that one need only check separable convex functions in order to verify majorization. In fact it suffices to check only a particularly simple subclass of the separable convex functions. In what follows we use the notation $a^+ = \max\{0, a\}$.

Theorem 2.2.5 (HLP, Karamata) *$\underline{x} \le_M \underline{y}$ if and only if $\sum_{i=1}^n h(x_i) \le \sum_{i=1}^n h(y_i)$ for every (continuous) convex function $h : \mathbb{R} \to \mathbb{R}$.*

Theorem 2.2.6 (HLP) *$\underline{x} \le_M \underline{y}$ if and only if $\sum_{i=1}^n x_i = \sum_{i=1}^n y_i$ and $\sum_{i=1}^n (x_i - c)^+ \le \sum_{i=1}^n (y_i - c)^+$ for every $c \in \mathbb{R}$.*

It is evidently sufficient to prove Theorem 2.2.6, since $h(x) = x$ and $h(x) = (x - c)^+$ are continuous convex functions. It is also then evident that Theorem 2.2.5 remains valid with or without the parenthetical word "continuous."

Proof One implication is trivial since the functions $\sum_{i=1}^n x_i$ and $\sum_{i=1}^n (x_i - c)^+$ are separable convex and hence Schur convex. The proof of the converse is also straightforward. Merely set c successively equal to $y_{(1:n)}, y_{(2:n)}, \ldots, y_{(n:n)}$. Thus, in terms of decreasing order statistics,

$$\begin{aligned}
\left(\sum_{i=1}^k x_{(i:n)} - k y_{(k:n)}\right) &\le \sum_{i=1}^k (x_{(i:n)} - y_{(k:n)})^+ \\
&\le \sum_{i=1}^n (x_i - y_{(k:n)})^+ \\
&\le \sum_{i=1}^n (y_i - y_{(k:n)})^+ \quad \text{by hypothesis}
\end{aligned}$$

$$= \sum_{i=1}^{k} (y_{(i:n)} - y_{(k:n)})^+$$

$$= \sum_{i=1}^{n} y_{(i:n)} - k y_{(k:n)}.$$

Adding $k y_{(k:n)}$ to each side, the desired result follows. ■

Theorem 2.2.5 is the more commonly quoted of the two theorems. It is sometimes called Karamata's theorem. Karamata's (1932) proof of this result antedated HLP's proof which appeared in their 1959 book. However HLP had stated the result sans proof in their ear1ier brief note in the Messenger of Mathematics (in 1929). To them (HLP) the fact that linear combinations of angles (functions of the form $g(x) = (x - a)^+$) were dense in the set of continuous convex functions was intuitively clear. Perhaps they consequently felt it unnecessary to overburden their note with elementary proofs.

Another important class of Schur convex functions are those introduced by Muirhead (1903), subsequently dubbed symmetrical means by HLP (1959). For a given vector $(a_1, \ldots, a_n)$ with each $a_i > 0$ we define the $\underline{x}$'th symmetrical mean of $\underline{a}$ for some set of n real numbers $x_1, \ldots, x_n$ to be

$$[\underline{x}]_{\underline{a}} = \frac{1}{n!} \sum_{\pi} a_{\pi(1)}^{x_1} a_{\pi(2)}^{x_2} \cdots a_{\pi(n)}^{x_n} \tag{2.18}$$

where the sum is over all permutations of the integers $1, 2, \ldots, n$.

Muirhead (1903) showed that majorization could be verified merely by checking that all symmetrical means were appropriately ordered. He restricted attention to integer valued x_i's, but his proof carries over to the case of more general values for the x_i's, as noted by HLP (1959).

Theorem 2.2.7 (Muirhead, HLP) *$\underline{x} \leq_M \underline{y}$ if and only if for every $\underline{a} > \underline{0}$,*

$$[\underline{x}]_{\underline{a}} \leq [\underline{y}]_{\underline{a}}.$$

Proof A symmetrical mean is evidently symmetric and is easily verified to be convex. From Exercise 4 we then conclude that it is Schur convex, i.e., $[\underline{x}]_{\underline{a}} \leq [\underline{y}]_{\underline{a}}$ whenever $\underline{x} \leq_M \underline{y}$.

Conversely, for a fixed $k < n$, define $\underline{a}^{(k)}(u)$ by $a_1 = a_2 = \cdots = a_k = u$ and $a_{k+1} = \cdots = a_n = 1$. The corresponding symmetric means $[\underline{x}]_{\underline{a}^{(k)}(u)}$ and $[\underline{y}]_{\underline{a}^{(k)}(u)}$ are (generalized) polynomials in u with the indices of their highest powers being, respectively, $\sum_{i=1}^k x_{(i:n)}$ and $\sum_{i=1}^k y_{(i:n)}$. In order to have $[\underline{x}]_{\underline{a}^{(k)}(u)} \leq [\underline{y}]_{\underline{a}^{(k)}(u)}$ for u large, we must have $\sum_{i=1}^k x_{(i:n)} \leq \sum_{i=1}^k y_{(i:n)}$. Finally, if we set $\underline{a}^{(n)}(u) = (u, u, \ldots, u)$, we can only have $[\underline{x}]_{\underline{a}^{(n)}(u)} \leq [\underline{y}]_{\underline{a}^{(n)}(u)}$ for u both large and small if $\sum_{i=1}^k x_{(i:n)} = \sum_{i=1}^k y_{(i:n)}$. ■

The proof of the converse given above is, modulo notation changes, exactly that provided by HLP (1959). The symmetric means are of considerable historical interest. They assume a prominent role in both Muirhead's (1903) paper and in HLP's (1959) Chap. 2. Incidently, at the present juncture we may quickly illustrate Muirhead's development of the arithmetic-geometric mean inequality. Both the arithmetic and the geometric mean of $\underline{a} > \underline{0}$ are symmetric means corresponding, respectively, to the choices $(1, 0, \ldots, 0)$ and $(\frac{1}{n}, \frac{1}{n}, \ldots, \frac{1}{n})$, respectively, for $\underline{x}$. Since evidently $(1, 0, 0, \ldots, 0) \geq_M (\frac{1}{n}, \ldots, \frac{1}{n})$, the arithmetic-geometric mean inequality follows from Theorem 2.2.7.

Marshall et al. (2011, Chapter 4, Section B) describe other classes of Schur convex functions which may be used to determine majorization. The symmetric means and the separable convex functions remain the classical examples.

2.3 Exercises

1. Evidently $\underline{x} = (2, 2, 2) \leq_M (1, 2, 3) = \underline{y}$. Verify that although $\underline{x}$ may be obtained from $\underline{y}$ by a single Robin Hood operation, it requires a countable number of elementary Robin Hood operations to obtain $\underline{x}$ from $\underline{y}$.

2. Give an example to show that the doubly stochastic matrix P alluded to in Theorem 2.1.1 is not necessarily unique.

3. Give an example of a discontinuous Schur convex function defined on $\mathbb{R}$.
   [*Hint*: If $\sum_{i=1}^{n} x_i = 1$ and $\sum_{i=1}^{n} y_i = 2$, then $g(\underline{x})$ and $g(\underline{y})$ do not have to be related in any way]. [More pathologically, can you construct a non-measurable Schur convex function on $\mathbb{R}_n$?].

4. Suppose that g is symmetric on A and convex on A. Prove that g is Schur convex in A.
   [We really only need convexity on sets of the form $\{\underline{x} : \sum_{i=1}^{n} x_i = c\}$.]

5. Prove that separable convex functions on $\mathbb{R}_n$ are Schur convex without assuming differentiability.

6. Verify that the functions defined in (2.11) and (2.13) are indeed Schur convex but not separable convex.

7. Consider the function $g(\underline{x}) = \sum_{i=1}^{n}(x_i - \bar{x})^2$. Is it Schur convex? Is it separable convex?

8. Suppose $g(\underline{x}) = \sum_{i=1}^{n} a_i x_{i:n}$. Supply suitable conditions on the vector $\underline{a}$ to guarantee that g will be Schur convex.

9. If h is convex on I, then $g(\underline{x}) = \sum_{i=1}^n h(x_i)$ is Schur convex on I^n. Prove the converse, i.e., if $g(\underline{x}) = \sum_{i=1}^n h(x_i)$ is Schur convex on I^n, then h is convex on I.

10. (Easy but useful). If g is Schur convex on A and if we define, for $t \in \mathbb{R}$,

$$\chi_t(\underline{x}) = \begin{cases} 1, \text{ if } & g(\underline{x}) \geq t, \\ 0, \text{ otherwise,} \end{cases}$$

then $\chi_t(\underline{x})$ is Schur convex on A.

11. If g is a non-decreasing Schur convex function defined on I^n and h is a convex function on $\mathbb{R}$, verify that

$$\phi(\underline{x}) = g(h(x_1), h(x_2), \ldots, h(x_n))$$

is Schur convex on I^n.

12. Suppose h is convex on $\mathbb{R}$ and define

$$g(x) = \sum_{\pi} e^{\sum_{i=1}^n h(c_{\pi(i)} x_i)}$$

where the sum is over all permutations of $(1, \ldots, n)$. Verify that g is Schur convex (e.g., Muirhead's symmetric means). [Marshall and Olkin (1979) point out that the result is also true when the summation is extended only over the k largest of the $n!$ summands.]

13. Suppose that $\sum_{i=1}^n x_i = \sum_{i=1}^n y_i$ and $x_{i:n}/y_{i:n}$ is a non-increasing function on i. Show that $\underline{x} \leq_M \underline{y}$.

14. Suppose the P is a doubly stochastic matrix. Verify that the matrix $I - P^\top P$ is positive semi-definite. Use this observation to prove that $g(\underline{x}) = \sum_{i=1}^n x_i^2$ is Schur convex.

15. Suppose that $\underline{x} \leq_M \underline{y}$ and that $\underline{y} \leq_M \underline{z}$. Prove that $\underline{x} \leq_M \underline{z}$.

Chapter 3
The Lorenz Order in the Space of Distribution Functions

The graphical measure of inequality proposed by Lorenz (1905) in an income inequality context is intimately related to the concept of majorization. The Lorenz curve, however, can be meaningfully used to compare arbitrary distributions rather than distributions concentrated on n points, as is the case with the majorization partial order. The Lorenz order can, thus, be thought of as a useful generalization of the majorization order. While extending our domain of definitions in one direction, to general rather than discrete distributions, we find it convenient to add a restriction which was not assumed in Chap. 2, a restriction that our distributions be supported on the non-negative reals and have positive finite expectation. In an income or wealth distribution context the restriction to non-negative incomes is often acceptable. The restriction to distributions with finite means is potentially more troublesome. Any real world (finite) population will have a (sample) distribution with finite mean. However, a commonly used approximation to real world income distributions, the Pareto distribution, only has a finite mean if the relevant shape parameter is suitably restricted. See Arnold (2015b) for a detailed discussion of Pareto distributions in the income modelling context. To avoid distorted Lorenz curves (as alluded to in Exercise 1 and illustrated in Wold (1935)), we will hold fast to our restriction that all distributions to be discussed will be supported on $\mathbb{R}^+$ and will have positive finite means. In terms of random variables our restriction is that they be non-negative with positive finite expectations. We will speak interchangeably of our Lorenz (partial) order as being defined on the class of distributions (supported on $\mathbb{R}^+$ with positive finite means) or as being defined on the class of positively integrable non-negative random variables.

Figure 3.1 shows a photo of the American economist Max Otto Lorenz.

B. C. Arnold, J. M. Sarabia, *Majorization and the Lorenz Order with Applications in Applied Mathematics and Economics*, Statistics for Social and Behavioral Sciences,
https://doi.org/10.1007/978-3-319-93773-1_3

Fig. 3.1 Max Otto Lorenz

3.1 The Lorenz Curve

First we need to recall Lorenz's original definition of his inequality curve, and then describe a version of it suitable for our purposes, i.e. one which can be used to (partially) order non-negative integrable random variables. After rotating and rescaling Lorenz's diagram, we may describe one of his curves as follows. The Lorenz curve corresponding to a particular population of individuals is a function, say $L(u)$, defined on the interval $[0, 1]$ such that for each $u \in [0, 1]$, $L(u)$ represents the proportion of the total income of the population which accrues to the poorest $100u$ percent of the population. Associated with such a finite population of n individual incomes is a sample distribution function say $F_n(x)$ where, by definition,

$$F_n(x) = \{\text{number of individuals with income } \leq x\}/n. \tag{3.1}$$

How is Lorenz's curve related to this sample distribution function? After a little thought we realize that something is lacking in Lorenz's original definition. The curve is not well defined for every $u \in [0, 1]$, only for $u = 0, \frac{1}{n}, \frac{2}{n}, \ldots, \frac{n-1}{n}, 1$ where n is the size of the population. It is reasonable to complete the curve by linear interpolation, and that is what we shall do. If we denote the ordered individual incomes in the population by $x_{1:n}, x_{2:n}, \ldots, x_{n:n}$, then for $i = 0, 1, 2, \ldots, n$

$$L\left(\frac{i}{n}\right) = \left(\sum_{j=1}^{i} x_{j:n} \Big/ \sum_{j=1}^{n} x_{j:n}\right). \tag{3.2}$$

The points $\left(\frac{i}{n}, L(\frac{i}{n})\right)$ are then linearly interpolated to complete the corresponding Lorenz curve. Such a curve does approximate the bow shape alluded to by

Lorenz. Obvious modifications are required in (3.2) if the x_i's are not all distinct. The Lorenz curve is well defined at 0 and at a number of points equal to the number of distinct values among the x_i's. It is then completed by linear interpolation. Now any distribution function can be approximated arbitrarily closely by discrete distributions. Thus (3.2) must essentially determine a functional on the space of all distribution functions which will associate a "Lorenz curve" with each distribution in a manner consistent with Lorenz's definition of the curve for sample distribution functions. The most convenient mathematical description of that functional is of recent provenance. It was implicitly known and explicitly available in parametric form before, but seems to have not been clearly enunciated until Gastwirth (1971) supplied the following definition.

For any distribution function F we define the corresponding "inverse distribution function" or Quantile function by

$$F^{-1}(y) = \sup\{x : F(x) \leq y\}, \quad 0 < y < 1. \tag{3.3}$$

With this definition the mean, μ_F of the distribution (assumed supported on $[0, \infty)$ is given by

$$\mu_F = \int_0^1 F^{-1}(y)dy, \tag{3.4}$$

in the sense that the mean exists if and only if the Riemann integral in (3.4) converges. With this definition of F^{-1}, Gastwirth defines the Lorenz curve corresponding to the distribution F by

$$L(u) = \left[\int_0^u F^{-1}(y)dy\right] \Big/ \left[\int_0^1 F^{-1}(y)dy\right], \quad 0 \leq u \leq 1. \tag{3.5}$$

It is a straightforward matter to verify that the definition (3.5) does indeed coincide with Lorenz's original definition, i.e., (3.2) with linear interpolation, in the case of a sample distribution function corresponding to n numbers (individual incomes), since

$$F_n^{-1}(y) = x_{i:n}, \quad \frac{i-1}{n} \leq y < \frac{i}{n}.$$

The form (3.5) is especially useful since it makes transparent several important properties of Lorenz curves. A Lorenz curve is a continuous function on $[0, 1]$ with $L(0) = 0$ and $L(1) = 1$. It is non-decreasing and differentiable almost everywhere. Convexity of the Lorenz curve is obvious since the function F^{-1} is non-decreasing. Thus the general definition provided by (3.5) does give us "bow-shaped" curves, as promised by Lorenz. A Lorenz curve will always lie below the 45° line joining (0, 0) to (1, 1) (by convexity). It will coincide with the 45° line in the case of a degenerate distribution. It is evident that the Lorenz curve determines the distribution up to a scale transformation ($L'(u) = cF^{-1}(u)$ a.e. and F^{-1} determines F).

If the Lorenz curve is twice differentiable in some interval say (u_1, u_2), then the corresponding distribution has a finite positive density in the interval $(\mu_F L'(u_1+), \mu_F L'(u_2-))$ where μ_F is defined in (3.4). The density in that interval is given by

$$f(x) = \left[\mu_F L''(F(x))\right]^{-1} \tag{3.6}$$

For an example of a Lorenz curve, consider the classical Pareto distribution defined by

$$F(x) = 1 - (x/\sigma)^{-\alpha}, \quad x > \sigma \tag{3.7}$$

where $\sigma > 0$ and $\alpha > 1$ (to ensure the existence of the mean). One finds

$$F^{-1}(u) = \sigma(1-u)^{-1/\alpha}, \quad 0 < u < 1$$

and consequently, from (3.5),

$$L(u) = 1 - (1-u)^{(\alpha-1)/\alpha}, \quad 0 \le u \le 1. \tag{3.8}$$

Figure 3.2 shows the original Lorenz curve proposed by M.O. Lorenz and published in his 1905 paper (Lorenz 1905, with permission of the American Statistical Association).

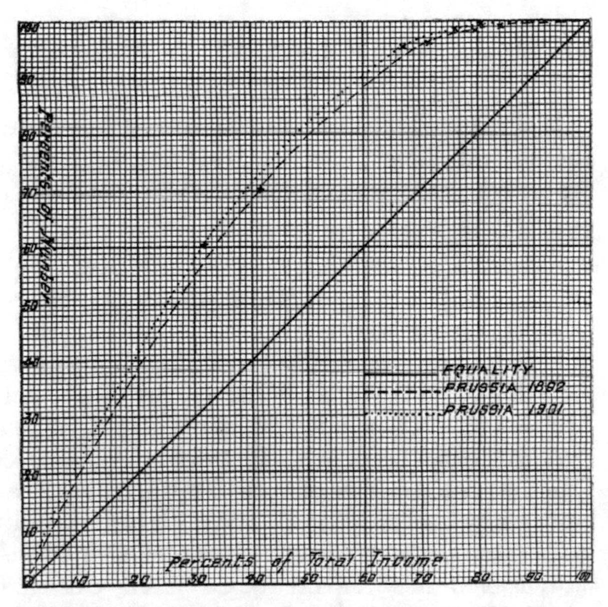

Fig. 3.2 Original Lorenz curve proposed by M.O. Lorenz in 1905

It is evident at a glance that the figures for 1901 show a greater concentration than those for 1892.

3.2 The Lorenz Order

Lorenz proposed ordering income distributions by the degree with which the Lorenz curve bow is bent; or more prosaically in terms of nested Lorenz curves. One associates a high level of inequality with a severely bent bow. The case of complete equality corresponds to the unbent bow or 45° line. We can associate an integrable non-negative random variable with any distribution function supported on $[0, \infty)$ with positive finite mean. We thus can and will discuss a Lorenz partial order on the class of all non-negative integrable random variables rather than on the class of distribution functions. For any non-negative random variable X with positive finite expectation we will denote its corresponding distribution function by F_X and its corresponding Lorenz curve (i.e., the Lorenz curve corresponding to F_X via (3.5)) by L_X. With this notation, we define the Lorenz partial order by

Definition 3.2.1 $X \leq_L Y$ (i.e., X does not exhibit more inequality in the Lorenz sense than does Y), if $L_X(u) \geq L_Y(u)$ for every $u \in [0, 1]$.

For example, suppose that X and Y have classical Pareto distributions with parameters (α_1, σ_1) and (α_2, σ_2) respectively (refer to Eq. (3.7)). Assume that $\alpha_1 < \alpha_2$. By referring to the calculated Lorenz curve (3.8), it is evident that $L_X(u) < L_Y(u)$, $\forall\ u \in (0, 1)$ and hence $X \geq_L Y$. Thus, in the case of the classical Pareto distribution an increase in the parameter α corresponds to a decrease in inequality as measured by the Lorenz ordering.

Although the Lorenz inequality ordering has achieved a remarkable degree of acceptance, especially in the economic arena, the development of related analytic theory for popular parametric families of densities has been slow. For any finite population there is no problem evaluating the Lorenz curve. For a continuous distribution an analytic expression for the Lorenz curve will rarely be available. This is true because one must first get an analytic expression for the inverse function, and then hope one can integrate it. Fortunately, it is sometimes possible to determine that Lorenz curves arc nested without explicitly deriving the Lorenz curves in question (see the example following Theorem 3.2.3 below, and also the material in Chap. 4).

Another possibility involves use of a parametric representation of the Lorenz curve. Most of the early work on Lorenz curves was done in terms of such a representation. Corresponding to a distribution function F (supported on $[0, \infty)$ with positive finite mean) we define its first moment distribution function to be

$$F^{(1)}(x) = \left[\int_0^x y dF(y)\right] \Big/ \left[\int_0^\infty y dF(y)\right]. \tag{3.9}$$

A parametric representation of the Lorenz curve is then possible, as follows. The Lorenz curve corresponding to the distribution F is the set of points $(F(x), F^{(1)}(x))$ in the unit square where x ranges from 0 to ∞ completed, if necessary, by linear interpolation. By using this representation a graphical comparison of the Lorenz curves of two distributions F and G is clearly possible. Analytically the comparison involves the quantities $F^{(1)}(x)$ and $G^{(1)}(G^{-1}(F(x)))$. A slight computational gain is observable here in that we only have to invert one of the distribution functions. Exercise 4 illustrates this technique. In this exercise it is verified that if $X \sim \Gamma(1, 1)$ and $Y \sim \Gamma(2, 1)$, then $X \geq_L Y$. In this example the distribution function for X is readily invertible while that for Y is not. The more general conclusion that if $X \sim \Gamma(k_1, 1)$ and $X \sim \Gamma(k_2, 1)$ with $k_1 < k_2$, then $X \geq_L Y$ is included in Taillie (1981).

A succinct expression for the Lorenz curve can be formed utilizing (3.9). One may write

$$L(u) = F^{(1)}(F^{-1}(u)). \tag{3.10}$$

To use this result, closed form expressions for both F^{-1} and $F^{(1)}$ are needed.

The list of parametric families of distributions for which closed form expressions are available for the corresponding Lorenz curve is remarkably short. It includes the family of classical Pareto distributions (Eq. (3.8) above), distributions uniform on finite intervals, exponential distributions (Exercise 9), and arc-sin distributions. An alternative approach is to propose parametric families of Lorenz curves (whose corresponding distribution functions are usually not simple or even available in closed form) to be used in fitting observed Lorenz curves. A simple example is

$$L(u) = 1 - (1 - u^{\beta})^{\alpha}$$

where $\alpha \in (0, 1]$ and $\beta \geq 1$. An alternative family is

$$L(u) = [1 - (1 - u)^{\alpha}]^{\beta}$$

where $0 < \alpha \leq 1 \leq \beta$.

Extensive discussion of parametric families of Lorenz curves, including the two families just described, will be found in Chap. 6.

The lognormal distribution has a remarkable representation for it Lorenz curve namely,

$$L_X(u) = \Phi(\Phi^{-1}(u) - \sigma) \tag{3.11}$$

(see Exercise 10) where Φ is the standard normal distribution function and σ is the scale parameter for $\log X$. The expression (3.11) can be used to generate other parametric families of Lorenz curves by replacing Φ by some other distribution function. It turns out that a sufficient condition for (3.11) to represent a family of Lorenz curves is that the distribution function Φ be strongly unimodal (Arnold et al. 1987). A judicious non-normal choice for Φ in (3.11) will yield the family of classical Pareto Lorenz curves, (3.8) (see Exercise 11). No matter which strongly unimodal Φ is used in (3.11), the resulting family is Lorenz ordered by σ.

A surprisingly good parametric family of Lorenz curves for fitting observed income distributions are the general quadratic curves. In many situations the observed graph $\{(u, L(u)) : 0 \leq u \leq 1\}$ is well approximated by a segment of an ellipse. Such Lorenz curves are remarkable in that it is possible to derive closed form (albeit not aesthetically pleasing) expressions for the corresponding distribution and density functions (Arnold and Villaseñor 1984). See Chap. 6 for more detailed discussion of such general quadratic curves.

Functionals of the Lorenz curve have been proposed as simple summary measures of inequality. The most popular such measure is the Gini index. The Gini index is conveniently defined as twice the area between the Lorenz curve and the 45° line (which is the Lorenz curve corresponding to an egalitarian distribution in which all individuals have identical incomes). The Pietra index is the maximal vertical deviation between the Lorenz curve and the egalitarian line. A third proposal is an index defined to be simply the length of the Lorenz curve (proposed, for example, by Amato (1968) and Kakwani (1980a)). Some alternative representations of these inequality measures are investigated in the exercises at the end of this chapter. Further discussion of inequality indices may be found in Chap. 5.

At this point it is convenient to relate the majorization partial order to the Lorenz order described in this chapter. Majorization is a partial order on n-tuples of real numbers. In the present context we restrict its domain to the non-negative orthant, i.e., sets of n non-negative numbers. Any such set of n numbers can have a (sample) distribution function associated with it (cf. Eq. (3.1)). If we use Gastwirth's definition of the Lorenz curve (3.5), we may see (using (3.2) that for two vectors $\underline{x}, \underline{y} \in \mathbb{R}_n^+$ we have $\underline{x} \leq_M \underline{y}$ if and only $\sum_{i=1}^n x_i = \sum_{i=1}^n y_i$ and $X \leq_L Y$ where the random variables X and Y are defined by

$$
\begin{aligned}
P(X = x_i) &= \frac{1}{n}, \quad i = 1, 2, \ldots, n \\
P(Y = y_i) &= \frac{1}{n}, \quad i = 1, 2, \ldots, n.
\end{aligned} \tag{3.12}
$$

Returning to Gastwirth's general definition of the Lorenz curve (3.5), it is immediately apparent that $X \leq_L Y$ if and only if $X/E(X) \leq_L Y/E(Y)$ (we have assumed non-negativity and integrability for our random variables and we explicitly exclude random variables degenerate at 0 so we can divide by expectations). Thus, the Lorenz order actually relates equivalence classes of random variables where two random variables are considered equivalent if one is a scalar multiple of the other. If we use (3.12) as a device to define a partial order on $\mathbb{R}_n^+$, conveniently called the Lorenz order and denoted by $\leq_L$, we see that $\underline{x} \leq_L \underline{y}$ if and only if the normalized vectors $\tilde{\underline{x}}, \tilde{\underline{y}}$ satisfy $\tilde{\underline{x}} \leq_L \tilde{\underline{y}}$ (where $\tilde{x}_i = x_i / \sum_{i=1}^n x_i$, $\tilde{y}_i = y_i / \sum_{i=1}^n y_i$). The relationship between the two orders $\leq_L$ and $\leq_M$ on $\mathbb{R}_n^+$ is evidently intimate but their distinct nature is exemplified by the observation that if $\sum_{i=1}^n x_i \neq \sum_{i=1}^n y_i$ then $\underline{x}$ and $\underline{y}$ cannot be related by $\leq_M$ but might be related by $\leq_L$.

In Chap. 2, we encountered some remarkable equivalent conditions for majorization in $\mathbb{R}_n$. Specifically consider Theorems 2.1.1, 2.2.5, and 2.2.6. To what extent can we extend these results to deal with the Lorenz partial order of non-negative integrable random variables (which can legitimately be viewed as an extension of majorization)? Theorems 2.2.5 and 2.2.6 extend readily. Thus

Theorem 3.2.1 *Suppose* $X \geq 0$, $Y \geq 0$ *and* $E(X) = E(Y)$. *We have* $X \leq_L Y$ *if and only if* $E(h(X)) \leq E(h(Y))$ *for every continuous convex function* $h : \mathbb{R}^+ \to \mathbb{R}$.

We have the obvious

Corollary 3.2.1 $X \leq_L Y$ *iff* $E[g(X/E(X))] \leq E[g(Y/E(Y))]$ *for every continuous convex* g.

or the sometimes more convenient

Corollary 3.2.2 $X \leq_L Y$ *iff* $E[g(E(Y)X)] \leq E[g(E(X)Y)]$ *for every continuous convex* g.

Theorem 3.2.2 *Suppose* $X \geq 0$, $Y \geq 0$ *and* $E(X) = E(Y)$. *we have* $X \leq_L Y$ *if and only if* $E[(X - c)^+] \leq E[(Y - c)^+]$ *for every* $c \in \mathbb{R}^+$.

It is not easy to track down explicit proofs of these theorems in the literature. Of course, if X and Y have only n equally likely possible values, the theorems reduce to the earlier proved Theorems 2.2.5 and 2.2.6. The general proof then follows easily by a limiting argument. Actually HLP (1929) includes the more general result. To make the correspondence one has to rewrite $E(h(X))$ as $\int_0^1 h(F_X^{-1}(u))du$ and $E(h(Y))$ as $\int_0^1 h(F_Y^{-1}(u))du$. The HLP result involves monotone rearrangements of functions. But since F_X^{-1} and F_Y^{-1} are already monotone the HLP (1929) theorem is indeed equivalent to Theorem 3.2.1 provided that, as assumed, $E(X) = \int_0^1 F_X^{-1}(u)du = \int_0^1 F_Y^{-1}(u)du = E(Y)$. See also Marshall et al. (2011, p. 22).

Theorem 3.2.1 and its corollaries suggest that a reasonable summary measure of inequality will be provided by an index of the form $E(g(X/E(X)))$ for any continuous convex g. The choice $g(x) = x^2$ leads to an ordering equivalent to that based on the coefficient of variation (see Exercise 20).

Now we turn to the possibility of extending Theorem 2.1.1 to the more general context of the Lorenz order on integrable non-negative random variables. To this end, let us first look at Theorem 2.1.1 from a slightly different perspective. Recall that the theorem states that $\underline{x} \geq_M \underline{y}$ if and only if $\underline{y} = P\underline{x}$ for some doubly stochastic matrix P. We may rewrite this as a statement involving random variables X and Y as defined in (3.12). But then what is the role of the matrix P? For discussion purposes it is convenient to assume that the coordinates of $\underline{x}$ and of $\underline{y}$ are distinct (i.e., $i \neq i'$ implies $x_i \neq x_i'$ and $y_i \neq y_i'$). On some convenient probability space construct a bivariate random variable (X, Z) where X has possible values $x_1, x_2, \ldots, x_n$ and Z has possible values $1, 2, \ldots, n$. Suppose that the joint distribution of (X, Z) is described by:

$$P(Z = i) = \frac{1}{n}, \quad i = 1, 2, \ldots, n$$

and

$$P(X = x_j|Z = i) = p_{ij}, \quad i, j = 1, 2, \ldots, n$$

where the p_{ij}'s are elements of a doubly stochastic matrix P. If denote $E(X|Z = i)$ by y_i, it is really verified that

$$y_i = \sum_{i=1}^{n} p_{ij} x_j.$$

Now $P(E(X|Z) = y_i) = P(Z = i) = \frac{1}{n}$ so that $E(X|Z) \stackrel{d}{=} Y$ (where Y is as defined in (3.12)). What we have shown is that in the context of random variables X and Y each with n equally likely distinct possible values, $X \geq_L Y$ if and only if there exist jointly distributed random variables X', Z' with $X \stackrel{d}{=} X'$ and $Y \stackrel{d}{=} E(X'|Z')$. This theorem is true in much more general settings. Restating a theorem of Strassen (1965) in terms of the Lorenz order we have:

Theorem 3.2.3 *Let X and Y be non-negative integrable random variables with $E(X) = E(Y)$. $Y \leq_L X$ if and only if there exist jointly distributed random variables X', Z' such that $X \stackrel{d}{=} X'$ and $Y \stackrel{d}{=} E(X'|Z')$.*

Proof We will prove only the easy part, referring the reader to Strassen's paper for the more difficult converse. Suppose that $Y \stackrel{d}{=} E(X'|Z')$, we claim that $Y \leq_L X'$. Obviously, $E(Y) = E(X')$ so that by Theorem 3.2.1 it will suffice to verify that $E(h(Y)) \leq E(h(X'))$ for every continuous convex h. This however is true since

$$\begin{aligned} E(h(X')) &= E(E(h(X')|Z')) \\ &\geq E(h(E(X'|Z'))) \quad \text{(Jensen's inequality)} \\ &= E(h(Y)) \end{aligned}$$

■

The conditional expectation of X given Z is an averaging of X, and our Theorem 3.2.3 merely quantifies the plausible statement that, in general, averaging will decrease inequality (as measured by the Lorenz ordering). The reverse operation is known as balayage or sweeping out. We could then phrase our theorem in the form: balayages increase inequality.

Suppose X has an exponential (λ) distribution and Y has a distribution function of the form

$$F_Y(y) = 1 - \left(1 + \frac{y}{\sigma}\right)^{-\alpha}, \quad y > 0 \tag{3.13}$$

where $\alpha > 1$ and $\sigma > 0$. Y can be said to have a Pareto (II) distribution (a translated classical Pareto distribution). We claim that $X \leq_L Y$. One approach would involve direct computation and comparison of the corresponding Lorenz curves. However, a simple observation allows us to draw the conclusion as a consequence of Theorem 3.2.3.

Suppose that X and Z are independent random variable with $X \sim \Gamma(1, \lambda^{-1})$ and $Z \sim \Gamma(\alpha, 1)$. If we define $Y = X/Z$, then by direct computation we find that Y has the Pareto II distribution given by (3.13). However, by construction $E(Y|X) = XE(1/Z)$ where X has an exponential (λ) distribution. Thus, $X = E(Y/E(Z^{-1})|X)$, and so by Theorem 3.2.3, $X \leq_L Y/E(Z^{-1})$, and by the scale invariance of the Lorenz order, $X \leq_L Y$.

Theorem 3.2.3 would be especially useful if one could derive some algorithm which, for a given pair of random variables X, Y ordered by $Y \leq_L X$, would generate the distribution of the appropriate bivariate random variable (X', Z') referred to in the theorem. Put more bluntly, how does one recognize a balayage? We will return to this problem in Chap. 4 when we discuss inequality attenuating transformations.

3.3 Exercises

1. Suppose we were to use Eq. (3.5) for a random variable which assumes negative values. How will the resulting "Lorenz curve" differ from that usually encountered? Illustrate with the case of a random variable distributed uniformly over the interval $(-1, 2)$.

2. Verify (3.4).

3. Let $L(u)$ be a Lorenz curve which for some $u^* \in (0, 1)$ satisfies $L(u^*) = u^*$. Prove that $L(u) = u, \forall\, u \in [0, 1]$.

4. Suppose $X \sim \Gamma(1, 1)$ and $Y \sim \Gamma(2, 1)$. Prove that $X \geq_L Y$. [*Hint*: Compare $F_Y^{(1)}(y)$ and $F_X^{(1)}(F_X^{-1}(F_Y(y)))$].

5. Provide a careful proof of Theorem 3.2.1 using the monotone convergence theorem and Theorem 2.2.5.

6. For any j, the function $g_j(\underline{x}) = \sum_{i=1}^{j} x_{i:n} / \sum_{i=1}^{n} x_{i:n}$ is Schur concave. Verify that $\underline{x} \leq_M \underline{y}$ iff $g_j(\underline{x}) \geq g_j(\underline{y})$, $j = 1, 2, \ldots, n$.

7. Verify that convex combinations of Lorenz curves are again Lorenz curves. Will it be easy to identify the distribution function corresponding to the combined curve?

8. Let $Z = \mu + \sigma X$ where $\mu \geq 0$ and $\sigma > 0$. Assume $X \geq 0$ and $0 < E(X) < \infty$.

   (a) Let $g(\mu, \sigma, u) = L_Z(u)$. Show that for fixed μ, $g(\mu, \sigma, u)$ is a non-decreasing function of σ. Thus, if $\sigma_1 < \sigma_2$, we have $\mu + \sigma_1 X \geq_L \mu + \sigma_2 X$.

   (b) Assume $\sigma = 1$. Investigate the relationship between the Lorenz curves of X and Z (*Hint*: $F_Z^{(-1)}(u) = \mu + E(X)L'_X(u)$).

9. (Lorenz curve for exponential distribution.) Suppose X has density $f_X(x) = \lambda e^{-\lambda x}$, $x > 0$. Determine the form of the corresponding Lorenz curve of X.

10. Suppose X has a lognormal distribution, i.e., $\log X \sim N(0, \sigma^2)$. Verify that the Lorenz curve of X has the form $L_X(u) = \Phi(\Phi^{-1}(u) - \sigma)$ where Φ is the standard normal distribution function.

11. Show that (3.8) can be thought of as being a special case of (3.11) for a suitable strongly unimodal distribution function Φ.

12. The Lorenz curve corresponding to a particular random variable X is itself a continuous distribution function with support $[0, 1]$. The mean of this distribution function, i.e., $\int_0^1 u dL_X(u)$, can be used as a summary measure of inequality of X. How is this measure related to the Gini index of X?

13. Uniform record values. Consider repeated sampling from a uniform distribution on the interval $[0, 1]$. Let $Y_1, Y_2, \ldots$ be the sequence of upper record values. Prove that $Y_i \leq_L Y_{i+1}$, $i = 1, 2, \ldots$ (see Arnold and Villaseñor 1998 for related material)

14. A Lorenz curve is symmetric if $L(1 - L(u)) = 1 - u$ for every $u \in (0, 1)$. Suppose a random variable X has mean μ and density $f(x)$. Show that its Lorenz curve is symmetric if and only if

$$\frac{f(\mu^2/x)}{f(x)} = \left(\frac{x}{\mu}\right)^3$$

   for every x for which $f(x) > 0$ (Taguchi 1968).

Certain well-known inequality indices are introduced in the next four exercises. More detailed discussion of these indices may be found in Chap. 5.

15. (Gini Index). The Gini index is defined by

$$G(X) = 2\int_0^1 (u - L(u))du.$$

   Let X_1, X_2 be i.i.d. copies of X and denote their minimum by $X_{1:2}$. Verify that

$$G(X) = E(|X_1 - X_2|)/2E(X)$$

and

$$G(X) = 1 - [E(X_{1:2})/E(X)].$$

16. (Pietra Index). The Pietra index is defined by

$$P(X) = \max_{u \in (0,1)} [u - L_X(u)].$$

Assume F_X is strictly increasing on its support and verify that the maximum is achieved when $u = F_X(E(X))$ and that the Pietra index can be expressed as

$$P(X) = E(|X - E(X)|)/2E(X).$$

17. Verify that the Pietra index is twice the area of the largest triangle which can be inscribed between the Lorenz curve and the egalitarian line.

18. Since the length of the Lorenz curve must be in the interval $(\sqrt{2}, 2)$, Kakwani (1980a), see also Amato (1968), proposed the following index of inequality.

$$K(X) = \frac{\ell_X - \sqrt{2}}{2 - \sqrt{2}}$$

where ℓ_X is the length of the Lorenz curve corresponding to F_X. Verify that

$$\ell_X = \frac{1}{E(X)} E\left(\sqrt{[E(X)]^2 + X^2}\right).$$

19. Let X have a classical Pareto distribution (Eq. (3.7)). Determine the corresponding Gini, Pietra, and Amato-Kakwani indices (defined in Exercises 15–18).

20. The coefficient of variation of X is defined by $\text{c.v.}(X) = \sqrt{\text{var}(X)}/E(X)$. Verify that if $X \leq_L Y$ then $\text{c.v.}(X) \leq \text{c.v.}(Y)$. Is the converse true?

21. Instead of considering the maximum vertical deviation between the Lorenz curve and the egalitarian line (as Pietra did), one might consider the maximum horizontal deviation. Discuss this summary measure of inequality. A third possibility is to consider the maximal distance of the Lorenz curve from the egalitarian line.

22. Using (3.5), determine the density of a random variable X whose mean is 2 and whose Lorenz curve is given by

$$L(u) = \frac{u}{9 - 8u}, \quad 0 \leq u \leq 1.$$

(Aggarwal and Singh 1984).

Chapter 4
Transformations and Their Effects

We may most easily motivate the material in the present chapter by setting it in the context of income distributions. Income distributions which exhibit a high degree of inequality (as indicated by their Lorenz curves) are generally considered to be undesirable. Consequently, there are frequent attempts to modify observed income distributions by means of intervention in the economic process. Taxation and welfare programs are obvious examples. Essentially then, we replace the original set of incomes (or, more abstractly, a vector in $\mathbb{R}_n^+$) by some function of the set of incomes. Interest centers on characterizing inequality preserving and inequality attenuating transformations. We will consider both deterministic and stochastic transformations.

4.1 Deterministic Transformations

Having motivated our task by considering vectors in $\mathbb{R}_n^+$, we make the natural extension suggested by the material in Chap. 3 and consider functions of non-negative integrable random variables with positive expectations. Thus, we seek more insight into two classes of functions mapping $\mathbb{R}^+$ into $\mathbb{R}^+$:

(1) Inequality preserving functions. g is inequality preserving if $X \leq_L Y$ implies $g(X) \leq_L g(Y)$.
(2) Inequality attenuating functions. g is inequality attenuating if for every non-negative random variable X with $0 < E(X) < \infty$ we have $g(X) \leq_L X$.

An obvious example of an inequality preserving transformation is the function $g(x) = cx$ for some $c > 0$. Are there others? The answer is yes. But there are not very many interesting ones. One may verify that the only inequality preserving transformations are those of the following three forms:

B. C. Arnold, J. M. Sarabia, *Majorization and the Lorenz Order with Applications in Applied Mathematics and Economics*, Statistics for Social and Behavioral Sciences,
https://doi.org/10.1007/978-3-319-93773-1_4

$$g_{1c}(x) = cx, \quad x \geq 0 \text{ for some } c > 0, \tag{4.1}$$

$$g_{2c}(x) = c, \quad x \geq 0 \text{ for some } c > 0, \tag{4.2}$$

and

$$g_{3c}(x) = \begin{cases} 0, & x = 0, \\ c, & x > 0 \quad \text{for some } c > 0. \end{cases} \tag{4.3}$$

The classes (4.2) and (4.3) preserve the Lorenz order by mapping the class of non-negative random variables into very restricted classes. As such, they may be considered to be of only academic interest. The proof that (4.1)–(4.3) constitutes a complete enumeration of the inequality preserving transformations involves a rather tiresome enumeration of cases (see Arnold and Villaseñor (1985) for details). A key observation in the arguments is that if $g(0) \neq 0$ then, to preserve inequality, g must be non-decreasing on $(0, \infty)$. Consequently, any inequality preserving function must be measurable. In fact, one can show they must be linear on $(0, \infty)$. Marshall et al. (2011, p. 166) show that functions on $\mathbb{R}$ which preserve majorization if measurable must be linear, but in their context, they cannot rule out nonmeasurable solutions. By restricting attention to functions on $\mathbb{R}^+$ and by asking for preservation of the Lorenz order we are able to avoid such anomalies.

What about inequality attenuating transformations? Sufficient conditions for inequality attenuation have repeatedly been discovered in the literature. In fact, the commonly quoted conditions are essentially necessary and sufficient as we shall now show. Earlier references to the conditions as necessary conditions are to be found in Marshall et al. (1967), Fishburn (1976), Kakwani (1980a), and Nygard and Sandstrom (1981). Marshall et al. (1967) are concerned with $*$-ordering (see Chap. 9) which implies Lorenz ordering. The Lorenz order is defined on random variables which are non-negative and whose expectations exist and are positive. As was remarked earlier, the positivity requirement is needed to rule out the random variable degenerate at 0. We denote this class of random variables by $\mathcal{L}$. Our theorem can then be stated as follows:

Theorem 4.1.1 *Let $g : \mathbb{R}^+ \to \mathbb{R}^+$. The following are equivalent*

(i) $g(X) \leq_L X, \forall X \in \mathcal{L}$
(ii) $g(x) > 0, \forall x > 0$, $g(x)$ *is non-decreasing on* $[0, \infty)$ *and* $g(x)/x$ *is non-increasing on* $(0, \infty)$.

Proof $(ii) \Rightarrow (i)$. Assume g satisfies (ii). Now if $X \in \mathcal{L}$ we need to verify that $g(X) \in \mathcal{L}$. Since $g(x) > 0 \, \forall x > 0$ and $E(X) > 0$, it follows that $E(g(X)) > 0$. Next, since $g(x)$ is non-decreasing on $[0, \infty)$, we have $g(X) \leq g(1)$ when $X \leq 1$. Since $g(x)/x$ is non-increasing on $(0, \infty)$, we have $g(x)/x \leq g(1)/1$ or $g(X) \leq Xg(1)$ when $X \geq 1$. Thus $g(X) \leq (X + 1)g(1)$, and hence $E(g(X)) < \infty$. Thus $g(X) \in \mathcal{L}$. To compare the Lorenz curves of X and $Y = g(X)$, it suffices to consider conveniently chosen random variables X' and Y' with $X \stackrel{d}{=} X'$ and $Y \stackrel{d}{=} Y'$. Let U

be a random variable uniformly distributed on the interval (0, 1) and let F_X be the distribution function of the random variable X. Define $X' = F_X^{-1}(U)$ and $Y' = g(F_X^{-1}(U))$ (where F_X^{-1} is defined in Eq. (3.3)). Note that both F_X^{-1} and g are non-decreasing. It follows that

$$\begin{aligned} L_Y(u) - L_X(u) &= L_{Y'}(u) - L_{X'}(u) \\ &= \int_0^u g(F_X^{-1}(v))dv \Big/ \int_0^1 g(F_X^{-1}(v))dv \\ &\quad - \int_0^u F_X^{-1}(v)dv \Big/ \int_0^1 F_X^{-1}(v)dv \\ &= \int_0^u \left[g(F_X^{-1}(v)) - F_X^{-1}(v)\frac{E(X')}{E(Y)} \right] \frac{dv}{E(Y')}. \end{aligned}$$

Since $g(x)/x$ is non-increasing on $(0, \infty)$, the last integrand is first positive then negative as v ranges from 0 to 1. Thus, the integral assumes its smallest value when $u = 1$. Since $L_Y(1) - L_X(1) = 0$, it follows that $L_Y(u) \geq L_X(u)\ \forall u \in [0, 1]$, i.e. $g(X) = Y \leq_L X$.

$(ii) \Rightarrow (i)$ Suppose g is such that there exists $x^* > 0$ with $g(x^*) = 0$. Then consider a random variable X such that $P(X = x^*) = 1$. Obviously $X \in \mathcal{L}$, but $P(g(X) = 0) = 1$, so $g(X) \notin \mathcal{L}$, and thus $g(X) \nleq_L X$.

Suppose $g(x) > 0$ for $x > 0$ but g is not non-decreasing on $[0, \infty)$. Thus, there exist x and y with $0 \leq x < y$ and $g(y) < g(x)$. Consider a random variable X such that $P(X = x) = p$, $P(X = y) = 1 - p$. There are two cases to consider.

Case 1: $x = 0$, $g(y) > 0$. Here, see Exercise 3, $g(X) \nleq_L X$ provided $p < (g(0) - g(y))/(2g(0) - g(y))$.

Case 2: $x > 0$, $g(y) > 0$. Here, see Exercise 4, $g(X) \nleq_L X$ provided $p < (\frac{y}{x} - 1)/(\frac{g(x)}{g(y)} + \frac{y}{x} - 2)$.

Finally, suppose g is non-decreasing, $g(x) > 0$ for $x > 0$ and $g(x)/x$ is not non-increasing on $(0, \infty)$. Thus, there exist x and y such that $0 < x < y$ with $0 < g(x)/x < g(y)/y$. Let X be a random variable defined by $P(X = x) = P(X = y) = 1/2$. One finds $L_{g(X)}(1/2) < L_X(1/2)$, so that $g(X) \nleq_L X$. ■

The above theorem has an attractive interpretation in terms of taxation policies. If we think of X as representing the distribution of income before taxes and $g(X)$ as representing income after taxes, then in order for our taxation policy g to be guaranteed to reduce inequality (for any pre-tax income distribution), g must satisfy conditions (ii) of Theorem 4.1.1. It must satisfy $g(x) > 0\ \forall x > 0$, i.e., everyone with some income before taxes should still have some money left after taxes. It must be monotone, i.e., if Sally earned more than Joe before taxes, her after tax income must also be more than Joe's after tax income. Finally, we must have $g(x)/x \downarrow$. But this just says that it must be a "progressive" tax which takes proportionally more from the rich than it does from the poor. Most taxation policies do satisfy these conditions and are, thus, inequality attenuating.

Another way to change income distributions is to mandate salary increases which depend on the current level of salary. Thus, if any individual with salary x is given a $100\gamma(x)\%$ increase in salary, his new salary will be $g(x) = x(1 + \gamma(x))$. We can refer to Theorem 4.1.1 in order to determine conditions on γ under which such a policy will reduce inequality. We assume $\gamma(x) \geq 0$ (i.e., we are indeed speaking of salary increases and not cuts). In order to have $g(x)/x \downarrow$, we must have $\gamma(x) \downarrow$ (i.e. bigger percentage increases go to the poorer individuals). Finally, to have $g(x) \uparrow$, we need $x(1 + \gamma(x)) \uparrow$. This means that γ cannot decrease too fast. For example, if we assume differentiability of γ (note that up to this point we have studiously avoided putting smoothness conditions on g or γ), then to have $g \uparrow$ we need

$$\gamma'(x) \geq -\frac{1 + \gamma(x)}{x}.$$

Although most taxation policies in vogue do qualify as being inequality attenuating, such is not always the case for policies dealing with salary increases. A not uncommon policy for salary increases which is not necessarily inequality attenuating is of the following type. All employees receiving less than $10,000 will be given a 15% increase, all employees earning between $10,000 and $20,000 will be given a 12% increase, and all others will be given a 10% increase. Although this is a policy involving generous increases, one can expect to hear from disgruntled employees whose previous salary was $10,050.

Analogous arguments to those used in Theorem 4.1.1 allow one to characterize inequality accentuating transformations. One finds

Theorem 4.1.2 *Let* $g : \mathbb{R}^+ \to \mathbb{R}^+$. *The following are equivalent*

(i) $X \leq_L g(X), \forall X \in \mathcal{L}$
(ii) $g(x) > 0 \; \forall x > 0$, $g(x)$ *is non-decreasing on* $[0, \infty)$ *and* $g(x)/x$ *is non-decreasing on* $(0, \infty)$.

4.2 Stochastic Transformations

What happens when we allow our transformation g to have random components? Two basic insights can guide us here. First, the introduction of extraneous randomness or noise should increase inequality and, second, averaging should decrease inequality. The second of these insights is illustrated by Theorem 3.2.3 which includes the observation that if $Y = E(X|Z)$ (or if $Y \stackrel{d}{=} E(X|Z)$), then $Y \leq_L X$.

Thus, if Y can be identified as a conditional expectation of X given some random variable Z, then $Y \leq_L X$. If Y is a conditional expectation of X, then X is said to be a balayage, dilation or sweeping out of Y. The first insight, regarding the introduction of noise, is illustrated in the misreported income example following Eq. (4.5) below. But, as a caveat, see Exercise 9.

We may rephrase our question about possibly random transformations which attenuate or accentuate inequality as follows. Let X be a non-negative random variable, and suppose that

$$Y = \psi(X, Z) \geq 0 \tag{4.4}$$

where Z is random and ψ is deterministic. Under what conditions on ψ, Z and on the joint distribution of Z and X can we conclude that necessarily $Y \leq_L X$ (attenuation) or $Y \geq_L X$ (accentuation)? We can see immediately that inequality accentuation rather than attenuation is most likely to result from transformations of the form (4.4). This is because such transformations usually involve additional randomness or noise. For example, if X is degenerate, then Y, in (4.4), is typically not degenerate so $Y \geq_L X$. A simple example in which inequality attenuation occurs is the following. If $Z \equiv X$ and $\psi(x, z) = g((x + z)/2)$ where g satisfies conditions (ii) of Theorem 4.1.1, then $Y \leq_L X$, for any $X \in \mathcal{L}$.

Let us consider a non-trivial example in which inequality accentuation obtains. Several authors have considered the following model for misreporting of income (on tax returns, for example)

$$Y = UX. \tag{4.5}$$

Here Y represents reported income, X is true income, and U is the misreporting factor. A common assumption is that U and X are independent (non-negative random variables). It follows immediately that $E(Y|X) = E(U)X$, and so, by Theorem 3.2.3, $Y \geq_L E(U)X$, whence $Y \geq_L X$. Thus, in this case, misreporting increases inequality. In fact we do not even need to assume U and X are independent. It suffices that $E(U|X) = c$. In practice, it seems reasonable to assume that, in (4.5), $U \leq 1$ (i.e., people underreport their income on their tax returns). A progressive tax (in the sense of Theorem 4.1.1), of course, is applied to reported income (not actual income). Does it still attenuate inequality? Now, post tax income is of the form

$$Y = (1 - U)X + g(UX). \tag{4.6}$$

Here $Y|U = u$ is a non-random function of X which satisfies conditions (ii) of Theorem 4.1.1. Thus, $Y|U = u \leq_L X$ for every $u \in [0, 1]$. It is tempting to conclude that this result still holds unconditionally, i.e., $Y \leq_L X$. Such an argument is a snare, however. We cannot, in general, expect to have $Y \leq_L X$ when X and Y are related by (4.6). The case of a degenerate X again provides a fly in the ointment. For even though X is degenerate, say $X \equiv 1$, Y is decidedly not degenerate, and so $Y \nleq_L X$. Continuity arguments allow us to conclude that non-degenerate counterexamples must also exist. Consequently, we cannot be sure that a progressive tax on reported income will necessarily attenuate the inequality of actual income, which is sad but true and, retrospectively, quite obvious.

It is possible to identify random transformations which necessarily accentuate inequality. For example, if U and X are independent non-negative random variables, and if g satisfies condition (ii) of Theorem 4.1.2, then $Y = g(UX) \geq_L X$. This is a direct consequence of Theorem 4.1.2, the fact that misreporting accentuates inequality and the fact that the Lorenz order is transitive. We have in this situation $X \leq_L UX \leq_L g(UX)$. A slight generalization of this observation is provided by the following theorem due to Arnold and Villaseñor (1985).

Theorem 4.2.1 *Suppose $g : \mathbb{R}_2^+ \to \mathbb{R}^+$ is such that $g(z, x)/x$ is non-decreasing in x for every z, and $g(z, x)$ is non-decreasing in x for every z. Assume that X and Z are independent non-negative random variables with $X \in \mathcal{L}$ and $g(Z, X) \in \mathcal{L}$. It follows that $X \leq_L g(Z, X)$.*

Proof Exercise 6. ■

In Theorem 4.2.1 in order to have $X \leq_L g(Z, X)$, we really only require that $g(z, x)/x$ and $g(z, x)$ be non-decreasing in x as x ranges over the set of possible values of X, and we only require this to hold for any z that is a possible value of Z (x is a possible value of X if for every $\epsilon > 0$, $P(x - \epsilon < X < x + \epsilon) > 0$). Similar "extensions" of Theorems 4.1.1 and 4.1.2 are possible. For example, in the setting of Theorem 4.1.1, we have $g(X) \leq_L X$ provided conditions (ii) hold as x ranges over the set of possible values of X.

In the misreported income scenario discussed above, instead of observing X, we observed a transformed version of X. In many scientific fields, the random variable X of interest is also not observed. What is observed is not a transformation of X but, rather, a weighted version of X. The basic reference is Rao (1965). Mahfoud and Patil (1982) provide a more recent survey of the area. Instead of observing random variables with density $f(x)$, because of the method of ascertainment (the way the data are collected), we actually observe random variables with a density proportional to $g(x)f(x)$. The function $g(x)$ is the weighting function. The special case $g(x) = x$, called size biased sampling, occurs when bigger units are more likely to be sampled than small ones. How do such weightings affect inequality as measured by the Lorenz order?

Suppose that $X \in \mathcal{L}$ and that g is a suitably measurable nonnegative function. The g-weighted version of X, denoted X_g is defined to be a random variable such that

$$P(X_g \leq x) = \int_0^x g(y)dF_X(y)/E[g(X)] \tag{4.7}$$

provided $0 < E[g(X)] < \infty$. Note that if $X \in \mathcal{L}$ then in order to have $X_g \in \mathcal{L}$ we require both $0 < E[g(X)] < \infty$ and $0 < E[Xg(X)] < \infty$.

Inequality preserving weightings will correspond to functions g for which $X \leq_L Y \Rightarrow X_g \leq_L Y_g$. Obviously a homogeneous function of the form $g(x) \equiv c > 0$ will preserve inequality. Using the basic Lemmas described in Exercises 1 and 2, it is not difficult to verify that there is very little scope for variation from homogeneity. In fact (Arnold 1986a) the only inequality preserving weightings are of the form

$$g(0) = 0, \quad x = 0$$

$$g(x) = \beta, \quad x > 0$$

where $\alpha \geq \beta > 0$ (the first step in the proof is Exercise 14).

In a similar fashion we can seek a "weighting" version of Theorem 4.1.1. Again the basic lemmas of Exercises 1 and 2 are helpful. Very few weightings are inequality attenuating. One may verify (Arnold 1986a) that $X_g \leq_L X$ for every $X \in \mathcal{L}$ if and only if

$$g(x) = \begin{cases} \alpha, \ x = 0, \\ \beta, \ x > 0 \end{cases}$$

where $\beta > 0$ and $0 \leq \alpha \leq \beta$ (the first step in the proof is Exercise 15).

4.3 Exercises

1. Suppose $0 < x_1 < x_2$ and that random variables X and Y are defined by

$$P(X = x_1) = p, \quad P(X = x_2) = 1 - p,$$
$$P(Y = x_1) = p', \quad P(Y = x_2) = 1 - p'.$$

   Show that X and Y are not Lorenz ordered except in the trivial cases when $p = p'$, $pp' = 0$ or $(1 - p)(1 - p') = 0$.

2. Suppose $0 < x$ and that random variables X and Y are defined by

$$P(X = 0) = p, \quad P(X = x) = 1 - p,$$
$$P(Y = 0) = p', \quad P(Y = x) = 1 - p'.$$

   Show that $p \leq p' \Rightarrow X \leq_L Y$.

3. Assume g is such that there exists $y > 0$ with $0 < g(y) < g(0)$. Let X be a random variable such that $P(X = 0) = p$, $P(X = y) = 1 - p$. Show that $g(X) \not\leq_L X$ for small values of p (specifically, for $p < [g(0) - g(y)]/[2g(0) - g(y)]$).

4. Assume g is such that there exist x and y with $0 < x < y$ and $0 < g(y) < g(x)$. Let X be a random variable such that $P(X = x) = p$, $P(X = y) = 1-p$. Show that $g(X) \not\leq_L X$ for large values of p (specifically, for $p > [\frac{y}{x} - 1]/[\frac{g(y)}{g(x)} + \frac{y}{x} - 2]$).

5. (Inequality attenuation in the sense of majorization.) Show that g satisfies conditions (ii) of Theorem 4.1.1 if and only if for any n and for any $\underline{x} \in \mathbb{R}_n^+$ we have

$$\left(\frac{g(x_1)}{\sum g(x_i)}, \ldots, \frac{g(x_n)}{\sum g(x_i)}\right) \leq_M \left(\frac{x_1}{\sum x_i}, \ldots, \frac{x_n}{\sum x_i}\right) \qquad (*)$$

[Note that it is possible to have a function satisfy $(*)$ for a fixed n without conditions (ii) of Theorem 4.1.1 being satisfied. For example with $n = 2$, consider $g(0) = 2$, $g(x) = 1, x \neq 0$.]

6. Prove Theorem 4.2.1.

7. Let $X \in \mathcal{L}$ and define $Y = \mu_1 + \sigma_1 X$ and $Z = \mu_2 + \sigma_2 X$ where $\mu_1, \mu_2 > 0$ and $\sigma_1, \sigma_2 > 0$. Under what circumstances can we claim $Y \leq_L Z$? This result, when X is a classical Pareto random variable, is discussed in Samuelson (1965). The Lorenz curves of Y and Z were discussed in Chap. 3, Exercise 8. Here we can use Theorems 4.1.1 and 4.1.2.

8. Suppose $X \leq_L Y$, $a > 0$, $b \geq 0$ and $E(X) = E(Y)$. Show that $aX + b \leq_L aY + b$. What happens if $E(X) \neq E(Y)$?

9. Does the addition of noise increase inequality? Suppose X, $U \in \mathcal{L}$ are independent random variables. Can we conclude that $X + U \geq_L X$? Can we conclude $X + U \leq_L X$?

10. Suppose X is a random variable with finite α'th and β'th moments. Prove that $X^\alpha \leq_L X^\beta$ if and only if $\alpha \leq \beta$.

11. Supply an example in which $X \leq_L Y$ yet $X + 1 \not\leq_L Y + 1$.

12. (Deterministic underreporting). Suppose that $U = \frac{1}{2}$ with probability one and that g is an inequality attenuating transformation. Can we conclude that for any $X \in \mathcal{L}$ we have

$$Y = (1 - U)X + g(XU) \leq_L X?$$

13. (Strong Lorenz Order). We write $X <_L Y$ if $X \leq_L Y$ and $Y \not\leq_L X$. State and prove a strong Lorenz order version of Theorem 4.1.1, i.e. give necessary and sufficient conditions on g to ensure that $g(X) <_L X$ for every non-degenerate $X \in \mathcal{L}$.

14. Suppose that for some $x_1, x_2 > 0$ we have $g(0) = \alpha$, $g(x_1) = \beta$ and $g(x_2) = \gamma$ where $\gamma > \beta$. Show that such a weighting g does not preserve the Lorenz order. (*Hint*: consider two random variables $X \leq_L Y$ where

$$P(X = 0) = P(X = x_1) = \frac{1}{2}$$

and

$$P(Y = 0) = \frac{1}{2} + \epsilon$$
$$P(Y = x_2) = \frac{1}{2} - \epsilon$$

in which $\epsilon = (\gamma - \beta)/4(\gamma + \beta)$).

15. Suppose that for $0 < x_1 < x_2$ we have $g(x_1) = \gamma_1 \neq \gamma_2 = g(x_2)$. Show that such a weighting g does not attenuate inequality. (*Hint*: consider a random variable X such that $P(X = x_1) = P(X = x_2) = \frac{1}{2}$).

Chapter 5
Inequality Measures

5.1 Introduction

In this chapter we discuss inequality measures, emphasizing their relationships with the Lorenz curve. Many of the early writers and some other more recent papers about inequality do not clearly distinguish between sample and population statistics. A "distribution" for them might refer to some genuine random variable or to the sample distribution of a finite number of observations from some population.

We however will distinguish results relative to theoretical distributions from results related to sample distributions. When we discuss population measures, we will speak of a single non-negative random variable X with cumulative distribution function $F_X(x)$ and survival function $\bar{F}_X(x) = P(X > x)$. A random variable in $\mathcal{L}$ refers to a non-negative random variable with positive and finite expectation.

When we are speaking about sample inequality measures, we will be dealing with a set of n quantities $X_1, \ldots, X_n$, together with their corresponding sample distribution function defined by

$$F_n(x) = \frac{1}{n} \sum_{i=1}^{n} I(X_i \leq x).$$

Moreover, when we discuss the distribution of sample measures of inequality, we will assume that $X_1, \ldots, X_n$ are independent and identically distributed, that is, they constitute a random sample of size n from some distribution F. In these situations, our main interest centers on inferring properties of F based on properties of the sample.

B. C. Arnold, J. M. Sarabia, *Majorization and the Lorenz Order with Applications in Applied Mathematics and Economics*, Statistics for Social and Behavioral Sciences,
https://doi.org/10.1007/978-3-319-93773-1_5

5.2 Common Measures of Inequality

5.2.1 *Seven Basic Inequality Measures*

Following Arnold (2015b), we present a list of seven of the most commonly used measures for quantifying the inequality exhibited by a random variable X. The first inequality measure in the list is often called the mean deviation though more precisely it should be labelled the mean absolute deviation,

$$I_1(X) = E(|X - \mu|) = \int_0^\infty |x - \mu| dF_X(x), \tag{5.1}$$

where $\mu = E(X)$. This measure is translation invariant,

$$I_1(X + \lambda) = I_1(X),$$

but, in contrast, $I_1(\lambda X) = \lambda I_1(X)$. This measure is frequently not easy to compute analytically.

Now, if we standardize (5.1), by dividing by the mean, we obtain the relative mean deviation

$$I_2(X) = \frac{I_1(X)}{\mu} = \frac{E(|X - \mu|)}{\mu} = \frac{1}{\mu}\int_0^\infty |x - \mu| dF(x). \tag{5.2}$$

Since, clearly, $I_2(\lambda X) = I_2(X)$ this measure is scale invariant, but not location invariant. This measure, the relative mean deviation, is actually directly related to the Lorenz curve, $L_X(u)$, since it can be shown to be proportional to the Pietra index, which will be defined in Sect. 5.3.5.

The next two measures which are frequently used to quantify inequality are the standard deviation,

$$I_3(X) = \sigma_X = \sqrt{E([X - \mu]^2)} = \left\{\int_0^\infty [x - \mu]^2 dF_X(x)\right\}^{1/2} \tag{5.3}$$

and the coefficient of variation,

$$I_4(X) = \frac{I_3(X)}{\mu} = \frac{\sqrt{E([X - \mu]^2)}}{\mu}. \tag{5.4}$$

The above four measures describe inequality or variability in terms of the average difference between one observation and the population mean and are ubiquitous in the physical sciences (physics, engineering, etc.). However, in economics and in the social sciences in general, a list of measures proposed by the Italian school are more popular and prevalent. These kinds of measures have their origin in the work of Corrado Gini and his co-workers (see David (1968) for a historical discussion). These measures quantify the variability by considering the average difference between two independent observations from the distribution.

If X is a random variable in $\mathcal{L}$ with mean μ, and if X_1, X_2 are two independent and identically distributed copies of X, then Gini's mean difference is defined as

$$I_5(X) = E(|X_1 - X_2|) = E(X_{2:2}) - E(X_{1:2}), \tag{5.5}$$

where we have used standard order statistics notation, so that $X_{1:2} = \min\{X_1, X_2\}$ and $X_{2:2} = \max\{X_1, X_2\}$.

The standardized version of (5.5) (obtained by dividing by twice the mean) is the Gini index or ratio of concentration (usually denoted by the letter G),

$$I_6(X) = G(X) = \frac{I_5(X)}{2\mu} = \frac{E(|X_1 - X_2|)}{2\mu}.$$

The Gini index has several possible interpretations and alternative ways in which it can be expressed, as we shall see as this chapter develops. Perhaps, the most popular description of this measure is one related to the area between the population Lorenz curve and the egalitarian line. However, other alternative expressions are possible. We may, for example, write

$$G(X) = 1 - \frac{E(X_{1:2})}{E(X_{1:1})} = \frac{E(X_{2:2})}{E(X_{1:1})} - 1, \tag{5.6}$$

which relates the Gini index to expectations of minimas or maximas of samples of sizes one and two. Equation (5.6) can be found in Arnold and Laguna (1977) and Dorfman (1979). Since

$$E(X_{1:2}) \leq E(X_{1:1}) \leq E(X_{2:2}),$$

Eq. (5.6) provides a quick proof that the Gini index is always in the interval [0, 1].

The next basic inequality measure presented in this section is the variance of the logarithm of X,

$$I_7(X) = \text{var}(\log X) = E([\log X]^2) - [E(\log X)]^2,$$

which is scale invariant (since $\log(\lambda X) = \log\lambda + \log X$ so that $I_7(\lambda X) = I_7(X)$, where $\lambda > 0$. For example, if $X \sim LN(\mu, \sigma^2)$ is a lognormal variable, then $\log X \sim N(\mu, \sigma^2)$ and consequently, for this distribution,

$$I_7(X) = \text{var}(\log X) = \sigma^2.$$

Values of the seven inequality measures, for the classical Pareto and lognormal distributions, respectively, are displayed in Tables 5.1 and 5.2.

Table 5.1 Seven basic inequality measures for the classical Pareto distribution

Inequality measure	Expression
Absolute mean deviation	$\frac{2\sigma(\alpha-1)^{\alpha-2}}{\alpha^{\alpha-1}}, \quad \alpha > 1$
Relative mean deviation	$\frac{2(\alpha-1)^{\alpha-1}}{\alpha^{\alpha}}, \quad \alpha > 1$
Standard deviation	$\frac{\sqrt{\alpha}\sigma}{(\alpha-1)\sqrt{\alpha-2}}, \quad \alpha > 2$
Coefficient of variation	$\frac{1}{\sqrt{\alpha(\alpha-2)}}, \quad \alpha > 2$
Gini mean difference	$\frac{2\alpha\sigma}{(\alpha-1)(2\alpha-1)}, \quad \alpha > 1$
Gini index	$\frac{1}{2\alpha-1}, \quad \alpha > 1$
Variance of logarithms	$\frac{1}{\alpha^2}, \quad \alpha > 0$

Table 5.2 Seven basic inequality measures for the lognormal distribution

Inequality measure	Expression
Absolute mean deviation	$\exp\left(\mu + \frac{\sigma^2}{2}\right)\left\{4\Phi\left(\frac{\sigma^2}{2}\right) - 2\right\}$
Relative mean deviation	$4\Phi\left(\frac{\sigma^2}{2}\right) - 2$
Standard deviation	$\exp\left(\mu + \frac{\sigma^2}{2}\right)\sqrt{\exp(\sigma^2) - 1}$
Coefficient of variation	$\sqrt{\exp(\sigma^2) - 1}$
Gini mean difference	$2\exp\left(\mu + \frac{\sigma^2}{2}\right)\left\{2\Phi\left(\frac{\sigma}{\sqrt{2}}\right) - 1\right\}$
Gini index	$2\Phi\left(\frac{\sigma}{\sqrt{2}}\right) - 1$
Variance of logarithms	σ^2

A general class of inequality measures are those of the form,

$$I_g(X) = E\left[g\left(\frac{X}{\mu}\right)\right], \tag{5.7}$$

where $E(X) = \mu$ and $g(\cdot)$ is a continuous convex function on $(0, \infty)$. This family was initially proposed by Ord et al. (1978).

Some properties of this family are:

- Scale invariance: $I_g(\lambda X) = I_g(X)$, if $\lambda > 0$
- Monotonicity with respect to the Lorenz order
- Additive decomposability, for some choices of g (see Sect. 5.4.2).

A subfamily of (5.7) corresponds to the choice,

$$g_\theta(x) = \frac{x^\theta - 1}{\theta(\theta - 1)},$$

with $\theta \in \mathbb{R}$. Such measures are known as generalized entropy indices and will be discussed in some detail in Sect. 5.4.2.

5.2.2 *Inequality Measures Based on the Concept of Entropy*

In addition to the large family of measures encompassed by (5.7), we can make use of the whole spectrum of available entropy measures as measures of inequality (see Cowell (2011), for a discussion of these measures in economic analysis). These measures are typically expressed in terms of the density function of the variable in question.

The Shannon (1948) entropy measure of a random variable X with distribution function F_X is defined by

$$H(X) = H(f_X) = -\int \log f_X(x) dF_X(x) = -E[\log f_X(X)],$$

where $f_X(x) = dF_X(x)$ is the probability density (mass) function for the absolutely continuous (discrete) distribution F_X. In particular, for a discrete random variable X we have

$$H(X) = H(f_X) = -\sum_{j=1}^{\infty} p_j \log(p_j) = -E[\log f_X(X)], \tag{5.8}$$

with $p_j = P(X = x_j)$, $j = 1, 2, \ldots$

The entropy can be viewed as a measure of disparity of the density $f_X(x)$ from the uniform density. It measures uncertainty in the sense of the utility of using $f_X(x)$ in place of the ultimate uncertainty of the uniform distribution (Good 1968). In the discrete case, the entropy is non-negative and is invariant under one-to-one transformations. In the continuous case, the entropy takes values in $(-\infty, \infty)$ but is not invariant under one-to-one transformations.

We consider two examples.

Example 5.2.1 Let X be a shifted geometric distribution with probability mass function $p_k = P(X = k) = p(1 - p)^{k-1}$, $k = 1, 2, \ldots$, with $0 < p < 1$. The Shannon entropy for this distribution is

$$H(X) = -\frac{p \log(p) + (1 - p) \log(1 - p)}{p},$$

since

$$\begin{aligned}H(X) = -E(\log p(X)) &= -E(\log p + (X-1)\log(1-p)) \\ &= -\log p - \log(1-p)(E(X)-1) \\ &= -\log p - \log(1-p)(1/p-1),\end{aligned}$$

and we obtain the result, taking into account the fact that $E(X) = \frac{1}{p}$.

Example 5.2.2 Next, let us consider a continuous random variable that has a uniform distribution in the interval $[a, b]$ with pdf $f_X(x) = \frac{1}{b-a}$ if $a \leq x \leq b$. The Shannon entropy in this case is

$$H(X) = -E(\log f(X)) = -\int_a^b \frac{1}{b-a} \log \frac{1}{b-a} dx = \log(b-a).$$

We remark that among all absolutely continuous distributions with support $[a, b]$, the maximal value of the Shannon entropy is attained by the uniform distribution in the interval $[a, b]$.

A frequently cited extension of the Shannon entropy was provided by Rényi (1961),

$$H_\lambda(X) = H_\lambda(f_X) = \frac{1}{1-\lambda} \log \left(\sum_{j=1}^{\infty} p_j^\lambda \right), \tag{5.9}$$

with, in the discrete case, $p_j = P(X = x_j)$, $j = 1, \ldots$, and in the continuous case,

$$H_\lambda(X) = H_\lambda(f_X) = \frac{1}{1-\lambda} \log \left(\int_{-\infty}^{\infty} f_X^\lambda(x) dx \right), \tag{5.10}$$

with $\lambda > 0$ and $\lambda \neq 1$. The Rényi entropy $H_\lambda(f)$ is monotonically decreasing in λ and the Shannon entropy is obtained by taking the limit as $\lambda \uparrow 1$ in (5.10).The following properties may be verified:

- The Shannon and the Rényi entropies are non-negative in the discrete case but not in the continuous case.
- The two measures of entropy are different in terms of additivity properties.

Finally we mention an alternative entropy measure that was proposed by Ord et al. (1981),

$$H_\gamma(X) = \frac{1}{\gamma} \int_0^\infty f_X(x)[1 - f_X^\gamma(x)]dx, \quad -1 < \gamma < \infty. \tag{5.11}$$

If $\gamma \to 0$ in (5.11), we obtain the Shannon entropy.

5.3 Inequality Measures Derived from the Lorenz Curve

5.3.1 The Gini Index

The Gini index is the most popular and important inequality measure. This index has a long history, dating back to Gini (1914), if not earlier. The Gini index is based on the area between the egalitarian line and the Lorenz curve. This quantity is multiplied by 2, in order to have a range of values in the interval [0, 1]. The Italian statistician, demographer and sociologist Corrado Gini is pictured in Fig. 5.1

Definition 5.3.1 The Gini index is defined as twice the area between the egalitarian line and the Lorenz curve.

Thus, if X is a random variable in $\mathcal{L}$ with Lorenz curve L_X, a formula for its Gini index, $G(X)$ or simply G if the random variable is known from the context, is

$$G(X) = 2\int_0^1 [u - L_X(u)]du = 1 - 2\int_0^1 L_X(u)du. \tag{5.12}$$

Figure 5.2 represents the egalitarian line $y = u$, the Lorenz curve $y = L(u)$, and the Gini index for a hypothetical distribution.

Example 5.3.1 Consider the family of power Lorenz curves,

$$L(u; \alpha) = u^\alpha, \quad 0 \leq u \leq 1,$$

with $\alpha \geq 1$. Using (5.12),

Fig. 5.1 Corrado Gini

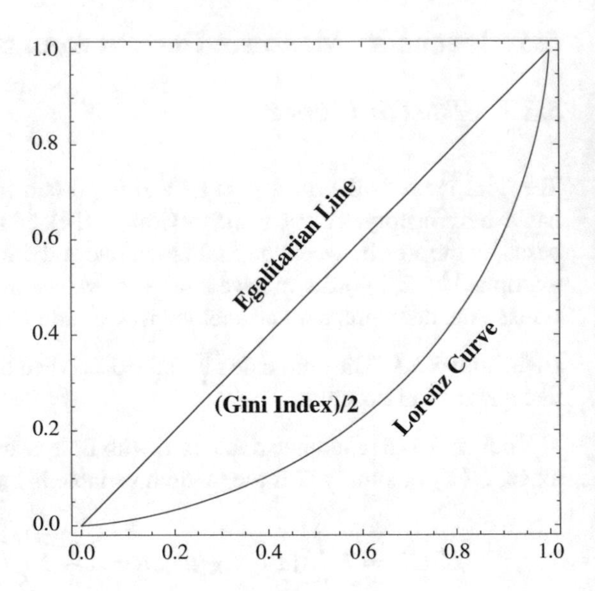

Fig. 5.2 Egalitarian line $y = u$, Lorenz curve $y = L(u)$, and Gini index

$$G = 1 - 2\int_0^1 u^\alpha du = 1 - \frac{2}{\alpha + 1} = \frac{\alpha - 1}{\alpha + 1}.$$

Consequently, if $\alpha = 1$ the Gini index is $G = 0$ and we have perfect equality while if $\alpha \to \infty$, $G \to 1$, which corresponds to perfect inequality.

The next result is a simple formula in terms of the first derivative of the Lorenz curve

Theorem 5.3.1 *The Gini index can be expressed as*

$$G(X) = 2\int_0^1 uL'_X(u)du - 1.$$

Proof The proof involves integrating by parts in (5.12). ∎

The following result is an expression for the Gini index in terms of moments of order statistics, as mentioned earlier without proof in Eq. (5.6).

Theorem 5.3.2 *The Gini index can be written as*

$$G(X) = 1 - \frac{E(X_{1:2})}{\mu} = 1 - \frac{1}{\mu}\int_0^\infty [1 - F_X(x)]^2 dx, \tag{5.13}$$

where $X_{1:2}$ is the smaller of a sample of size 2 with the same distribution as X.

Proof Using (5.12),

$$\begin{aligned}G(X) &= 1 - 2\int_0^1 L_X(u)du \\ &= 1 - \frac{2}{\mu}\int_0^1 \int_0^u F_X^{-1}(y)dydu \\ &= 1 - \frac{2}{\mu}\int_0^1 \left[\int_y^1 du\right] F_X^{-1}(y)dy \\ &= 1 - \frac{1}{\mu}E(X_{1:2}).\end{aligned}$$

■

The following result provides an expression for the Gini index in terms of a covariance.

Theorem 5.3.3 *Let X be a random variable in $\mathcal{L}$ with cdf F_X and mean μ. Then, the Gini index can be computed as*

$$G(X) = \frac{2}{\mu}cov(X, F_X(X)). \tag{5.14}$$

Proof Using formula (5.12),

$$\begin{aligned}G(X) &= 1 - 2\int_0^1 L_X(u)du \\ &= \frac{2}{\mu}\int_0^\infty xF_X(x)f_X(x)dx - 1 \\ &= \frac{2}{\mu}\left(\int_0^\infty xF_X(x)f_X(x)dx - \frac{\mu}{2}\right),\end{aligned}$$

and since $F_X(X) \sim U[0, 1]$ we obtain the result. ■

The next results provide alternative ways to compute the Gini index from the cdf and the pdf of the random variable.

Theorem 5.3.4 *Let X be a random variable in $\mathcal{L}$ with cdf F_X, pdf f_X, and mean μ. Then, the Gini index can be computed as*

$$G(X) = -1 + \frac{2}{\mu}\int_0^\infty xf_X(x)F_X(x)dx, \tag{5.15}$$

or

$$G(X) = \frac{1}{\mu}\int_0^\infty F_X(x)[1 - F_X(x)]dx.$$

5.3.2 *Generalizations of the Gini Index*

There are several generalizations and extensions of the Gini index that have been proposed in the literature.

Mehran (1976) suggested the general class of linear measures of the form

$$I_w(X) = \int_0^1 [u - L_X(u)]dw(u),$$

where $w(u)$ is some increasing function which allows value judgements about inequality to be incorporated in the measure. Note that $I_w(X)$ is always monotone with respect to the Lorenz order. If we take $w(u) = 2u$, with $0 \leq u \leq 1$, we obtain the Gini index.

An alternative generalization of the Gini index was proposed by Donaldson and Weymark (1980) and Kakwani (1980a) and was studied in detail by Yitzhaki (1983). These authors proposed a generalized Gini index defined as

$$G_\nu(X) = 1 - \nu(\nu - 1) \int_0^1 (1 - u)^{\nu-2} L_X(u)du, \tag{5.16}$$

where $\nu > 1$. If $\nu = 2$, we obtain the usual Gini index. When ν increases, higher weights are attached to small incomes. The limiting case when ν goes to infinity depends on the lowest income, and as such it would be considered to be appropriate if we accept the judgement introduced by Rawls, that social welfare depends only on the poorest society member.

For positive integer values of ν, these measures can be expressed in terms of the expectation of the minimum of a sample of size n,

$$m_n = E(X_{1:n}), \quad n = 1, 2, \ldots \tag{5.17}$$

The elements in the sequence $\{m_n\}$, $n = 1, 2, \ldots$, can be called absolute Gini indices. It can be proved that (Muliere and Scarsini 1989)

$$G_n(X) = 1 - \frac{m_n}{\mu_X} = 1 - \frac{E(X_{1:n})}{\mu_X}, \tag{5.18}$$

which can be viewed as a generalization of formula (5.13).

Arnold (1983, p. 109) has proposed the following generalization of the Gini index,

$$\tilde{G}_n(X) = 1 - \frac{E(X_{1:n+1})}{E(X_{1:n})}, \quad n = 1, 2, \ldots \tag{5.19}$$

If we set $n = 1$ in (5.19), we obtain the usual Gini index.

Each of the three sequences (5.17), (5.18), and (5.19) characterizes the underlying distribution. We have the following result.

Theorem 5.3.5 *Any $F \in \mathcal{L}$ is characterized by its sequences of:*

(i) Absolute Gini indices $m_n = E(X_{1:n})$, $n = 1, 2, \ldots$
(ii) Relative Gini indices $G_n(X)$, $n = 1, 2, \ldots$ up to a scale factor
(ii) Relative Gini indices $\tilde{G}_n(X)$, $n = 1, 2, \ldots$ up to a scale factor

Additional results about characterization of income distributions in terms of generalized Gini indices can be found in Kleiber and Kotz (2002).

Example 5.3.2 Consider $X \sim P(I)(\sigma, \alpha)$ a classical Pareto distribution. The sequences of absolute Gini indices are

$$m_n = \frac{\alpha n \sigma}{\alpha n - 1}, \quad n = 1, 2, \ldots$$

if $\alpha n > 1$, since the distribution of the minimum is again of the Pareto form, i.e., $X_{1:n} \sim \mathcal{P}(\alpha n, \sigma)$. The sequences of relative Gini indices are

$$G_n = \frac{n - 1}{\alpha n - 1}, \quad n = 2, 3, \ldots$$

if $\alpha n > 1$ and

$$\tilde{G}_n = \frac{1}{n(\alpha n + \alpha - 1)}, \quad n = 1, 2, \ldots$$

if $\alpha(n + 1) > 1$.

Example 5.3.3 Let $X \sim U[0, 1]$ a uniform distribution on the interval $[0, 1]$. It is well known that the kth order statistic of a sample of size n from such a uniform distribution has a beta distribution, i.e., $X_{k:n} \sim B(k, n - k + 1)$. Moreover, the $U[0, 1]$ distribution is characterized by the sequence of expected minima:

$$m_n = E(X_{1:n}) = \frac{1}{n + 1}, \quad n = 1, 2, \ldots$$

In addition, the sequences of relative Gini indices

$$G_n = 1 - \frac{1/(n + 1)}{1/2} = \frac{n - 1}{n + 1}, \quad n = 2, 3, \ldots$$

and

$$\tilde{G}_n = 1 - \frac{1/(n + 2)}{1/(n + 1)} = \frac{1}{n + 2}, \quad n = 1, 2, \ldots$$

characterize the $U[0, 1]$ distribution.

5.3.3 Decomposition of the Gini and Yitzhaki Indices

The decomposition of the various inequality measures is a key issue in current economic analysis. In this chapter we work with two kinds of decompositions of the target random variable X:

(a) Decompositions by subpopulations,
(b) Decompositions by factors (or sources of income).

The decomposition of type (a) based on subgroups in populations is based on the assumption that the pdf of the random variable X can be written in the form,

$$f_X(x) = p_1 f_1(x) + \cdots + p_k f_k(x),$$

where $f_i, i = 1, \ldots, k$ are k pdfs corresponding to k subgroups with weights $p_i \geq 0, i = 1, \ldots, k$ and $\sum_{i=1}^{k} p_i = 1$. That is, the whole population is composed of k groups with weights $p_1, \ldots, p_k$.

On the other hand, the decomposition by factors of the type (b) is based on the assumption that the random variable can be written as

$$X = \sum_{i=1}^{k} X_i,$$

or

$$X = \prod_{i=1}^{k} X_i,$$

where $X_1, \ldots, X_k$ are independent (or, in some cases, dependent) random variables.

Decomposition of the Gini Index by Subgroups of a Population

Here we consider decomposition of the Gini index by subgroups or subpopulations. As an example, with an economic interpretation, of this decomposition we consider modeling a regional income distribution involving several countries. This problem has been studied by Chotikapanich et al. (2007).

Assume that we have k countries and we know the income distribution in each country defined in terms of the pdf's $f_j(x)$, $j = 1, 2, \ldots, k$ and we also know the population proportions $p_1, \ldots, p_k$, where $p_j > 0$ and $\sum_{j=1}^{k} p_j = 1$.

The pdf of the regional income distribution is then given by the finite mixture,

$$f(x) = \sum_{j=1}^{k} p_j f_j(x), \quad x \geq 0. \tag{5.20}$$

Similarly, the regional cdf is

$$F(x) = \sum_{j=1}^{k} p_j F_j(x), \quad x \geq 0, \tag{5.21}$$

where $F_j(x)$ is the cdf of the jth country. The mean income is then expressible as

$$\mu = \sum_{j=1}^{k} p_j \mu_j,$$

where $\mu_j = \int_0^\infty x f_j(x)dx$ is the mean income of country j.

The regional first moment distribution (reflecting the regional cumulative income shares) is given by

$$\begin{aligned} F^{(1)}(x) &= \frac{1}{\mu} \int_0^x z f(z)dz \\ &= \frac{1}{\mu} \sum_{j=1}^{k} p_j \int_0^x z f_j(z)dx \\ &= \sum_{j=1}^{k} \frac{p_j \mu_j}{\mu} F_j^{(1)}(x), \end{aligned} \tag{5.22}$$

The regional Lorenz curve is given by (in implicit form)

$$(F(x), F^{(1)}(x)) = \left(\sum_{j=1}^{k} p_j F_j(x), \sum_{j=1}^{k} \frac{p_j \mu_j}{\mu} F_j^{(1)}(x) \right),$$

which can be numerically graphed for a grid of values of x.

The following theorem provides the expression for the regional Gini index.

Theorem 5.3.6 *The regional Gini index can be written as*

$$G = -1 + \frac{1}{\mu} \left\{ \sum_{i=1}^{k} p_i^2 m_{ii} + \sum_{i \neq j} p_i p_j m_{ij} \right\}, \tag{5.23}$$

where

$$m_{ii} = \int_0^\infty x F_i(x) f_i(x)dx = \frac{\mu_i}{2}(G_i + 1), \quad i = 1, 2, \ldots, k, \tag{5.24}$$

and

$$m_{ij} = \int_0^\infty x F_j(x) f_i(x) dx = E[X(i) F_j(X(i))], \quad i \neq j,$$

where $X(i)$ has density f_i, $i = 1, 2, \ldots, k$, and G_i, $i = 1, 2, \ldots, k$ is the Gini index of the ith country.

Proof Using formula (5.15) for the Gini index and expressions (5.20) and (5.21) we obtain

$$\begin{aligned} G &= -1 + \frac{2}{\mu} \int_0^\infty x F(x) f(x) dx \\ &= -1 + \frac{2}{\mu} \int_0^\infty x \left(\sum_{j=1}^k p_j F_j(x) \right) \left(\sum_{i=1}^k p_i f_i(x) \right) dx \\ &= -1 + \frac{2}{\mu} \sum_{j=1}^k \sum_{i=1}^k p_j p_i \int_0^\infty x F_j(x) f_j(x) dx, \end{aligned}$$

which, after rearrangement is (5.23). ■

In practice, formula (5.24) can be computed directly using the mean and the Gini index of the ith country. However, m_{ij} with $i \neq j$ is in general difficult to compute. However, it can be estimated in the following manner. We draw observations $x_i^{(h)}$, $h = 1, 2, \ldots, H$ from the pdf of the ith country, $f_i(x)$, and then compute the values $x_i^{(h)} F_j(x_i^{(h)})$, $i = 1, 2, \ldots, n$ for each draw and then compute the averages,

$$\hat{m}_{ij} = \frac{1}{H} \sum_{h=1}^H x_i^{(h)} F_j(x_i^{(h)}).$$

For large values of H (we can choose $H = 10{,}000$), the $\hat{m}_{ij}$'s will be accurate estimates of the m_{ij}'s. See Chotikapanich et al. (2007) for details.

Remark (More General Mixtures) The above discussion dealt with what were essentially k-component mixture distributions. Instead, we can consider a general mixture of distributions from a parametric family of distributions, $\{F_0(x; \alpha) : \alpha \in (-\infty, \infty)\}$, with a quite arbitrary mixing distribution denoted by G. Thus our mixed distribution is of the form

$$F(x) = \int_{-\infty}^\infty F_0(x; \alpha) dG(\alpha).$$

The corresponding first moment distribution will then be obtainable as follows:

$$\begin{aligned} F^{(1)}(x) &= \frac{1}{\mu}\int_0^x zf(z)dz \\ &= \frac{1}{\mu}\int_0^x z\int_{-\infty}^{\infty} f_0(z;\alpha)dG(\alpha) \\ &= \int_{-\infty}^{\infty}\frac{1}{\mu}\int_0^x zf_0(z;\alpha)dzdG(\alpha) \\ &= \int_{-\infty}^{\infty}\frac{\mu_\alpha}{\mu}F_0^{(1)}(x;\alpha)dG(\alpha), \end{aligned}$$

where, for each α, μ_α and $f_0(x;\alpha)$are the mean and the density, respectively, of the distribution $F_0(x;\alpha)$, and where μ and f correspond in parallel fashion to the distribution F.

The corresponding Gini index of the mixed distribution F is then available, as an expression parallel to (5.23) involving a triple integral, thus

$$G = -1 + \frac{2}{\mu}\int_0^x\int_{-\infty}^{\infty}\int_{-\infty}^{\infty} xF(x;\alpha)f(x;\beta)dG(\beta)dG(\alpha)dx.$$

Decomposition by Factors or Sources of Income

Let X be a random variable in $\mathcal{L}$, with cdf F_X and mean $\mu = E(X)$, which represents income. We have proved in (5.14) that

$$G(X) = \frac{A}{\mu} = \frac{2cov(X, F_X(X))}{\mu},$$

where A is one half of Gini's mean difference.

The next theorem by Lerman and Yitzhaki (1985) provides a decomposition of the Gini index of X, when X can be written as sum of income components or sources of income. Thus we assume that the overall income X can be written as the sum of k components $X_1, \ldots, X_k$, i.e.,

$$X = \sum_{j=1}^{k} X_j, \tag{5.25}$$

where each X_j belongs to $\mathcal{L}$ with cdf F_j and mean $\mu_j = E(X_j)$, $j = 1, 2, \ldots, k$. The X_j's may or may not be independent.

Theorem 5.3.7 *Under the previous hypothesis, (5.25), the Gini index of X can be decomposed as*

$$G(X) = \sum_{j=1}^{k} \left\{ \frac{cov(X_j, F_X(X))}{cov(X_j, F_{X_j}(X_j))} \cdot \frac{2cov(X_j, F_{X_j}(X_j))}{\mu_j} \cdot \frac{\mu_j}{\mu} \right\} \tag{5.26}$$

$$= \sum_{j=1}^{k} R_j \cdot G(X_j) \cdot S_j, \tag{5.27}$$

where R_j is the "Gini correlation" between income component j and total income, $G(X_j)$ is the usual Gini index of component j, and S_j is component j's share of total income.

Proof Since X can be written as (5.25), we can write

$$2cov(X, F_X(X)) = 2cov\left(\sum_{j=1}^{k} X_j, F_X(X)\right) = 2\sum_{j=1}^{k} cov(X_j, F_X(X)),$$

from which we obtain (5.26). ■

The next result provided by Lerman and Yitzhaki (1984) is a new decomposition for the index proposed by Yitzhaki (1983) defined in (5.16). We write this index as

$$G_\nu(X) = 1 - \nu(\nu - 1) \int_0^1 (1 - u)^{\nu-2} L_X(u) du,$$

with $\nu > 1$ and where $L_X(u)$ is the Lorenz curve of X and ν is the parameter that reflects a relative preference for equality.

Theorem 5.3.8 *If the overall income X can be written as the sum of k components $X_1, \ldots, X_k$, the Yitzhaki (1983) index of X can be decomposed as*

$$G_\nu(X) = \sum_{j=1}^{k} \left\{ \frac{cov(X_j, (1 - F_X(X))^{\nu-1})}{cov(X_j, (1 - F_{X_j}(X_j))^{\nu-1})} \cdot \frac{-\nu cov(X_j, (1 - F_{X_j}(X_j)^{\nu-1})}{\mu_j} \cdot \frac{\mu_j}{\mu} \right\}$$

$$= \sum_{j=1}^{k} R_j \cdot G_j(\nu) \cdot S_j, \tag{5.28}$$

where R_j is the "general correlation" between income component j and total income, $G_j(\nu)$ is the usual Yitzhaki index of component j, and S_j is component j's share of total income.

Proof If we substitute $(1/\mu)\int_0^u F_X^{-1}(y)dy$ for $L_X(u)$ in the definition of $G_\nu(X)$, and then interchange the order of integration, we may verify the following alternative expression for the Yitzhaki index:

$$G_\nu(X) = -\frac{\nu}{\mu} cov(X, [1 - F_X(X)]^{\nu-1}). \tag{5.29}$$

Then write $cov(X, [1 - F_X(X)]^{\nu-1}) = \sum_{j=1}^{k} cov(X_j, [1 - F_X(X)]^{\nu-1})$ in order to verify (5.28). ■

5.3.4 Inequality Indices Related to Lorenz Curve Moments

Note that the Lorenz curve can be considered to be a cumulative distribution function on [0, 1]. We can exploit this fact and employ the moments of the Lorenz curve to develop new measures of inequality.

Definition 5.3.2 Let X be a random variable in $\mathcal{L}$ with Lorenz curve $L_X(u)$. The kth Lorenz curve moment for X is defined as

$$\tilde{D}_k(X) = \int_0^1 u^k dL_X(u), \quad k = 1, 2, \ldots$$

The set of all such Lorenz curve moments uniquely determines the Lorenz curve. In addition it can be verified that all members of the family $\{\tilde{D}_k(X) : k = 1, 2, \ldots\}$ satisfy the principles of transfer and scale invariance.

Note that the range of $\tilde{D}_k(X)$ varies with k. To avoid this drawback, Aaberge (2000) has defined the modified family,

$$\begin{aligned} D_k(X) &= \frac{k+1}{k}\tilde{D}_k(X) - \frac{1}{k} \\ &= \frac{1}{k}\left\{(k+1)\tilde{D}_k(X) - 1\right\}. \end{aligned} \tag{5.30}$$

With this definition,$\{D_k(X) : k = 1, 2, \ldots\}$ is a new family of inequality measures, each having values which range over the interval [0, 1]. We have the following result (Aaberge 2000).

Theorem 5.3.9 *Let X be a random variable in $\mathcal{L}$ with Lorenz curve $L_X(u)$. Then,*

(i) The Lorenz measures of inequality $D_k(X)$, $k = 1, 2, \ldots$ exist.
(ii) The Lorenz curve L_X is characterized by the Lorenz measures of inequality $D_k(X)$, $k = 1, 2, \ldots$
(iii) The distribution of the random variable X is characterized by its mean μ and its Lorenz measures $D_k(X)$, $k = 1, 2, \ldots$

An alternative expression for $D_k(X)$ is

$$D_k(X) = (k+1)\int_0^1 u^{k-1}(u - L_X(u))du,$$

which proves that $D_1(X) = G(X)$, the Gini index.

Aaberge (2000) has also proved that D_k can be expressed in terms of the income gaps $\{g_k(x)\}_{k=1}^{\infty}$ between a unit with income x and the expected maximum income of a random sample of size $k+1$ from incomes lower than x. If $X_1, \ldots, X_{k+1}$ are iid with common distribution F_X, we have

$$g_k(x) = x - E(X_{k+1:k+1}|X_{k+1:k+1} \leq x),$$

and it can be proved that

$$D_k(X) = \frac{E(g_k(X))}{\mu},$$

verifying that $D_k(X)$ is the ratio of the mean of the income gap $g_k(X)$ to the overall mean.

Finally, it is interesting to investigate the relation between $G_k(X)$ and $D_k(X)$. It can be shown that (with k an integer)

$$G_k(X) = 1 + (k+1)\sum_{i=1}^{k}(-1)^i\binom{k}{i}\frac{i}{i+1}(1 - D_i(X)), \quad k = 1, 2, \ldots, \tag{5.31}$$

since

$$\begin{aligned} G_k(X) &= 1 - k(k+1)\int_0^1 (1-u)^{k-1}L_X(u)du \\ &= 1 + k(k+1)\sum_{i=1}^{k}(-1)^i\binom{k-1}{i-1}\int_0^1 u^{i-1}L_X(u)du, \end{aligned}$$

from which we obtain Eq. (5.31).

Aaberge (2001) has also considered the sequence

$$A_k(X) = 1 - (k+1)\int_0^1 [L_X(u)]^k du, \quad k = 1, 2, \ldots, \tag{5.32}$$

which corresponds to the moment sequence of the inverse Lorenz curve viewed as a distribution function on (0, 1). If we set $k = 1$ in (5.32), we obtain the Gini index.

Example 5.3.4 Let X be a classical Pareto distribution with Lorenz curve,

$$L_X(u) = 1 - (1-u)^{\delta}, \quad 0 \leq u \leq 1, \tag{5.33}$$

where $0 < \delta \leq 1$, and $\delta = 1 - \frac{1}{\alpha}$, where $\alpha > 1$ is the shape parameter of the classical Pareto distribution. The corresponding value for $\tilde{D}_k(X)$ is

$$\begin{aligned}\tilde{D}_k(X) &= \int_0^1 u^k dL_X(u)\\ &= \int_0^1 u^k \delta(1-u)^{\delta-1} du\\ &= \frac{\Gamma(k+1)\Gamma(\delta+1)}{\Gamma(k+\delta+1)},\end{aligned}$$

consequently the $D_k(X)$ indices are given by

$$D_k(X) = \frac{1}{k}\left\{\frac{\Gamma(k+2)\Gamma(\delta+1)}{\Gamma(k+\delta+1)} - 1\right\}. \tag{5.34}$$

If we set $k = 1$ in (5.34) we obtain

$$D_1(X) = \frac{\Gamma(3)\Gamma(\delta+1)}{\Gamma(\delta+2)} - 1 = \frac{1-\delta}{1+\delta},$$

the Gini index corresponding to (5.33), as expected.

The $A_k(X)$ indices (5.32) for this distribution are given by

$$A_k(X) = 1 - (k+1)\sum_{j=0}^{k} \frac{(-1)^j \binom{k}{j}}{\delta j + 1}, \quad k = 1, 2, \ldots.$$

Figure 5.3 presents the graphs of some indices $D_k(X)$ defined in Eq. (5.34) as functions of δ.

5.3.5 The Pietra Index

The index suggested by Pietra is based on a simple geometric characteristic of the Lorenz curve. We begin with the basic definition and two alternative representations (Pietra 1915)

Definition 5.3.3 Let X be a non-negative random variable in $\mathcal{L}$. The Pietra index is defined as the maximal vertical deviation between the Lorenz curve and the egalitarian line, that is,

$$P(X) = \max_{0 \leq u \leq 1} \{u - L_X(u)\}.$$

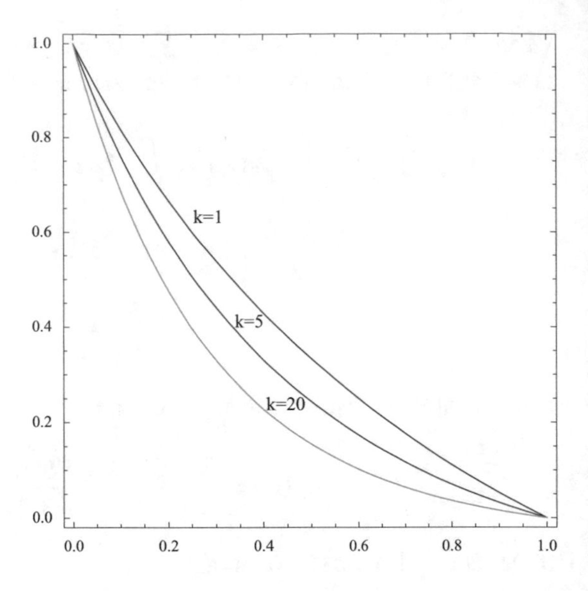

Fig. 5.3 Graphs of the indices $D_k(X)$ (Eq. (5.34)) as functions of δ for some selected values of k

We have available the basic representation of the Pietra index in terms of the first moment and the mean absolute deviation.

Lemma 5.3.1 *If $X \in \mathcal{L}$, the Pietra index can be written as*

$$P(X) = \frac{E(|X - \mu|)}{2\mu}. \tag{5.35}$$

Proof If we assume that F_X is strictly increasing on its support, the function $u - L_X(u)$ will be differentiable everywhere on $(0, 1)$ and its maximum will be reached when $1 - F_X^{-1}(x)/\mu$ is zero, that is, when $x = F_X(\mu)$. The value of $u - L_X(u)$ at this point is given by

$$P(X) = F_X(\mu) - \frac{1}{\mu}\int_0^{F_X(\mu)} [\mu - F_X^{-1}(y)]dy = \frac{1}{2\mu}\int_0^{\infty} |z - \mu| dF_X(z),$$

which is Eq. (5.35). If F_X is not strictly increasing, a limiting argument may be used. ■

It is generally accepted that an inequality measure should be monotone with respect to the Lorenz order. From this viewpoint, a wide range of possible inequality indices are given by the representation $E[g(X/\mu_X)]$, where $g(\cdot)$ is a continuous convex function such that the expectation exists. Note that from Eq. (5.35) the Pietra index admits such a representation with $g(x) = \frac{|x-1|}{2}$ (see Arnold 1987, 2012).

The following results permit one to obtain the Pietra index in relatively simple ways.

Lemma 5.3.2 *If $X \in \mathcal{L}$ with $E(X) = \mu$, the Pietra index can be written as*

$$P(X) = F_X(\mu) - L_X(F_X(\mu)). \tag{5.36}$$

Proof This follows directly since the maximum value of $u - L_X(u)$ occurs when $x = F_X(\mu)$. ■

The following lemma can also be used for the computation of the Pietra indices.

Lemma 5.3.3 *If $X \in \mathcal{L}$ with $E(X) = \mu$, the Pietra index can be written as*

$$P(X) = F_X(\mu) - F_X^{(1)}(\mu), \tag{5.37}$$

where $F_X^{(1)}$ is the first moment distribution of X.

Proof The proof is direct taking into account Lemma 5.3.2 and the following alternative expression for a Lorenz curve,

$$L_X(u) = F_X^{(1)}(F_X^{-1}(u)),$$

where $0 \leq u \leq 1$. ■

To illustrate the use of (5.36) consider

Example 5.3.5 For the classical Pareto distribution with cdf,

$$F_X(x) = 1 - \left(\frac{x}{\sigma}\right)^{-\alpha}, \quad x \geq \sigma$$

the mean is $\mu = \alpha\sigma/(\alpha - 1)$ if $\alpha > 1$ and the Lorenz curve is $L_X(u) = 1 - (1 - u)^{1-1/\alpha}$ for $0 \leq u \leq 1$. We have $F_X(\mu) = 1 - (\alpha/(\alpha - 1))^{-\alpha}$ and thus the Pietra index is

$$P(X) = F_X(\mu) - L_X(F_X(\mu)) = \frac{1}{\alpha}\left(1 - \frac{1}{\alpha}\right)^{\alpha-1}.$$

As a consequence of previous results in order to obtain the Pietra index for a non-negative random variable with finite mathematical expectation using formula (5.36), we basically need three ingredients:

- The cumulative distribution function $F_X(\cdot)$.
- The value of the mathematical expectation μ.
- The first moment distribution. $F_X^{(1)}(\cdot)$ or, alternatively, an expression of the Lorenz curve $L_X(\cdot)$.

Finite mixtures of distributions are common in economics. Assume that the data come from a mixture of distributions with cdf,

$$F(x; \boldsymbol{\pi}, \boldsymbol{\theta}) = \pi_1 F_1(x; \theta_1) + \pi_2 F_2(x; \theta_2) + \cdots + \pi_k F_k(x; \theta_k), \tag{5.38}$$

where $\pi_i \geq 0$, $i = 1, 2, \ldots, k$, $\sum_{i=1}^k \pi_i = 1$ and F_i, $i = 1, 2, \ldots, k$ are genuine cdf's. We denote by $F_i^{(1)}(x; \theta_i)$, $i = 1, 2, \ldots, k$ the first moment distributions of the component distributions. We then have

$$F^{(1)}(x;\boldsymbol{\pi},\boldsymbol{\theta}) = \pi_1 \frac{\mu_1(\theta_1)}{\mu(\boldsymbol{\pi},\boldsymbol{\theta})} F_1^{(1)}(x;\theta_1) + \pi_2 \frac{\mu_2(\theta_2)}{\mu(\boldsymbol{\pi},\boldsymbol{\theta})}$$

$$F_2^{(1)}(x;\theta_2) + \cdots + \pi_k \frac{\mu_k(\theta_k)}{\mu(\boldsymbol{\pi},\boldsymbol{\theta})} F_k^{(1)}(x;\theta_k).$$

The following theorem provides a closed expression for the Pietra index of a finite mixture of distributions.

Theorem 5.3.10 *The Pietra index of the finite mixture (5.38) is given by*

$$P(\boldsymbol{\pi},\boldsymbol{\theta}) = \sum_{i=1}^{k} \pi_i \left\{ F_i(\mu(\boldsymbol{\pi},\boldsymbol{\theta});\theta_i) - \frac{\mu_i(\theta_i)}{\mu(\boldsymbol{\pi},\boldsymbol{\theta})} F_i^{(1)}(\mu(\boldsymbol{\pi},\boldsymbol{\theta});\theta_i) \right\}, \tag{5.39}$$

where

$$\mu(\boldsymbol{\pi},\boldsymbol{\theta}) = \sum_{i=1}^{k} \pi_i \mu_i(\theta_i),$$

in which $\mu_i(\theta_i)$ the mean of the component distribution F_i, $i = 1, 2, \ldots, k$.

Proof The proof is direct using Lemma 5.3.3 and the result $\mu(\boldsymbol{\pi},\boldsymbol{\theta}) = \sum_{i=1}^{k} \pi_i \mu_i(\theta_i)$. ■

Note that this result does not quite provide us with the kind of decomposition that we would like to have. It does not provide an expression for the Pietra index of the mixture as a mixture of component Pietra indices.

Remark More general mixtures can of course be considered, of the form $F(x; G, \boldsymbol{\theta}) = \int_{-\infty}^{\infty} F_\alpha(x;\theta_\alpha) dG(\alpha)$.

5.3.6 The Palma Index and Income Share Ratios Inequality Indices

The Palma inequality index is defined as the ratio of the richest 10% of the population's share of total income divided by the poorest 40 percent's share. This index was introduced recently by the Chilean economist Gabriel Palma (2011) and can be written in terms of the Lorenz curve as

$$P_L = \frac{1 - L(1 - 0.1)}{L(0.4)}, \tag{5.40}$$

where $L(\cdot)$ represents the Lorenz curve of the income distribution.

This definition can be extended to one defined in terms of other income share ratios. For two values α, β with $0 < \alpha, \beta < 0.5$, the $(1-\alpha, \beta)$ income share ratio is defined by

$$R(1-\alpha, \beta) = \frac{1 - L(1-\alpha)}{L(\beta)}. \quad (5.41)$$

A possible reasonable choice for (α, β) is $(0.2, 0.2)$, which represents the ratio of the total income accruing to the richest 20% of the population to the total income accruing to the poorest 20%. The Palma index (5.40) corresponds to the choice $R(0.9, 0.4)$ in (5.41). See also Cobham and Sumner (2014).

5.3.7 *The Amato Index*

The Amato index describes another geometric feature of the Lorenz curve, namely its length.

Definition 5.3.4 Let X be a random variable in $\mathcal{L}$ with $E(X) = \mu$ and with Lorenz curve $L_X(u)$. The Amato inequality index, $A(X)$ is defined as the length of the Lorenz curve.

The Amato index can be written in three alternative ways,

- In terms of the Lorenz curve

$$A(X) = \int_0^1 \sqrt{1 + [L'_X(u)]^2 du} \quad (5.42)$$

- As the mathematical expectation of a convex function (see Arnold 2012)

$$A(X) = E\left[\sqrt{1 + (X/\mu)^2}\right] \quad (5.43)$$

- As an infinite numerical series. If $E(X^{2n}) < \infty, \forall n$, expanding formula (5.43) we have

$$A(X) = \sum_{n=0}^{\infty} \binom{1/2}{n} \frac{E(X^{2n})}{\mu^{2n}} \quad (5.44)$$

Formula (5.43) has been recently discussed by Arnold (2012), and it permits one to obtain the Amato index from the pdf of the random variable X.

The following example was taken from Arnold and Sarabia (2017).

Example 5.3.6 Let X_δ be a classical Pareto distribution with Lorenz curve $L_{X_\delta}(u) = 1 - (1-u)^\delta$ for $0 \leq u \leq 1$, where $\delta = 1 - 1/\alpha$, so $\delta \in (0, 1)$. Using formula (5.42), the Amato index is

Table 5.3 The Amato index for the classical Pareto distribution for some representative values of δ

δ	Amato index	δ	Amato index
0.01	1.95167	0.5	1.47894
0.05	1.84214	0.6	1.45267
0.1	1.75441	0.7	1.43440
0.2	1.64056	0.8	1.42258
0.3	1.56695	0.9	1.41615
0.4	1.51566	0.99	1.41423

$$\begin{aligned} A(X_\delta) &= \int_0^1 \sqrt{1+\delta^2(1-u)^{2(\delta-1)}}du \\ &= {}_2F_1\left[-\frac{1}{2}, \frac{\delta}{2(1-\delta)}; \frac{2-\delta}{2(1-\delta)}; -\frac{1}{\delta^2}\right], \end{aligned} \tag{5.45}$$

where ${}_2F_1[a, b; c; z]$ is the Gauss hypergeometric function, which is defined as

$${}_2F_1[a, b; c; z] = \frac{\Gamma(c)}{\Gamma(b)\Gamma(c-b)} \int_0^1 t^{b-1}(1-t)^{c-b-1}(1-tz)^{-a}dt, \tag{5.46}$$

with $c > b$. Expression (5.45) is decreasing in δ, with $lim_{\delta\to 0}A(X_\delta) = 2$ and $lim_{\delta\to 1}A(X_\delta) = \sqrt{2}$. Table 5.3 includes some values of Amato's index for the Pareto distribution as a function of $\delta = 1 - 1/\alpha$.

Other explicit expressions for the Amato index for different parent distributions can be found in Arnold and Sarabia (2017).

5.3.8 The Elteto and Frigyes Inequality Measures

Here we present three measures of inequality proposed by Elteto and Frigyes (1968), obtained by relating the average income of individuals above and below average to the overall average income.

Let X be a random variable in $\mathcal{L}$ and define

$$\begin{aligned} \mu &= \int_0^\infty xdF(x), \\ \mu_1 &= \frac{\int_0^\mu xdF(x)}{F(\mu)}, \\ \mu_2 &= \frac{\int_\mu^\infty xdF(x)}{1-F(\mu)}. \end{aligned}$$

The quantity μ is the average income, μ_1 is the average income of the poorer that average individuals, and μ_2 is the average income of the richer than average individuals. Based on these means, Elteto and Frigyes (1968) proposed the following three inequality measures

$$(U, V, W) = \left(\frac{\mu}{\mu_1}, \frac{\mu_2}{\mu_1}, \frac{\mu_2}{\mu}\right).$$

it is easily verified that the Elteto and Frigyes vector is scale invariant. In addition it admits an interpretation in terms of the Lorenz curve. Note that $V = UW$, so that we need only to interpret U and W. The component U represents the reciprocal of the slope of the line joining $(0, 0)$ to $(F(\mu), L(F(\mu)))$ on the Lorenz curve. The component W provides the slope of the line joining $(F(\mu), L(F(\mu)))$ to $(1, 1)$. Since $F(\mu)$ is the point where $u - L(u)$ is maximized, we are actually dealing with the slopes of the sides of the maximal triangle which can be inscribed within the Lorenz curve.

A simple relation between the Pietra index and the Elteto and Frigyes vector was discovered by Kondor (1971). This author observed that the Pietra index P, which is equal to twice the area of the maximal inscribed triangle, can be written as

$$P = \frac{2(U-1)(W-1)}{V-1}.$$

Example 5.3.7 For the classical Pareto distribution $P(I)(\sigma, \alpha)$ the Elteto and Frigyes inequality indices are (provided that $\alpha > 1$),

$$U = \frac{\beta^\alpha - 1}{\beta^\alpha - \beta},$$

$$V = \frac{\beta^\alpha - 1}{\beta^{\alpha-1} - 1},$$

$$W = \beta,$$

where $\beta = \alpha/(\alpha - 1)$.

5.4 The Atkinson and the Generalized Entropy Indices

Now we present three inequality indices that are very popular in modern economic analysis: the Atkinson, the generalized entropy, and the Theil indices.

5.4.1 The Atkinson Indices

The Atkinson (1970) inequality indices are defined as

$$A_\epsilon(X) = 1 - \left[E\left(\frac{X}{\mu}\right)^{1-\epsilon}\right]^{1/(1-\epsilon)}$$
$$= 1 - \left[\int_0^\infty \left(\frac{x}{\mu}\right)^{1-\epsilon} dF_X(x)\right]^{1/(1-\epsilon)}, \qquad (5.47)$$

where $\epsilon > 0$ is a parameter that controls the inequality aversion. The limiting cases when $\epsilon \to 1$ and $\epsilon \to \infty$ are

$$A_1(X) = 1 - \frac{1}{\mu}\exp\{E(\log X)\}$$
$$= 1 - \frac{1}{\mu}\exp\left\{\int_0^\infty \log(x) dF_X(x)\right\},$$

and

$$A_\infty(X) = 1 - \frac{F_X^{-1}(0)}{\mu},$$

respectively. The Atkinson indices can be written in terms of the Lorenz curve as

$$A_\epsilon(X) = 1 - \left\{\int_0^1 [L'_X(u)]^{1-\epsilon} du\right\}^{1/(1-\epsilon)}.$$

An example of the Atkinson indices for the lognormal distribution will be provided in the next section.

5.4.2 The Generalized Entropy Indices and the Theil Indices

In this section we introduce some inequality measures, which are very popular in the analysis of income inequality.

The Gini index is undoubtedly the most popular inequality index. However, quite subtle features of the income distribution can have noticeable effects on the Gini index. In fact, this index is more sensitive to transfers around the middle of the income distribution.

As an alternative, the generalized entropy indices provide a class of inequality measures which can be more sensitive to transfers near the bottom or the top of the distribution, depending on the value of the parameter which indexes these indices.

Definition 5.4.1 Let X be a positive random variable with $E(X) = \mu < \infty$ and $E(X^\theta) < \infty$, for some $\theta \in (-\infty, \infty)$, where appropriate. The generalized entropy (GE) indices are defined by

$$\mathrm{GE}_\theta(X) = \frac{1}{\theta(\theta-1)}\left[E\left(\frac{X}{\mu}\right)^\theta - 1\right], \tag{5.48}$$

if $\theta \neq 0, 1$. The limiting cases $\theta = 0$ and $\theta = 1$ are

$$\mathrm{GE}_0(X) = T_0(X) = -E\left(\log\frac{X}{\mu}\right), \tag{5.49}$$

and

$$\mathrm{GE}_1(X) = T_1(X) = E\left(\frac{X}{\mu}\log\frac{X}{\mu}\right), \tag{5.50}$$

respectively. The index (5.49) is the Theil 0 index or mean logarithmic deviation (MLD), and (5.50) is the Theil 1 coefficient.

The GE family (5.48) includes one half of the square of the coefficient of variation for $\theta = 2$:

$$\mathrm{GE}_2(X) = \frac{\sigma^2}{2\mu^2} = \frac{cv^2(X)}{2},$$

where $cv(X) = \frac{\sigma}{\mu}$ is the coefficient of variation of X.

The $\mathrm{GE}_\theta(X)$ index, for large positive values of θ is sensitive to changes in the upper tail of the distribution, typically for $\theta > 2$. If $\theta < 0$ the $\mathrm{GE}_\theta(X)$ index is sensitive to changes in the lower tail of the distribution. According to Shorrocks (1980), for practical use of $\mathrm{GE}_\theta(X)$ in empirical work, the range of values of the parameter θ should be $[-1, 2)$.

Example 5.4.1 Let us consider a lognormal distribution $X \sim LN(\mu, \sigma^2)$. We have $E(X^r) = e^{r\mu + r^2\sigma^2/2}$, so that the Atkinson indices (5.47) are

$$A_\epsilon(X) = 1 - \exp(-\epsilon\sigma^2/2), \tag{5.51}$$

and the GE indices are

$$\mathrm{GE}_\theta(X) = \frac{\exp\{\theta(\theta-1)\sigma^2/2\} - 1}{\theta(\theta-1)}, \quad \theta \neq 0, 1. \tag{5.52}$$

The MLD and Theil indices are

$$T_0(X) = T_1(X) = \frac{\sigma^2}{2}.$$

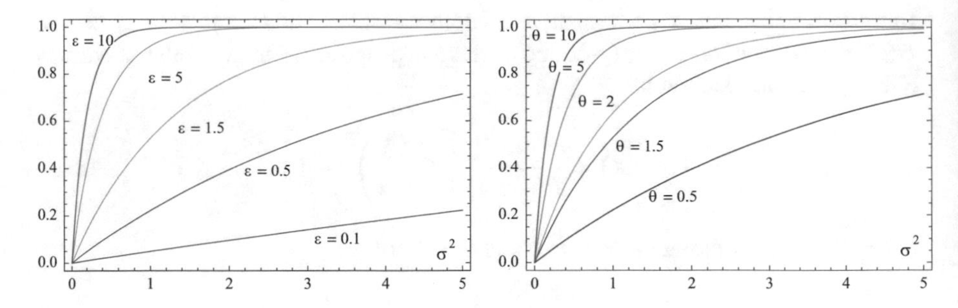

Fig. 5.4 Inequality indices for the lognormal distribution: Atkinson indices (left) and Generalized entropy indices (right) for some selected values of the parameters

Table 5.4 Inequality measures for the classical Pareto distribution

Inequality measure	Expression
Lorenz curve	$1-(1-u)^{1-1/\alpha},\quad \alpha>1$
Gini index	$\dfrac{1}{2\alpha-1},\quad \alpha>1$
Pietra index	$\dfrac{(\alpha-1)^{\alpha-1}}{\alpha^{\alpha}},\quad \alpha>1$
GE indices	$\dfrac{1}{\theta^2-\theta}\left\{\left(\dfrac{\alpha-1}{\alpha}\right)^{\theta}\dfrac{\alpha}{\alpha-\theta}-1\right\},\quad \alpha>\theta$
Atkinson indices	$1-\dfrac{\alpha-1}{\alpha}\left(\dfrac{\alpha}{\alpha+\epsilon-1}\right)^{1/(1-\epsilon)},\quad \alpha+\epsilon>0$
Theil T_0, MLD index	$-\dfrac{1}{\alpha}+\log\dfrac{\alpha}{\alpha-1},\quad \alpha>1$
Theil T_1 index	$\dfrac{1}{\alpha-1}-\log\dfrac{\alpha}{\alpha-1},\quad \alpha>1$

Figure 5.4 provides graphs of the Atkinson indices (Eq. (5.51)) and Generalized entropy indices (Eq. (5.52)) for the lognormal distribution.

Tables 5.4 and 5.5 summarize some inequality measures for the classical Pareto and lognormal distributions, respectively.

5.4.3 *Decomposability of Certain Indices*

First, we study the decomposition properties of the Theil index. Once again we consider two kinds of decompositions: by factors and by population subgroups. We begin with the decomposition by multiplicative factors (see Sarabia et al. 2017a)

Table 5.5 Inequality measures for the lognormal distribution

Inequality measure	Expression
Lorenz curve	$\Phi(\Phi^{-1}(u) - \sigma)$
Gini index	$2\Phi\left(\frac{\sigma}{\sqrt{2}}\right) - 1$
Pietra index	$2\Phi\left(\frac{\sigma^2}{2}\right) - 1$
GE indices	$\frac{\exp[\theta(\theta-1)\sigma^2/2] - 1}{\theta(\theta-1)}$
Atkinson indices	$1 - \exp(\epsilon\sigma^2/2)$
Theil T_0, MLD index	$\frac{\sigma^2}{2}$
Theil T_1 index	$\frac{\sigma^2}{2}$

Theorem 5.4.1 *Let $X_1, X_2, \ldots, X_k$ be independent positive random variables with finite expectations and consider the new random variable $X = \prod_{i=1}^{k} X_i$. Then, for the $T_0(X)$ and $T_1(X)$ indices we have*

$$T_i(X) = \sum_{j=1}^{k} T_i(X_j), \quad i = 0, 1. \tag{5.53}$$

Examples of this phenomenon include the following:

Example 5.4.2 The Pareto-lognormal distribution was defined by Colombi (1990) as the product of two independent random variables Pareto and lognormal, that is $X = X_1 \cdot X_2$, where $X_1 \sim P(I)(\sigma, \alpha)$ and $X_2 \sim LN(\mu, \sigma^2)$ are independent. The pdf of the Pareto-lognormal is

$$f(x; \alpha, \mu, \sigma) = \frac{\alpha}{x}\phi\left(\frac{\log x - \mu}{\sigma}\right) R\left(\alpha\sigma - \frac{\log x - \mu}{\sigma}\right), \quad x \geq 0,$$

where $R(z) = \frac{1-\Phi(z)}{\phi(z)}$ is Mills ratio in which $\phi(z)$ and $\Phi(z)$ denote the pdf and cdf of the standard normal distribution, respectively. Using (5.53) for $T_0(X)$ we have

$$T_0(X) = T_0(X_1) + T_0(X_2) = -\frac{1}{\alpha} + \log\frac{\alpha}{\alpha - 1} + \frac{\sigma^2}{2}.$$

A similar result holds for $T_1(X)$ using (5.53).

Example 5.4.3 Let X be a GB2 distribution with shape parameters (a, p, q) and unit scale. It is well known that X can be written as $X = \frac{X_1}{X_2}$, where X_1 and X_2 are independent generalized gamma distributions, with unit scale parameters and shape parameters (a, p) and (a, q), respectively (see chapter 6 Kleiber and Kotz 2003). The random variable X_2^{-1} is distributed as an inverted generalized gamma distribution with Theil parameter $T_0(X_2^{-1}) = \frac{\psi(q)}{a} - \log \frac{\Gamma(q-1/a)}{\Gamma(q)}$, if $q > 1/a$. Consequently,

$$\begin{aligned} T_0(X) &= T_0(X_1) + T_0(X_2^{-1}) \\ &= -\frac{\psi(p)}{a} + \log \frac{\Gamma(p+1/a)}{\Gamma(p)} + \frac{\psi(q)}{a} - \log \frac{\Gamma(q-1/a)}{\Gamma(q)}, \end{aligned}$$

which is the $T_0(X)$ index of the GB2 distribution. In a similar way we obtain an analogous result for the $T_1(X)$ index of the GB2 distribution.

The generalized entropy indices GE_θ have the important property of being additively decomposable by population groups.

According to Shorrocks (1980), an additively decomposable measure is one which can be written as a weighted sum of the inequality values computed for population subgroups plus the contribution arising from differences between subgroups means. This property responds to a question frequently discussed in the analysis of income inequality concerning the extent to which inequality in the total population can be attributed to income differences between major population subgroups.

We describe the problem in the following terms, using five assumptions. The inequality index value for a population of n individuals with distribution $\mathbf{x} = (x_1, \ldots, x_n)$ will be denoted by $I(\mathbf{x}, n)$. We assume that $I(\mathbf{x}, n)$ satisfies the following assumptions (Shorrocks 1980):

- Assumption 1: $I(\mathbf{x}, n)$ is continuous and symmetric in $\mathbf{x}$.
- Assumption 2: $I(\mathbf{x}, n) \geq 0$, with equality holding if and only if $x_i = \mu$ for all i.
- Assumption 3: (additive decomposability): Given a population of any size $n \geq 2$ and a partition into k non-empty subgroups, there exists a set of coefficients $w_i^k(\boldsymbol{\mu}, \mathbf{n})$ such that

$$I(\mathbf{x}^1, \ldots, \mathbf{x}^k, n) = \sum_{i=1}^{k} w_i^k(\boldsymbol{\mu}, \mathbf{n}) I(\mathbf{x}^i; n_i) + I(\mu_1 \mathbf{u}_{n_1}, \ldots, \mu_k \mathbf{u}_{n_k}; n), \qquad (5.54)$$

for all $\mathbf{x}^1, \ldots, \mathbf{x}^k$, with $\mathbf{x}^i = (x_{i1}, \ldots, x_{in_i})$, $i = 1, 2, \ldots, k$, and where $\boldsymbol{\mu} = (\mu_1, \ldots, \mu_k)$ is the vector of subgroup means and $w_i^k(\boldsymbol{\mu}, \mathbf{n})$ is the weight attached to subgroup i in a decomposition of k subgroups. The term

$$I(\mu_1 \mathbf{u}_{n_1}, \ldots, \mu_k \mathbf{u}_{n_k}; n),$$

is the between-group term, which we assumed to be independent of inequality within the individual subgroups. The term $\mathbf{u}_n$ represents the unit vector $(1, 1, \ldots, 1)$ with n components. The coefficients $w_i^k(\boldsymbol{\mu}, \mathbf{n})$ may vary with the vector of subgroup means $\boldsymbol{\mu}$ and subgroup populations $n = (n_1, \ldots, n_k)$, but are independent of the level of inequality within groups.

- Assumption 4: $I(\mathbf{x}, n)$ has continuous second derivatives.
- Assumption 5: $I(c\mathbf{x}, n) = I(\mathbf{x}, n)$ for all $c > 0$.

Remark In the identity (5.54), the term,

$$I(W) = \sum_{i=1}^{k} w_i^k(\boldsymbol{\mu}, \mathbf{n}) I(\mathbf{x}^i; n_i), \tag{5.55}$$

represents the inequality within-groups and the term

$$I(B) = I(\mu_1 \mathbf{u}_{n_1}, \ldots, \mu_k \mathbf{u}_{n_k}; n), \tag{5.56}$$

the inequality between-groups.

The previous five assumptions characterize the class of generalized entropy measures. We have the following theorem due to Shorrocks (1980).

Theorem 5.4.2 *$I(\boldsymbol{x}, n)$ satisfies Assumptions 1 to 5 only if it has the form*

$$I(\boldsymbol{x}, n) = \frac{A_n}{\beta(\beta - 1)} \sum_{i=1}^{n} \left[\left(\frac{x_i}{\mu} \right)^{\beta} - 1 \right], \quad \beta \neq 0, 1, \tag{5.57}$$

or

$$I(\boldsymbol{x}, n) = A_n \sum_{i=1}^{n} \log \frac{\mu}{x_i}, \tag{5.58}$$

or

$$I(\boldsymbol{x}, n) = A_n \sum_{i=1}^{n} \frac{x_i}{\mu} \log \frac{x_i}{\mu}, \tag{5.59}$$

where $A_n > 0$.

In the following theorem we describe the decomposition (5.54) in the case of random variables. The whole population is described in terms of a finite mixture of densities.

Theorem 5.4.3 *Suppose that the population is divided into k mutually exclusive groups with probabilities p_j, $j = 1, 2, \ldots, k$, that is*

$$f_X(x) = p_1 f_{X_1}(x) + \cdots + p_k f_{X_k}(x), \tag{5.60}$$

with $p_i \geq 0$, $i = 1, \ldots, k$, $\sum_{i=1}^{k} p_i = 1$ *and* $E(X_i) = \mu_i$. *Then, the overall* GE_θ *of the population can be expressed as*

$$\mathrm{GE}_\theta(X) = \sum_{j=1}^{k} p_j \left(\frac{\mu_j}{\mu}\right)^\theta \mathrm{GE}_\theta(X_j) + \frac{1}{\theta(\theta-1)}\left(\sum_{j=1}^{k} p_j \left(\frac{\mu_j}{\mu}\right)^\theta - 1\right) \tag{5.61}$$

$$= \sum_{j=1}^{k} p_j^{1-\theta} s_j^\theta \mathrm{GE}_\theta(X_j) + \frac{1}{\theta(\theta-1)}\left(\sum_{j=1}^{k} p_j^{1-\theta} s_j^\theta - 1\right) \tag{5.62}$$

where the p_j*'s are the population shares,* $\mathrm{GE}_\theta(X_j)$ *the* GE_θ *index of group* j, s_j *stands for the proportion of mean income of the group* j *in the overall mean, i.e.,*

$$s_j = \frac{p_j \mu_j}{\mu} = \frac{p_j \mu_j}{\sum_{j=1}^{k} p_j \mu_j}, \tag{5.63}$$

and $\mu = \sum_{j=1}^{k} p_j \mu_j$ *is the overall mean.*

Proof See Exercise 15. ■

Now, we consider special cases of Eqs. (5.61)–(5.62) for the Theil indices. For the MLD index (Theil 0) we have

$$T_0(X) = \sum_{j=1}^{k} p_j T_0(X_j) - \sum_{j=1}^{k} p_j \log\left(\frac{\mu_j}{\mu}\right)$$

$$= \sum_{j=1}^{k} p_j T_0(X_j) - \sum_{j=1}^{k} p_j \log\left(\frac{s_j}{p_j}\right)$$

and for the Theil 1 index,

$$T_1(X) = \sum_{j=1}^{k} p_j \frac{\mu_j}{\mu} T_1(X_j) + \sum_{j=1}^{k} p_j \frac{\mu_j}{\mu} \log\left(\frac{\mu_j}{\mu}\right),$$

$$= \sum_{j=1}^{k} s_j T_1(X_j) + \sum_{j=1}^{k} s_j \log\left(\frac{s_j}{p_j}\right).$$

The Theil entropy measures are consistent with the concept of "income weighted decomposability", which defines the decomposition coefficients in terms of incomes shares (Bourguignon 1979). See also Cowell (1980, 2011) and Sarabia et al. (2017a).

Remark In the decomposition (5.61) we can write

$$\mathrm{GE}_\theta(X) = I_\theta(W) + I_\theta(B),$$

where

$$I_\theta(W) = \sum_{j=1}^{k} p_j \left(\frac{\mu_j}{\mu}\right)^\theta \mathrm{GE}_\theta(X_j),$$

is the inequality within-groups term and

$$I_\theta(B) = \frac{1}{\theta(\theta-1)}\left(\sum_{j=1}^{k} p_j \left(\frac{\mu_j}{\mu}\right)^\theta - 1\right)$$

the inequality between groups. These terms correspond to the elements (5.58) and (5.59), respectively. The weight coefficient w_i^k in (5.54) is

$$w_i^k = p_i \left(\frac{\mu_i}{\mu}\right)^\theta.$$

5.5 Estimation with Partial Information

In many practical situations we do not have full information about income distributions. In this section we consider the problem of specification or estimation of inequality measures when we only have some form of partial information.

5.5.1 Bounds on the Gini Index

First, we consider the problem of estimating the Gini index when we have partial information about the underlying cdf F. We have the following theorem due to Gastwirth (1972).

Theorem 5.5.1 *Let F be a distribution function with support $[\underline{M}, \overline{M}]$, with $0 < \underline{M} < \overline{M}$, and mean μ, then the corresponding Gini index satisfies*

$$0 \leq G \leq \frac{(\mu - \underline{M})(\overline{M} - \mu)}{\mu(\overline{M} - \underline{M})}. \quad (5.64)$$

On the other hand, if the distribution F has a support of the type $[\underline{M}, \infty)$, it is also possible to provide a bound for the Gini index.

Theorem 5.5.2 *Let F be a distribution function with support $[\underline{M}, \infty)$ for some $\underline{M} > 0$ and mean μ, then the corresponding Gini index satisfies*

$$0 \leq G \leq \frac{\mu - \underline{M}}{\mu}. \tag{5.65}$$

Proof Bound (5.65) can be obtained from bound (5.64) taking limit when $\overline{M}$ goes to infinity. To prove (5.65) directly we observe that

$$0 \leq G = 1 - \frac{E(X_{1:2})}{E(X_{1:1})} \leq 1 - \frac{\underline{M}}{\mu}.$$

■

The following result can be useful when we have a bound for the pdf.

Theorem 5.5.3 *Let F be an absolutely continuous distribution function with support $[0, 1]$ and mean μ, whose probability density function f is bounded a.e. by C. The corresponding Gini index satisfies*

$$G \geq 1 - \frac{2}{3}\sqrt{2\mu C}. \tag{5.66}$$

The proof of this result utilizes the representation $G = 1 - (\mu_{1:2}/\mu_{1:1})$. For details, see Arnold (2015b, p. 128).

5.5.2 Parameter Identification Using the Mean and the Gini Index

Here we consider a common practical problem in economics related to the identification of the income distribution with partial information. In economics analysis it is quite common to have a few pieces of information available for the identification of the complete income distribution. Typically we are willing to assume that the distribution in question is a member of some specific parametric family. This problem has been considered and studied by Chotikapanich et al. (1997) and Jordá et al. (2014).

Consider a scenario where for any given country only the mean income μ and a numerical value G of the Gini index are available.

More specifically, assume a random variable in X in $\mathcal{L}$, where $X \sim f(x; \underline{\theta})$, where $\underline{\theta} = (\theta_1, \theta_2)^\top$ depends on two parameters. We write

$$\mu = \mu(\theta_1, \theta_2),$$
$$G = G(\theta_1, \theta_2).$$

Then, if we solve these equations for θ_1 and θ_2 we get

$$\theta_1 = h_1(\mu, G),$$
$$\theta_2 = h_2(\mu, G).$$

In this fashion, we are able to identify the density

We consider two relevant cases: the lognormal and the Pareto II distributions. Let $X \sim LN(\mu_X, \sigma_X^2)$ be a lognormal distribution with parameters μ_X and σ_X. The mean and the Gini index are

$$\mu = \exp\left(\mu_X + \frac{\sigma_X^2}{2}\right),$$
$$G = 2\Phi\left(\frac{\sigma_X}{\sqrt{2}}\right) - 1.$$

If we solve for μ_X and σ_X we get the equations

$$\mu_X = \log\mu - \frac{\sigma_X^2}{2},$$
$$\sigma_X = \sqrt{2}\Phi^{-1}\left(\frac{G+1}{2}\right).$$

Now, let us consider a Pareto II distribution with cdf

$$F(x) = 1 - \frac{1}{(1 + x/\sigma)^{\alpha}}, \quad x \geq 0.$$

If $\alpha > 1$ the mean and the Gini index are

$$\mu = \frac{\sigma}{\alpha - 1},$$
$$G = \frac{\alpha}{2\alpha - 1},$$

and solving for α and σ we get the equations,

$$\alpha = \frac{G}{2G - 1},$$
$$\sigma = \frac{\mu(1 - G)}{2G - 1}.$$

5.6 Moment Distributions

The moment distributions plays an important role in many discussions about inequality.

An important survey about this topic is Hart (1975). See also Butler and McDonald (1987, 1989). Hart pointed out that many common measures of inequality have a simple interpretation in terms of moments of moment distributions.

We define the rth moment distribution of a non-negative random variable X with cumulative distribution function F_X by

$$F_X^{(r)}(x) = \frac{\int_0^x t^r dF_X(t)}{E(X^r)}, \tag{5.67}$$

if $E(X^r)$ exists.

The first moment distribution has already been introduced in Chap. 3. Further discussion of moment distributions (for bounded random variables) will be found in Chap. 9.

For the moments of the moment distribution we use the notation,

$$\mu_k^{(r)} = \int_0^\infty x^k dF_X^{(r)}(x), \tag{5.68}$$

$$(\sigma^{(r)})^2 = \mu_2^{(r)} - \{\mu_1^{(r)}\}^2. \tag{5.69}$$

If we set $r = 0$ in (5.68) and (5.69), we obtain the moments and the variance of the original distribution, respectively. Using the definition (5.67), we have the following simple formula,

$$\mu_k^{(r)} = \frac{E(X^{k+r})}{E(X^r)} = \frac{\mu_{k+r}^{(0)}}{\mu_r^{(0)}}. \tag{5.70}$$

Thus the moments of the moment distributions are simple functions of the moments of the parent distribution F_X.

For a non-negative random variable X, since $\log E(X^\gamma)$ is a convex function of γ, if $\alpha < \beta$ then any $\delta > 0$

$$\frac{\mu_{\alpha+\delta}^{(0)}}{\mu_\alpha^{(0)}} \le \frac{\mu_{\beta+\delta}^{(0)}}{\mu_\beta^{(0)}}.$$

The quantities $\mu_1^{(1)}$ and $\mu_1^{(1)}/\mu_1^{(0)}$ have been proposed as candidate measures of concentration. The quantity,

$$\tau_H = 1 - \frac{\mu_1^{(0)}}{\mu_1^{(1)}} = 1 - \frac{(EX)^2}{E(X^2)}, \tag{5.71}$$

is a linear function of the reciprocal of the Herfindahl index. Note that $0 \leq \tau_H \leq 1$ and $\tau_H = 0$ if the underlying distribution is degenerate, which corresponds to the case of complete equality. Moreover, formula (5.71) can also be written in terms of the coefficient of variation cv as

$$\tau_H = \frac{cv^2}{1 + cv^2}.$$

The Lorenz curve and various associated inequality measures can also be related to moment distributions.

We saw in Chap. 3 that the Lorenz curve can be written as

$$L_X(u) = F_X^{(1)}(F_X^{-1}(u)), \quad 0 \leq u \leq 1.$$

The Gini index is directly related to moment distributions since we can write it in the form

$$G(X) = 1 - 2\int_0^\infty F_X^{(1)}(x) dF_X(x).$$

Also, it will be recalled that the Pietra index can be expressed as $P(X) = F_X(\mu) - F_X^{(1)}(\mu)$. Other simple inequality measures admit formulas in terms of the moment distributions. For example, the first quintile can be written as

$$F_X^{(1)}(F_X^{-1}(0.2)).$$

Example 5.6.1 Let $X \sim P(I)(\sigma, \alpha)$, i.e., a classical Pareto distribution. The rth moment distribution is again of the Pareto form (provided that $r < \alpha$). If we denote by $X^{(r)}$ a random variable with the r'th moment distribution of X, we have

$$X^{(r)} \sim P(I)(\sigma, \alpha - r), \quad \alpha > r. \tag{5.72}$$

Using (5.70),

$$\mu_k^{(r)} = E([X^{(r)}]^k) = \frac{E(X^{k+r})}{E(X^r)} = \frac{\alpha\sigma^{k+r}}{\alpha - k - r} \cdot \frac{\alpha - r}{\alpha\sigma^r} = \frac{(\alpha - r)\sigma^k}{\alpha - k - r},$$

if $\alpha > k + r$, which can also be obtained directly using (5.72). Using (5.71), we have a simple expression for the Herfindal index of a Pareto distribution, specifically $\tau_H = (\alpha - 1)^{-2}$ if $\alpha > 2$.

5.7 Relations Between Inequality Measures

In general, it is difficult to interpret the significance of any relationships which exist between the different inequality measures. However, for completeness we will mention some of these relations.

First, consider the Pietra index P and the Gini index G, which are related in the sense that an inscribed triangle cannot have an area which exceeds that of the figure within which it is inscribed.

Taguchi (1968) extended this result. Using geometric reasoning he proved,

$$P \leq G \leq P(2 - P).$$

This result was rediscovered by Moothathu (1983).

On the other hand, Glasser (1961) observed that the Gini index G and the coefficient of variation I_4 are related by

$$G \leq \frac{I_4}{\sqrt{3}}.$$

An interesting relation exists between the Shannon entropy and the moments of the distribution. First, it can be shown that given $E|X|^k$,

$$H(X) \leq \frac{1}{k} \log \frac{2^k e \Gamma(1/k) E(|X|^k)}{k^{k-1}}, \quad k > 0, \tag{5.73}$$

(Wyner and Ziv 1969). The equality in (5.73) is attained by the maximum entropy distribution

$$f(x) = c(\eta, k) \exp(-\eta |x|^k),$$

where the parameter η is identifiable as the Lagrange multiplier corresponding to the constraint $E(|X|^k) < \infty$. If we set $k = 2$ in (5.73), we get

$$\frac{e^{2H(X)}}{2\pi e} \leq \mathrm{var}(X). \tag{5.74}$$

The ratio in Eq. (5.74) is the entropy power fraction proposed by Shannon (1948) for comparison of continuous random variables. Equality in Eq. (5.74) is attained if X is the normal distribution. Ebrahimi et al. (1999) have provided significant insights about entropy and its relation to higher moments (and the variance) by approximating the pdf using a Legendre series expansion.

Other interesting relations concern the effect of the shape of the income distribution (in terms of the skewness and kurtosis coefficients) on the Theil measures $T_0(X)$ and $T_1(X)$ (see Maasoumi and Theil 1979). We consider the log-income variable

$Y = \log X$, and we define $E(Y) = \mu$, $\mathrm{var}(Y) = \sigma^2$ and we consider the skewness and kurtosis coefficients of Y, i.e., $\gamma_1 = E(Y-\mu)^3/\sigma^3$ and $\gamma_2 = E(Y-\mu)^4/\sigma^4 - 3$.

Maasoumi and Theil (1979) have proved that

$$T_0(X) = \frac{\sigma^2}{2}\left[1 + \frac{1}{2}\gamma_1\sigma + \frac{1}{12}\gamma_2\sigma^2 + o(\sigma^2)\right],$$

and

$$T_1(X) = \frac{\sigma^2}{2}\left[1 + \frac{2}{3}\gamma_1\sigma + \frac{1}{4}\gamma_2\sigma^2 + o(\sigma^2)\right].$$

As a consequence, if the log-income distribution has positive skewness with a long tail on the right, both measures $T_0(X)$ and $T_1(X)$ exceed the lognormal Theil index $\frac{\sigma^2}{2}$ and the excess is twice as large for $T_1(X)$ as it is for $T_0(X)$. In a similar way (see Maasoumi and Theil 1979), when the log-income distribution is leptokurtic ($\gamma_2 > 0$), both measures $T_0(X)$ and $T_1(X)$ exceed the lognormal value $\frac{\sigma^2}{2}$.

5.8 Sample Versions of Analytic Measures of Inequality

In this section we study distributional properties of various sample measures of inequality. We will adopt a convention of using a subscripted T to denote the sample version of the corresponding population measure of inequality denoted by the correspondingly subscripted τ. Exceptions are the sample Gini, Amato, Bonferroni, Zenga, Atkinson, and Generalized Entropy indices, which will be denoted by G, A, B, Z, A_θ, and G_β.

5.8.1 Absolute and Relative Mean Deviation and the Sample Pietra Index

We begin with the mean absolute deviation

$$T_1(\underline{X}) = \frac{1}{n}\sum_{i=1}^{n}|X_i - \bar{X}|,$$

which can also be written in terms of the sample df as,

$$T_1^{(n)}(\underline{X}) = \int_0^\infty |x - \bar{X}|dF^{(n)}(x)$$

where $\bar{X}$ is the sample mean. A third convenient representation is provided by Gastwirth (1974)

$$T_1^{(n)}(\underline{X}) = \frac{2}{n}\left(N\bar{X} - \sum_{X_i<\bar{X}} X_i \right),$$

where N is the random number of observations which are less than $\bar{X}$. If the observations X_i's are i.i.d. from a distribution $F(x)$ with finite mean μ, $T_1^{(n)}(\underline{X})$ will provide an asymptotic unbiased estimator of $\tau_1(F)$. In the normal case, the exact distribution of $T_1^{(n)}$ has been obtained by Godwin (1945).

Provided that the parent distribution F has a continuous density in a neighborhood of its mean μ and a finite variance σ^2, $T_1^{(n)}(\underline{X})$ is asymptotically normally distributed as (Gastwirth 1974)

$$\sqrt{n}\left[T_1^{(n)}(\underline{X}) - \tau_1(F)\right] \xrightarrow{d} N(0, \sigma_1^2)$$

where

$$\sigma_1^2 = 4\left\{F^2(\mu)\int_{\mu}^{\infty}(x-\mu)^2 dF(x) - \bar{F}^2(\mu)\int_{-\infty}^{\mu}(x-\mu)^2 dF(x)\right\} - \tau_1^2(F). \tag{5.75}$$

The asymptotic variance (5.75) is not in general easy to compute, but for some parent distributions a closed form for it is available. If the X_i's are exponential variables, i.e., $X_i \sim \Gamma(1, \sigma)$, it may be verified that (5.75) takes the simple form

$$\sigma_1^2 = 4\sigma^2(2e^{-1} - 4e^{-2}).$$

In order to have a scale invariant inequality measure we consider the relative mean deviation statistic,

$$T_2(\underline{X}) = \frac{T_1(\underline{X})}{\bar{X}}.$$

The sample Pietra index is expressible as

$$P(\underline{X}) = \frac{1}{2}T_2(\underline{X}),$$

that is, one-half the relative mean deviation. The advantage of the Pietra index is that its range is in the interval [0, 1], but the value 1 is not achievable. In fact, we have

$$0 \leq P(\underline{X}) \leq \frac{n-1}{n}.$$

The small sample behavior of the relative mean deviation is intractable. However, large sample theory can be developed.

If the parent distribution has a finite variance σ^2 and a continuous density in a neighborhood of its mean μ, we may conclude that $T_2(\underline{X})$ is asymptotically normal with distribution (Gastwirth 1974),

$$\sqrt{n}\left[T_2(\underline{X}) - \tau_2(F)\right] \xrightarrow{d} N(0, \sigma_2^2)$$

where

$$\sigma_2^2 = \frac{\sigma_1^2}{\mu^2} + \frac{\sigma^2 \tau_2^2(F)}{\mu^2}\left[F(\mu)\sigma^2 - \int_{-\infty}^{\mu}(x-\mu)^2 dF(x)\right],$$

where σ_1^2 is given in (5.75).

5.8.2 *The Sample Amato and Bonferroni Indices*

The sample Amato index corresponds to the length of the sample Lorenz curve and admits the representation,

$$A_n(\underline{X}) = \frac{1}{n}\sum_{i=1}^{n}\sqrt{1 + \left(\frac{X_i}{\bar{X}}\right)^2}.$$

The extremal income configurations are $(1, 1, \ldots, 1)$ and $(0, 0, \ldots, 0, 1)$, and the range of possible values of the Amato index is

$$\sqrt{2} \leq A_n(\underline{X}) \leq 1 - \frac{1}{n}\sqrt{1 + \frac{1}{n^2}}.$$

The asymptotic normality of this index is readily confirmed. Thus (see Lombardo 1979),

$$\sqrt{n}\left[A_n(\underline{X}) - A(X)\right] \xrightarrow{d} N(0, \sigma_A^2)$$

where $A(X) = E(\sqrt{1 + (X/\mu)^2})$ and

$$\sigma_A^2 = \frac{1}{2}\left(\frac{\sigma_X^2}{\mu_X^2} - \frac{\sigma_X^4}{16\mu_X^4}\right).$$

The sample version of the Bonferroni index can be written as

$$B_n(\underline{X}) = 1 - \frac{1}{n-1}\sum_{i=1}^{n-1}\frac{m_i}{m},$$

where $m_i = (1/i)\sum_{j=1}^{i} X_{j:n}$ and $m = (1/n)\sum_{j=1}^{n} X_{j:n}$. Again, the asymptotic normality of this index can be confirmed using standard techniques.

The sample version of the Zenga II index is

$$Z_n^{II}(\underline{X}) = 1 - \frac{1}{n}\sum_{i=1}^{n}\frac{(1/i)\sum_{k=1}^{i} X_{k:n}}{(1/(n-i))\sum_{k=i+1}^{n} X_{k:n}}. \tag{5.76}$$

The asymptotic normality of the Zenga II index (5.76) has been verified by Greselin and Pasquazzi (2009).

5.8.3 *The Sample Standard Deviation and Coefficient of Variation*

The sample standard deviation is

$$T_3(\underline{X}) = \sqrt{\frac{1}{n-1}\sum_{i=1}^{n}(X_i - \bar{X})^2}.$$

This measure is location invariant and its sampling properties are well-known in the normal case.

If the parent distribution has a finite fourth moment, we have

$$\sqrt{n}\left[T_3(\underline{X}) - \tau_3(F)\right] \xrightarrow{d} N(0, \sigma_3^2),$$

where

$$\tau_3(F) = \sqrt{\text{var}(X)} \text{ and } \sigma_3^2 = \frac{E(X^4) - (E(X^2))^2}{E(X^2)}.$$

If we consider the corresponding standardized measure we obtain the sample coefficient of variation

$$T_4(\underline{X}) = \frac{T_3(\underline{X})}{\bar{X}}.$$

We may verify

$$0 \leq T_4(\underline{X}) \leq \sqrt{n-1}$$

Asymptotic normality of $T_4(\underline{X})$ occurs provided that the fourth moment exists. We have

$$\sqrt{n}\left[T_4(\underline{X}) - \tau_4(F)\right] \xrightarrow{d} N(0, \sigma_4^2),$$

where $\tau_4(F) = \tau_3(F)/\mu$ and

$$\sigma_4^2 = \frac{\mu^2\{E(X^4) - \mu^2\} - 4\{[E(X^2)]^3 - \mu E(X^2)E(X^3)\}}{4\mu^4\{E(X^2) - \mu^2\}},$$

in which, as usual, $\mu = E(X)$.

5.8.4 *Gini's Mean Difference*

Gini's mean difference is defined as

$$T_5(\underline{X}) = \frac{1}{n(n-1)} \sum_{i=1}^{n} \sum_{j \neq i} |X_i - X_j|, \tag{5.77}$$

which is the average of all the pairwise differences between the X_i's. A simple variant of (5.77) is

$$T_5'(\underline{X}) = \frac{1}{n^2} \sum_{i=1}^{n} \sum_{j \neq i} |X_i - X_j|. \tag{5.78}$$

One of the advantages of (5.78) is that when it is used to define a sample Gini index, the sample index will be two times the area of the concentration polygon defined by the sample Lorenz curve.

An alternative representation of (5.77) in terms of the spacings is (David 1968),

$$T_5(\underline{X}) = \frac{2}{n(n-1)} \sum_{i=1}^{n} i(n-i)\{X_{i+1:n} - X_{i:n}\}.$$

One relevant property of (5.77) is its unbiasedness, and in fact, it is a U statistics in the sense of Hoeffding (1948), and it is the unique unbiased estimate of the population Gini mean difference $\tau_5(X) = E(X_{2:2}) - E(X_{1:2})$. The variance of T_5 is (Nair 1936; Lomnicki 1952),

$$\mathrm{var}(T_5(\underline{X})) = \frac{1}{n(n-1)}[4\xi^2 + 4(n-2)\lambda - 2(2n-3)\tau_5^2], \tag{5.79}$$

where

$$\xi^2 = \frac{1}{2}E(X_1 - X_2)^2, \tag{5.80}$$

$$\lambda = E(|X_1 - X_2||X_1 - X_3|), \tag{5.81}$$

$$\tau_5 = E(|X_1 - X_2|). \tag{5.82}$$

We have the following special cases (Nair 1936)

- Exponential distribution, if $X_i \sim \Gamma(1, \sigma),\ i = 1, 2, \ldots, n$.

$$E(T_5(\underline{X})) = \sigma,$$

$$\text{var}(T_5(\underline{X}) = \frac{(4n-2)\sigma^2}{3n^2 - 2n}$$

- Uniform distribution, if $X_i \sim U(0, \sigma),\ i = 1, 2, \ldots, n$.

$$E(T_5(\underline{X})) = \frac{\sigma}{3},$$

$$\text{var}(T_5(\underline{X}) = \frac{(n+3)\sigma^2}{45(n^2 - n)}.$$

If we consider a classical Pareto case, that is, if $X_i \sim P(I)(\sigma, \alpha),\ i = 1, 2, \ldots, n$, the quantities (5.80)–(5.82) are given by

$$\xi^2 = \frac{\sigma^2\alpha}{(\alpha-1)^2(\alpha-2)},$$

$$\tau_5 = \frac{2\sigma\alpha}{(\alpha-1)(2\alpha-1)},$$

$$\lambda = \frac{\sigma^2(8\alpha^3 - 11\alpha^2 + 2\alpha)}{(\alpha-1)^2(\alpha-2)(2\alpha-1)(3\alpha-2)},$$

which when substituted in (5.79) provide the variance of $T_5(\underline{X})$. In an alternative approach, the quantity λ can instead be computed as (see Fraser 1957)

$$\lambda = \int_0^\infty \left[\int_0^\infty (u - v)dF(v)\right] dF(u).$$

In the exponential case the exact distribution of $T_5(\underline{X})$ can be derived by considering again the representation in terms of the spacings,

$$T_5(\underline{X}) = \frac{2}{n(n-1)} \sum_{i=1}^{n-1} i(n-i)Y_{i+1:n}, \tag{5.83}$$

where $Y_{i:n} = X_{i:n} - X_{i-1:n}$ represents the ith spacing. In the exponential case, the spacings are again exponentially distributed $Y_{i:n} \sim \Gamma(1, \frac{\sigma}{n-i+1})$, $i = 2, 3, \ldots, n$, and are independent. Consequently (5.83) can be seen to be a linear combinations of independent exponential variables. Then, using the result given in Feller (1971), if the X_i's are i.i.d. with a common exponential distribution with scale parameter σ, the probability density function of the Gini mean difference is

$$f_{T_5(\underline{X})}(x) = \frac{n}{2\sigma(n-2)!} \sum_{i=1}^{n-1} \left[\prod_{j \neq i} \left(\frac{1}{j} - \frac{1}{i} \right) \right] e^{-n(n-1)x/2i\sigma}, \tag{5.84}$$

with $x \geq 0$.

Alternatively, the asymptotic distribution of the Gini mean difference can be obtained taking into account that it is a U statistic and using the results on the limiting distribution of such statistics stated by Hoeffding (1948). Provided second moments exist, it follows that

$$\sqrt{n} \left[T_5(\underline{X}) - \tau_5(F) \right] \stackrel{d}{\rightarrow} N(0, \sigma_5^2),$$

where

$$\sigma_5^2 = 4(\lambda - \tau_5^2(F)),$$

and λ is defined in (5.81).

5.8.5 *The Sample Gini Index*

The sample Gini index is a scale invariant version of Gini's mean difference. The two most commonly used versions of the sample Gini index are

$$G_n = T_6(\underline{X}) = \frac{1}{2n(n-1)\bar{X}} \sum_{i=1}^{n} \sum_{j \neq i} |X_i - X_j|, \tag{5.85}$$

and

$$G_n' = \frac{1}{2n^2 \bar{X}} \sum_{i=1}^{n} \sum_{j=1}^{n} |X_i - X_j|. \tag{5.86}$$

The Gini index can also be written in terms of a ratio of linear functions of order statistics as

$$G_n = \frac{\sum_{i=1}^{n} (2i - n - 1) X_{i:n}}{\sum_{i=1}^{n} (n-1) X_{i:n}}.$$

Since the area between the sample Lorenz curve and the egalitarian line cannot exceed $(n-1)/2n$, and since the sample Gini index defined in (5.85) satisfies,

$$0 \leq G_n \leq 1,$$

G_n cannot be interpreted as twice the area between the egalitarian line and the sample Lorenz curve. The extremal data configurations are $(1, 1, \ldots, 1)$ and $(0, 0, \ldots, 1)$. In contrast, as mentioned following Eq. (5.78), G_n' can be so interpreted. Despite this observation, the form G_n remains the more commonly used of the two candidate versions of the sample Gini index.

We do have a closed form expression for the density of G_n, for a fixed value of n in the exponential case. If $Y_{i:n} = X_{i:n} - X_{i-1:n}$ are the spacings of the order statistics of an exponential sample, then $(n-i+1)Y_{i:n}$ are i.i.d. exponential random variables. If we define

$$T = \sum_{i=1}^{n} X_i = \sum_{i=1}^{n} (n-i+1) Y_{i:n},$$

it follows that G_n can be written as a linear combination of the coordinates of a symmetric Dirichlet $(1, 1, \ldots, 1)$ random vector. Specifically, we can write

$$G_n = \sum_{j=1}^{n} \frac{n-j}{n-1} D_j,$$

where $(D_1, \ldots, D_n)$ has a Dirichlet distribution. If $X_i \sim \Gamma(1, \sigma)$, $i = 1, 2, \ldots, n$, Dempster and Kleyle (1968) obtained the following expression for the distribution function of G_n

$$P(G_n \leq x) = \frac{x^{n-1}}{\prod_{i=1}^{n} c_i} - \sum_{j=1}^{n-1} \frac{[(x-c_j)_+]^{n-1}}{c_j \prod_{k \neq j}^{n-1} (c_k - c_j)}, \quad 0 \leq x \leq 1,$$

where $c_j = (n-j)/(n-1)$ and $(u)_+ = u$ if $u > 0$ and is zero otherwise. An alternative expression for this distribution is

$$P(G_n \leq x) = 1 - \sum_{j=1}^{n-1} \frac{[(c_j - x)_+]^{n-1}}{c_j \prod_{k \neq j}^{n-1} (c_j - c_k)}, \quad 0 \leq x \leq 1.$$

If we consider a general income distribution F with a finite second moment, the asymptotic normality of the sample Gini index (5.85) can be verified. The proof is based on the asymptotic joint normality of Gini's mean difference and the sample mean. We have

$$\sqrt{n}\,[G_n - G(F)] \xrightarrow{d} N(0, \sigma_6^2),$$

where

$$\sigma_6^2 = \frac{\lambda}{\mu^2} - \frac{\rho\tau_5}{\mu^3} + \frac{\xi^2}{2\mu^4},$$

and λ, τ_5 and ξ^2 are defined in (5.80)–(5.82), $\mu = E(X)$ and

$$\rho = \int_0^\infty \int_0^\infty x|x-y|dF(x)dF(y).$$

5.8.6 Sample Lorenz Curve

For the definition of the sample Lorenz curve we have two alternatives. The first one is

$$L_n(u) = \frac{\int_0^u F_n^{-1}(y)dy}{\int_0^1 F_n^{-1}(y)dy}, \tag{5.87}$$

with $0 \le u \le 1$, where F_n is the sample distribution function.. This formulation corresponds to a polygonal line joining the $n+1$ points (0, 0) and

$$\left(\frac{j}{m}, \frac{\sum_{i\le j} X_{i:n}}{\sum_{i=1}^n X_{i:n}}\right), \quad j = 1, 2, \ldots, n.$$

The second of the sample Lorenz curve definitions is the one considered by Taguchi (1968), Gail and Gastwirth (1978), and Chandra and Singpurwalla (1978). It is given by the expression,

$$L_n'(u) = \frac{\sum_{i\le[nu]} X_{i:n}}{n\bar{X}}, \quad 0 \le u \le 1,$$

where $[\cdot]$ denotes the integer part. This second definition is discontinuous at the points $X_{i:n}$ and flat between these points. Goldie (1977) has pointed out that asymptotically both definitions are equivalent because

$$\sup_{0\le u\le 1} |L_n(u) - L_n'(u)| = \frac{X_{n:n}}{n\bar{X}},$$

and the quotient $X_{n:n}/n\bar{X}$ converges almost surely to 0, provided that the first moment of the X_i's exists. Since both definitions are asymptotically indistinguishable and since L_n defined in (5.87) is consistent with the usual definition of Lorenz curve for a general distribution function F, Goldie recommended use of (5.87) as a definition.

If we assume finite second moments, then we can reasonably hope to have asymptotic normality of suitable scaled deviations $\{L_n(u) - L(u)\}$, since such is the case for deviations between sample and population distributions. Following Goldie's notation, we define the Lorenz process λ_n as

$$\lambda_n = \sqrt{n}(L_n - L) \tag{5.88}$$

Weak convergence of these processes to a limiting normal process is described in the following result of Goldie (1977). See the Goldie paper for several alternative sufficient conditions that might be invoked instead to guarantee the weak convergence of the process.

Theorem 5.8.1 *Let F be a continuous distribution function with a finite second moment and connected support S with $\sup S = \infty$. Assume there exists $a < 1$, $t_0 < \infty$, and $A < \infty$ such that*

$$\frac{F^{-1}(1-(vt)^{-1})}{F^{-1}(1-t^{-1})} \leq Av^a, \quad v \geq 1, \ t > t_0.$$

Then, if we let λ_n denote the sample Lorenz process (5.88) based on a sample of size n, we have λ_n converging weakly in the space of continuous functions on $[0, 1]$ to a normal process λ. The limiting normal process may be expressed in the form

$$\lambda(u) = \frac{1}{m}\int_0^1 [L(u) - I(t \leq u)]\beta(t)dF^{-1}(t), \quad 0 \leq u \leq 1,$$

where $I(t \leq u) = 1$ if $t \leq u$ and 0 otherwise and β is a Brownian bridge, that is, a normal process with $E[\beta(u)] = 0$, $\forall u$ and $cov(\beta(u), \beta(v)) = u(1-v)$ if $u \leq v$.

Gail and Gastwirth (1978) have proved the following result, for the variant definition of the Lorenz process. Note that the theorem provides the asymptotic distribution of $L_n'(u)$ for one fixed value of u.

Theorem 5.8.2 *Let F correspond to a positive random variable with finite second moment. Assume that F has a unique uth quantile x_u (i.e., $F(x_u) = u$) and that F is continuous at x_u. It follows that $[L_n'(u) - L(u)]/\sqrt{\mathrm{var} L_n'(u)}$ converges in distribution to a standard normal variable. Equivalently we may write*

$$\sqrt{n}[L_n'(u) - L(u)] \xrightarrow{d} N(0, \sigma^2_{\lambda(u)})$$

where

$$\sigma^2_{\lambda(u)} = \frac{\tau_1^2}{\mu^2} + \frac{\tau_2^2\eta^2}{\mu^4} - \frac{2\tau_{12}\eta}{\mu^3}, \tag{5.89}$$

in which

$$\mu = \int_0^\infty x dF(x),$$

$$\eta = \int_0^{x_u} x dF(x),$$

$$\tau_1^2 = 2\int_0^{F^{-1}(u)} \left[\int_0^{F^{-1}(y)} F(x)dx\right][1 - F(y)]dy, \tag{5.90}$$

$$\tau_2^2 = \int_0^{\infty} (x - m)^2 dF(x),$$

$$\tau_{12} = \tau^2 + \int_0^{F^{-1}(u)} F(x)dx \int_{F^{-1}(u)}^{\infty} [1 - F(y)]dy$$

Equation (5.89) for the asymptotic variance can be difficult to evaluate. In the exponential case with $X_i \sim \Gamma(1, \sigma),\ \ i = 1, 2, \ldots, n$, it can be evaluated. Gail and Gastwirth (1978) have obtained in the exponential case,

$$\sigma_{\lambda(u)}^2 = 2(1 - u)\log(1 - u) + u(2 - u) - [u + (1 - u)\log(1 - u)]^2.$$

5.8.7 Bias of the Sample Lorenz Curve and Gini Index

Inequality measures are often underestimated using sample data. It has been noted that the sample Lorenz curve often exhibits less inequality than does the population Lorenz curve. This fact suggests that the sample curve is a positively biased estimate of the population curve. In this context, Arnold and Villaseñor (2015) have provided several sufficient conditions for such positive bias.

If we have a sample $X_1, X_2, \ldots, X_n$ of size n from a distribution $F_X(x)$, recall that the corresponding sample Lorenz curve is defined to be a linear interpolation of the points $(0, 0)$ and $(j/n, \sum_{i=1}^{j} X_{i:n}/\sum_{i=1}^{n} X_{i:n}),\ j = 1, 2, \ldots, n$. As usual denote the sample Lorenz curve by $L_n(u)$.

The function

$$L^*(u) = E(L_n(u)),\ \ 0 \le u \le 1$$

is a valid Lorenz curve corresponding to a discrete random variable $\widetilde{X}$ defined by $P(\widetilde{X} = E(X_{j:n}/\sum_{i=1}^{n} X_{i:n})) = 1/n$. Note that the function $L^*(u)$ is a linear interpolation of the points $(0, 0)$ and $(j/n, E(\sum_{i=1}^{j} X_{i:n}/\sum_{i=1}^{n} X_{i:n})),\ j = 1, 2, \ldots, n$. Consequently if we wish to show that $\widetilde{X} \le_L X$, it will be sufficient to verify that $L^*(j/n) \ge L_X(j/n),\ j = 1, 2, \ldots, n$. This is true since $L_X(u)$ is a convex function and $L^*(u)$ is piecewise linear.

In general, it is quite challenging to derive an analytic expression for

$$L^*\left(\frac{j}{n}\right) = E\left(\frac{\sum_{i=1}^{j} X_{i:n}}{\sum_{i=1}^{n} X_{i:n}}\right).$$

There is one case in which the computation is straightforward. It is the case in which the common distribution of the X_i's is an exponential distribution, i.e., $X_i \sim \Gamma(1, \beta)$. Without loss of generality we may assume that $\beta = 1$. In this case, we do find that

$$L^*\left(\frac{j}{n}\right) > L_X\left(\frac{j}{n}\right),$$

confirming the positive bias of the sample Lorenz curve. In general, a sufficient condition for positive bias of the sample Lorenz curve is provided in the following theorem (Arnold and Villaseñor 2015).

Theorem 5.8.3 *Suppose that $X_1, X_2, \ldots, X_n$ are i.i.d. random variables with a common distribution F_X with the property that*

$$E\left(\frac{\sum_{i=1}^{j} X_{i:n}}{\sum_{i=1}^{n} X_{i:n}}\right) \geq \frac{\sum_{i=1}^{j} \mu_{i:n}}{\sum_{i=1}^{n} \mu_{i:n}}, \quad j = 1, 2, \ldots, n,$$

where $\mu_{i:n} = E(X_{i:n})$. In such a case, the corresponding sample Lorenz curve is a positively biased estimate of the population Lorenz curve, i.e., $L^*(u) = E(L_n(u)) \geq L_X(u)$ for all $u \in [0, 1]$.

Note: It is reasonable to ask whether sample Lorenz curves are *always* positively biased estimates of $L(u)$. Arnold and Villaseñor (2015) provide a simple example to show that indeed it is actually possible for the sample Lorenz curve to be negatively biased. The example that they provide deals with a sample of size 2 with common distribution given by

$$P(X = 1) = p \text{ and } P(X = c) = 1 - p,$$

where $p \in (0, 0.5)$ and $c > 1$. For many choices of c and p the resulting sample Lorenz curve is negatively biased (one particular choice that will lead to this phenomenon is $c = 3$ and $p = 0.1$).

Next, following Arnold and Villaseñor (2015), we study the bias of the sample Gini index. In the case of the sample Gini index, we consider the two frequently used versions

$$G_n = \frac{\sum_{i=1}^{n}(2i - n - 1)X_{i:n}}{\sum_{i=1}^{n}(n - 1)X_{i:n}}$$

and

$$G'_n = \frac{\sum_{i=1}^{n}(2i - n - 1)X_{i:n}}{\sum_{i=1}^{n} nX_{i:n}} = \frac{n - 1}{n} G_n.$$

The second version G'_n corresponds to the Gini index of the sample Lorenz curve. We would consequently expect that the sample Gini index, G'_n, will frequently be negatively biased. Thus we would expect to have

$$E(G'_n) \le G = \frac{E(X_{2:2}) - E(X_{1:2})}{E(X_{2:2}) + E(X_{1:2})},$$

i.e., that

$$E\left(\frac{\sum_{i=1}^{n}(2i-n-1)X_{i:n}}{\sum_{i=1}^{n} nX_{i:n}}\right) \le \frac{E(X_{2:2}) - E(X_{1:2})}{E(X_{2:2}) + E(X_{1:2})}.$$

For the exponential distribution, i.e., when the X_i's have a common $\Gamma(1, 1)$ distribution, we have, using Basu's lemma,

$$E(G'_n) = \frac{\sum_{i=1}^{n}(2i-n-1)E(X_{i:n})}{n^2}$$

and

$$G = \frac{E(X_{2:2}) - E(X_{1:2})}{E(X_{2:2}) + E(X_{1:2})} = \frac{1}{2}. \tag{5.91}$$

Substituting the well-known expressions for the expectations of exponential order statistics and simplifying, it may be verified that

$$E(G'_n) = \left(\frac{n-1}{n}\right)\frac{1}{2}, \tag{5.92}$$

which indeed is less than $\frac{1}{2}$.

Note that in this exponential case we have

$$E(G_n) = \frac{n}{n-1}E(G'_n) = \frac{n}{n-1}\left(\frac{n-1}{n}\right)\frac{1}{2} = \frac{1}{2},$$

that is, in this special case, G_n is unbiased. Note that, since G_n is not the Gini index of the sample Lorenz curve, we did not have strong justification for expecting it to be negatively biased in most cases.

On the other hand, we do have some reason to expect that G'_n will often be negatively biased. Will it always be negatively biased? Just as was the case for studying the bias of the sample Lorenz curve, consideration of samples of size 2 from a two-point distribution will be instructive.

Suppose that X_1 and X_2 are i.i.d. with $P(X_i = 1) = p$ and $P(X_i = c) = 1 - p$ where $c > 1$. We have in this case the sample Gini index, G'_2, given by

Table 5.6 Probabilities for computing the sample Gini indices in the discrete case in which $P(X_i = 1) = p$ and $P(X_i = c) = 1 - p$ (from Arnold and Villaseñor 2015)

	Variable			
Probability	$X_{1:2}$	$X_{2:2}$	G_2	G'_2
p^2	1	1	0	0
$2p(1-p)$	1	c	$\frac{c-1}{c+1}$	$\frac{c-1}{2(c+1)}$
$(1-p)^2$	c	c	0	0

$$G'_2 = \frac{1}{2}\frac{X_{2:2} - X_{1:2}}{X_{2:2} + X_{1:2}}$$

and the population Gini index given by

$$G = \frac{E(X_{2:2}) - E(X_{1:2})}{E(X_{2:2}) + E(X_{1:2})}.$$

For this sample we have the needed distributional information provided in Table 5.6.

In Table 5.6, for example, the probability that G_2 takes on the value $(c-1)/(c+1)$ is $2p(1-p)$. From this table, it follows that

$$\begin{aligned} G - E(G_2) &= \frac{(c-1)2p(1-p)}{2[p + c(1-p)]} - 2p(1-p)\frac{c-1}{c+1} \\ &= \frac{p(1-p)(c-1)^2}{[p + c(1-p)](c+1)}(2p-1). \end{aligned}$$

The sign of the difference $G - E(G_2)$ depends on the value of p. It is positive if $p > 0.5$, negative if $p < 0.5$ and equal to 0 if $p = 0.5$. However if we consider the bias of G'_2, we have

$$\begin{aligned} G - E(G'_2) &= \frac{(c-1)2p(1-p)}{2[p + c(1-p)]} - p(1-p)\frac{c-1}{c+1} \\ &= \frac{p(1-p)(c-1)}{[p + c(1-p)](c+1)}[1 + (c-1)p] > 0 \end{aligned}$$

for every $p \in (0, 1)$ and $c > 1$. Thus the expected negative bias of L'_2 is present, not just for values of p for which the corresponding sample Lorenz curve is positively biased, but even for values of p for which the curve is negatively biased.

5.8.8 *Asymptotic Distribution of Lorenz Ordinates and Income Shares*

In this section we consider distribution-free statistical inference for Lorenz curve ordinates and income shares. These results can be found in Beach and Davidson (1983). We will consider two kinds of results:

- The asymptotic distribution of a set of income quantiles
- The asymptotic distribution of a set of Lorenz curve ordinates

Let X be a positive random variable, which represents income for different individuals or families, and denote the population cdf of X by $F(x)$. The proportion of the total income received by those individuals with income less or equal to x is

$$L(F(x)) = \frac{1}{\mu}\int_0^x v dF(v), \quad 0 < x < \infty,$$

where all incomes are assumed to be positive, and the mean and variance of X, $E(X) = \mu$ and $\mathrm{var}(X) = \sigma^2$, exist and are finite. In this expression, L denotes, as usual, the Lorenz curve corresponding to the distribution function F.

Then, the objective is to perform statistical inference using a set of ordinates of the sample Lorenz curve, i.e., the set a set of K ordinates

$$L_n(\underline{u}) = (L_n(u_1), L_n(u_2), \ldots, L_n(u_K)),$$

corresponding to the abscissae $(u_1, \ldots, u_K)$, with $0 < u_1 < u_2 < \cdots < u_K < 1$.

An income quantile ξ_u corresponding to a distribution function F is defined implicitly by $F(\xi_u) = u$, where F is assumed to be strictly increasing. Thus, corresponding to a set of K values $u_1 < \cdots < u_K$, we have a set of K income quantiles $\xi_{u_1} < \cdots < \xi_{u_K}$ and a set of K population Lorenz curve ordinates $L(u_1) < \cdots < L(u_K)$, where

$$\begin{aligned} L(u_i) &= \frac{1}{\mu}\int_0^{\xi_{u_i}} v dF(v) = \frac{F(\xi_{u_i})}{\mu}\int_0^{\xi_{u_i}} \frac{v dF(v)}{F(\xi_{u_i})} \\ &= \frac{F(\xi_{u_i})}{\mu} E(X|X \le \xi_{u_i}) \\ &= u_i \frac{\gamma_i}{\mu}, \end{aligned}$$

where

$$\gamma_i = E(X|X \le \xi_{u_i}) \tag{5.93}$$

is the conditional mean of incomes less than or equal to ξ_{u_i}.

Now, let $X_1, \ldots, X_n$ be a random sample of size n from F and we denote the ordered sample by $X_{(1)} \leq X_{(2)} \leq \cdots \leq X_{(n)}$. Then, let the sample quantile $\hat{\xi}_u$ be defined as the r-th order statistics $X_{(r)}$, where $r = [nu]$ denotes the greatest integer less than or equal to nu. If F is strictly increasing, $\hat{\xi}_u$ has the property of strong or almost sure consistency as an estimate of ξ_u, and if F is differentiable then for any finite set $\{u_1, \ldots, u_K\}$, the $\hat{\xi}_{u_i}$'s are asymptotically multivariate normal, according the following lemma (see, Wilks 1962 or Arnold et al. 1992)

Lemma 5.8.1 *Suppose that for a set of proportions $\{u_1, \ldots, u_K\}$ with $u_1 < \cdots < u_K$, $\hat{\xi} = (\hat{\xi}_{u_1}, \ldots, \hat{\xi}_{u_K})^\top$ is a vector of K sample quantiles from a random sample of size n drawn from a continuous density $f(x)$ such that the ξ_{u_i}'s are uniquely defined and $f_i \equiv f(\xi_{u_i}) > 0$ for all $i = 1, \ldots, K$. Then, we have*

$$\sqrt{n}(\hat{\xi} - \xi) \xrightarrow{d} N^{(K)}(\underline{0}, \Lambda),$$

that is, there is convergence in distribution to a K-variate normal distribution with mean $\underline{0}$ and covariance matrix Λ with elements,

$$\Lambda_{ij} = \frac{u_i(1 - u_j)}{f_i f_j}, \quad i, j = 1, 2, \ldots, K.$$

The sample estimate of the Lorenz curve ordinate $L(u_i)$ is computed as

$$\begin{aligned} L_n(u_i) &= \frac{\sum_{j=1}^{r_i} X_{(j)}}{\sum_{j=1}^{n} X_{(j)}}, \quad \text{with } r_i = [nu_i] \\ &\doteq u_i \frac{\hat{\gamma}_i}{\hat{\mu}}, \end{aligned}$$

where $\hat{\mu} = (1/n) \sum_{j=1}^{n} X_{(j)}$ is the sample mean and

$$\hat{\gamma}_i = \frac{1}{r_i} \sum_{j=1}^{r_i} X_{(j)}.$$

The first main result is the following.

Theorem 5.8.4 *Consider the $(K + 1)$-dimensional random vector*

$$\hat{\theta} = (u_1 \hat{\gamma}_1, \ldots, u_K \hat{\gamma}_K, u_{K+1} \hat{\gamma}_{K+1})^\top$$

where the $\hat{\gamma}_i$'s, $i = 1, \ldots, K$ are the conditional means defined in (5.93), the proportions u_i are such that $0 < u_1 < \cdots < u_K < 1$ and where $u_{K+1} \equiv 1$ and $\hat{\gamma}_{K+1} \equiv \hat{\mu}$ the unconditional sample mean. Then, assuming that the population has a finite mean and a finite variance and that the cdf F is strictly increasing and twice differentiable, $\hat{\theta}$ is asymptotically normal, i.e.,

$$\sqrt{n}(\hat{\theta} - \theta) \xrightarrow{d} N^{(K+1)}(\underline{0}, \Omega),$$

with zero mean and covariance matrix $\Omega = \{w_{ij}\}$,

$$w_{ij} = u_i[\lambda_i^2 + (1 - u_j)(\xi_{u_i} - \gamma_i)(\xi_{u_j} - \gamma_j) + (\xi_{u_i} - \gamma_i)(\gamma_j - \gamma_i)], \tag{5.94}$$

for $i \leq j$.

In Eq. (5.94) λ_i^2 is defined as $\text{var}(X|X \leq \xi_{u_i})$, given by the equation

$$u_i(\lambda_i^2 + \gamma_i^2) = \int_0^{\xi_{u_i}} x^2 dF(x).$$

Note that setting $u_i = u_j$, the asymptotic variance of $u_i\hat{\gamma}_i$ is

$$\frac{u_i[\lambda_i^2 + (1 - u_i)(\xi_{u_i} - \gamma_i)^2]}{n}.$$

As well, setting $u_j = u_{K+1} = 1$ the asymptotic covariance of $u_i\hat{\gamma}_i$ and $\hat{\mu}$ is

$$\frac{u_i[\lambda_i^2 + (\xi_{u_i} - \gamma_i)(\mu - \gamma_i)]}{n}.$$

The variance of $u_{K+1}\hat{\gamma}_{K+1} = \hat{\mu}$ (in the bottom right-element of Ω) is $\frac{u_{K+1}\lambda_{K+1}^2}{n} = \frac{\sigma^2}{n}$, as usual.

In order to make practical use of Theorem 5.8.4 and render our analysis distribution free, we can substitute consistent estimates of the λ_i^2's, namely (Beach and Davidson 1983),

$$\hat{\lambda}_i^2 \equiv \frac{1}{r_i}\sum_{i=1}^{r_i}(X_{(k)} - \hat{\gamma}_i)^2,$$

and also insert consistent estimates of ξ_{u_i}, γ_i, i.e., $X_{(r_i)}$, $\hat{\gamma}_i$ respectively.

The second main result is the following.

Theorem 5.8.5 *Under the conditions of Theorem 5.8.4, the vector of sample Lorenz curve ordinates* $L_n(\underline{u}) = (L_n(u_1), \ldots, L_n(u_K))^\top$ *is asymptotically normal, i.e.,*

$$\sqrt{n}(L_n(u) - L(u)) \xrightarrow{d} N^{(K)}(\underline{0}, V_L), \tag{5.95}$$

with zero mean and covariance matrix $V_L = \{v_{ij}^L\}$ *specified by*

$$v_{ij}^L = \left(\frac{1}{\mu^2}\right) w_{ij} + \left(\frac{u_i\gamma_i}{\mu^2}\right)\left(\frac{u_j\gamma_j}{\mu^2}\right)\sigma^2 - \left(\frac{u_i\gamma_i}{\mu^3}\right) w_{j,K+1} - \left(\frac{u_j\gamma_j}{\mu^3}\right) w_{i,K+1},$$

for $i \le j = 1, \ldots, K$. In the case of $i = j$ the diagonal elements of V_L are

$$v_{ii}^L = \frac{u_i}{\mu^2}[\lambda_i^2 + (1-u_i)(\xi_{u_i} - \gamma_i)^2] + \left(\frac{u_i \gamma_i}{\mu^2}\right)^2 \sigma^2$$
$$-2\left(\frac{u_i^2 \gamma_i}{\mu^3}\right)[\lambda_i^2 + (\mu - \gamma_i)(\xi_{u_i} - \gamma_i)^2].$$

As a consequence of the previous formulation, the asymptotic standard errors for the sample estimates $\hat{L}_i$ are given by

$$\sqrt{\frac{v_{ii}^L}{n}}, \quad \text{for } i = 1, \ldots, K.$$

With the asymptotic joint distribution of $(L(u_1), L(u_2), \ldots, L(u_K))$ at hand, it is a simple operation to obtain the asymptotic joint distribution of the corresponding income shares, namely

$$\underline{S} = (L(u_1), L(u_2) - L(u_1), \ldots, L(u_K) - L(u_{K-1})). \tag{5.96}$$

If we define

$$\underline{\widetilde{S}} = (L_n(u_1), L_n(u_2) - L_n(u_1), \ldots, L_n(u_K) - L_n(u_{K-1})),$$

then we have

$$\sqrt{n}(\underline{S} - \underline{\widetilde{S}}) \xrightarrow{d} N^{(K)}(\underline{0}, \Delta V_L \Delta^\top), \tag{5.97}$$

where

$$\Delta_{K \times K} = \begin{pmatrix} 1 & 0 & 0 & \cdots\cdots & 0 \\ -1 & 1 & 0 & \cdots\cdots & 0 \\ 0 & -1 & 1 & \cdots\cdots & 0 \\ \vdots & \vdots & \ddots & \ddots & \vdots & \vdots \\ 0 & 0 & \cdots\cdots & & -1 & 1 \end{pmatrix},$$

and V_L is as defined following (5.95).

In order to make practical distribution free use of the asymptotic distributions given in (5.95) and (5.97), it is appropriate to substitute available consistent estimates of the parameters in V_L (defined following (5.95)).

5.8.9 The Elteto and Frigyes Indices

Now, we consider the sample versions of the Elteto and Frigyes indices studied in Sect. 5.3.8. These sample indices are defined by

$$U_n = \frac{N \sum_{i=1}^{N} X_i}{n \sum_{X_i<\bar{X}} X_i},$$

$$V_n = \frac{N \sum_{X_i>\bar{X}} X_i}{(n-N) \sum_{X_i<\bar{X}} X_i},$$

$$W_n = \frac{n \sum_{X_i>\bar{X}} X_i}{(N-n) \sum_{i=1}^{n} X_i},$$

where n is the sample size and N is the random number of observations X_i less than the mean $\bar{X}$. The asymptotic distributions of these indices were obtained by Gastwirth (1974). For the case of U_n we have

$$\sqrt{n}(U_n - U) \xrightarrow{d} N(0, \sigma_U^2),$$

where

$$\sigma_U^2 = \frac{1}{F^2(\mu)} \left\{ \frac{v_1}{\mu_1^2} + \frac{v_2 \mu^2}{\mu_1^4} - \frac{2c\mu}{\mu_1^3} \right\},$$

in which $\mu = E(X)$, $\xi^2 = \text{var}(X)$ and

$$\mu_1 = \frac{1}{F(\mu)} \int_0^{\infty} x dF(x),$$

$$v_1 = \mu^2 F(\mu) \bar{F}(\mu) + \xi^2 [F(\mu) + \mu f(\mu)]^2 - 2\mu[\mu f(\mu) + F(\mu)] F(\mu)(\mu - \mu_1),$$

$$v_2 = \mu^2 f^2(\mu) \xi^2 + \int_0^{\mu} x^2 dF(x) - F^2(\mu)\mu_1^2 + 2\mu f(\mu) \int_0^{\mu} x(x-\mu) dF(x),$$

$$c = \mu\mu_1 F(\mu)\bar{F}(\mu) - \mu^2 f(\mu) F(\mu)(\mu - \mu_1) + \mu f(\mu)[\mu f(\mu) + F(\mu)]\xi^2$$
$$+ [\mu f(\mu) + F(\mu)] \int_0^{\mu} x(x-\mu) dF(x).$$

See Gastwirth (1974) for the analogous expressions for V_n and W_n. There does not appear to be any finite sample size distributions theory for the Elteto and Frigyes indices in the literature.

5.8.10 Further Classical Sample Measures of Inequality

Here we include two inequality measures defined in terms of logarithms of the data points. If $Y_i = \log X_i$, Yntema (1933) considers the mean deviation of the logarithms,

$$T_7(\underline{X}) = \frac{1}{n} \sum_{i=1}^{n} |Y_i - \bar{Y}|, \tag{5.98}$$

as a suitable inequality measure.

The second similar measure corresponds to the standard deviation of the logarithms of the data points,

$$T_8(\underline{X}) = \sqrt{\frac{1}{n-1} \sum_{i=1}^{n} (Y_i - \bar{Y})^2}, \tag{5.99}$$

with $Y_i = \log X_i$. The asymptotic distribution of (5.98) and (5.99) can be obtained using standard techniques. The use of the logarithmic transformation (in preference to some other monotone transformation) has a long tradition in economics. This may or may not be because, if the parent distribution is lognormal (as it well might be for income data), the logarithms have a particularly well-understood distribution.

Bowley's interquartile ratio can be written as a function of order statistics as suggested by Dalton (1920). It is defined by

$$T_9(\underline{X}) = \frac{X_{3n/4:n} - X_{n/4:n}}{X_{3n/4:n} + X_{n/4:n}}. \tag{5.100}$$

The main attractive of this measure is its robustness. The measure ignores outlying observations. This aspect could be inappropriate in income studies, where emphasis is often on the upper tail of the distribution.

For obtaining the asymptotic distribution of (5.100) we begin with the distribution of $(X_{3n/4:n}, X_{n/4:n})$. If we denote the population quartiles by η_i, with $i = 1/4, 3/4$ and assuming that the population density f is continuous in neighborhoods of η_i, we have (see Arnold et al. 1992)

$$\sqrt{n}(X_{n/4:n}, X_{3n/4:n}) \xrightarrow{d} N^{(2)}((\eta_{1/4}, \eta_{3/3}), \Sigma),$$

where the elements of the covariance matrix are

$$\sigma_{11} = \frac{3}{16 f^2(\eta_{1/4})},$$

$$\sigma_{22} = \frac{3}{16 f^2(\eta_{3/4})},$$

$$\sigma_{12} = \sigma_{21} = \frac{1}{16 f(\eta_{1/4}) f(\eta_{2/4})}.$$

Now, if we define $\tau_9 = \frac{\eta_{3/4} - \eta_{1/4}}{\eta_{3/4} + \eta_{1/4}}$, using the δ-method we have

$$\sqrt{n}(T_9^{(n)} - \tau_9) \xrightarrow{d} N(0, \sigma_9^2),$$

where

$$\sigma_9^2 = 3\xi_{1/4}^2 - 2\xi_{1/4}\xi_{1/4} + 3\xi_{1/4}^2,$$

in which

$$\xi_{1/4} = \eta_{3/4}/[2f(\eta_{1/4})],$$
$$\xi_{3/4} = \eta_{1/4}/[2f(\eta_{3/4})].$$

Dalton (1920) also considered use of the ratio of the logarithms of the arithmetic and geometric means, as an inequality measure. Thus he investigated

$$T_{10}(\underline{X}) = \frac{\log \bar{X}}{\log \bar{X}_g},$$

where $\bar{X}_g = (\prod_{i=1}^n X_i)^{1/n}$ and where we assume $X_i > 1$, $i = 1, 2, \ldots, n$. This index satisfies $T_{10}(\underline{X}) \geq 1$, with equality if all X_i's are equal. Then, it is natural to consider the following index which takes on values in the interval (0, 1),

$$\tilde{T}_{10}(\underline{X}) = 1 - \frac{1}{T_{10}(\underline{X})}. \tag{5.101}$$

The asymptotic distribution of $T_{10}(\underline{X})$ is given by

$$\sqrt{n}\left[T_{10}(\underline{X}) - \frac{\log \eta}{\tilde{\eta}}\right] \xrightarrow{d} N(0, \sigma_{10}^2),$$

where

$$\eta = \int_1^\infty x dF(x), \quad \tilde{\eta} = \int_1^\infty \log x dF(x),$$

and

$$\sigma_{10}^2 = \frac{\xi_1^2}{\eta^2\tilde{\eta}^2} - \frac{2\xi_{12}\log\eta}{\eta\tilde{\eta}^3} + \frac{\xi_2^2(\log\eta)^2}{\tilde{\eta}^4},$$

in which

$$\xi_1^2 = \text{var}(X), \quad \xi_2^2 = \text{var}(\log X), \quad \text{and} \quad \xi_{12} = cov(X, \log X).$$

Several classes of indices which might be considered take the following forms,

$$T_g(\underline{X}) = c\sum_{i=1}^{n} g(X_i), \tag{5.102}$$

$$\tilde{T}_g(\underline{X}) = \frac{1}{n}\sum_{i=1}^{n} g\left(\frac{X_i}{\bar{X}}\right) \tag{5.103}$$

and

$$\tilde{\tilde{T}}_g(\underline{X}) = \frac{\frac{1}{n}\sum_{i=1}^{n} g(X_i)}{g(\bar{X})}, \tag{5.104}$$

where g is a convex function. Indices in these classes are asymptotically normal provided the needed moments exist. Classes (5.102) to (5.104) include some new inequality measures and some measures that have been already discussed in this book. Special cases include: (1) $g(x) = x^2$; (2) $g(x) = x\log x$; (3) $g(x) = -U(x)$, where $U(x)$ is a concave utility function (proposed by Dalton (1920) and Atkinson (1970)); (4) $g(x) = \sqrt{1+x^2}$, which corresponds to the Amato index. Many other choices merit consideration.

As an extension of the Gini index, Mehran (1976) proposed linear measures of inequality of the form,

$$T_W(\underline{X}) = \frac{\sum_{i=1}^{n} X_{i:n} W\left(\frac{i}{n+1}\right)}{n\bar{X}},$$

where $W(\cdot)$ is a smooth weight function. Asymptotic normality of these measures is a consequence of results due to Stigler (1974).

The sample Gini index is asymptotically equivalent to the Mehran linear measure with weight function $W(u) = 2u - 1$.

Weymark (1979) has considered generalized Gini indices of the form,

$$G_{\underline{a}}(\underline{X}) = \frac{\sum_{i=1}^{n} a_{i:n} X_{i:n}}{n\bar{X}}. \tag{5.105}$$

To ensure Schur convexity in (5.105) we need to have $a_{i:n} \uparrow$ as $i \uparrow$ for each n. In particular, Donaldson and Weymark (1980) considered the special case,

$$a_{i:n} = \frac{i^\delta - (i-1)^\delta}{n^\delta},$$

where the case $\delta = 2$ is closely related to the Gini index.

5.8.11 The Sample Atkinson and Generalized Entropy Indices

In this section we discuss the sample versions of the Atkinson and generalized entropy indices (which we considered in Sect. 5.4) and their corresponding asymptotic distributions.

Let X be a random variable in $\mathcal{L}$ with cdf F. The αth raw moment will be denoted by

$$\mu_\alpha = E(X^\alpha) = \int_0^\infty x^\alpha dF(x), \tag{5.106}$$

provided that the integral exists. The usual estimator of (5.106) is the sample moment, which we will denote as

$$m_\alpha = \frac{1}{n}\sum_{i=1}^{n} X_i^\alpha.$$

We will denote the variance of $\sqrt{n}m_\alpha$ by σ_α^2 and the covariance between two sample moments m_α and $m_{\alpha'}$ by

$$cov(\sqrt{n}m_\alpha, \sqrt{n}m_{\alpha'}) = \gamma_{\alpha,\alpha'}.$$

The values $\mu_1 = \mu$ and $\sigma_1^2 = \sigma^2$ are, as usual, the mean and variance of X, respectively.

We write the population Atkinson's (1970) index as (note that this is equivalent to the earlier definition which was written in terms of $\epsilon = 1 - \theta$),

$$A_\theta = 1 - \left[\int_0^\infty \left(\frac{x}{\mu}\right)^\theta dF(x)\right]^{1/\theta} = 1 - \frac{\mu_\theta^{1/\theta}}{\mu_1}, \tag{5.107}$$

where $-\infty < \theta < 1$. The cases $\theta = 0$ and $\theta = -\infty$ will be discussed below. The sample version of this index is given by

$$A_\theta(\underline{X}) = 1 - \frac{m_\theta^{1/\theta}}{m_1}$$
$$= 1 - \left[\frac{1}{n}\sum_{i=1}^{n}\left(\frac{X_i}{\bar{X}}\right)^\theta\right]^{1/\theta}.$$

The asymptotic distribution of the sample Atkinson index, provided that the needed moments exist is given by Thistle (1990),

$$\sqrt{n}\left[A_\theta(\underline{X}) - A_\theta\right] \xrightarrow{d} N(0, v_A^2),$$

with $n \to \infty$, where

$$v_A^2 = K(\theta)\left\{\sigma_\theta^2 - \frac{2\theta\mu_\theta}{\mu}\gamma_{\theta,1} + \left(\frac{\theta\mu_\theta}{\mu}\right)^2\sigma^2\right\},$$

and $K(\theta) = \{(1 - A_\theta)/\theta\mu_\theta\}^2$.

In empirical economic inequality analysis it is common to estimate the Atkinson index for several values of the parameter θ. In consequence, the estimates of the different values of A_θ will be correlated, because the sample moments m_θ are correlated. Let $\underline{\theta} = (\theta_1, \ldots, \theta_m)^\top$ be a selection of values of θ, and we denote $A_{\underline{\theta}} = (A_{\theta_1}, \ldots, A_{\theta_m})^\top$. The vector of the corresponding estimators will be denoted by $A_{\underline{\theta}}(\underline{X})$. Provided adequate moments exist, the asymptotic distribution of $A_{\underline{\theta}}(\underline{X})$ is

$$\sqrt{n}\left[A_{\underline{\theta}}(\underline{X}) - A_{\underline{\theta}}\right] \xrightarrow{d} N^{(m)}(0, \Sigma),$$

where

$$\sigma_{ij} = K_{ij}\left\{\gamma_{\theta_i,\theta_j} - K_j\gamma_{\theta_i,1} - K_i\gamma_{\theta_j,1} + K_iK_j\sigma^2\right\},\ i, j = 1, \ldots, m,$$

with $K_{ij} = (1 - A_{\theta_i})(1 - A_{\theta_j})/\theta_i\theta_j\mu_{\theta_i}\mu_{\theta_j}$ and $K_i = \theta_i\mu_{\theta_i}/\mu$.

There are two special extremal cases: $A_0 = 1 - \exp(E(\log X))$ and $A_{-\infty} = 1 - F_X^{-1}(0)/\mu$. The asymptotic distribution of $A_0(\underline{X})$ can be obtained as was done in the derivation of the distribution for finite negative values of θ. For the second situation, the sample estimate is

$$A_{-\infty}(\underline{X}) = 1 - \frac{X_{1:n}}{\bar{X}},$$

where $X_{1:n}$ is the minimum of the sample. Since the limiting distribution of the sample minimum depends on which domain of minimal attraction that F belongs to, one must consider three cases. See Arnold et al. (1992) for relevant details.

The generalized entropy index is (see Cowell and Kuga 1981 and Shorrocks 1984)

$$G_\beta = k \int_0^\infty \left[\left(\frac{x}{\mu} \right)^\beta - 1 \right] dF(x) = k \left(\frac{\mu_\beta}{\mu^\beta} - 1 \right), \tag{5.108}$$

where $k = 1/\beta(\beta - 1)$ and $\beta \neq 0, 1$. Compare with Eq. (5.107), with $\beta = \theta$. The sample versions of (5.108) are

$$\begin{aligned} G_\beta(\underline{X}) &= \frac{1}{\beta(\beta - 1)} \left(\frac{m_\beta}{m_1^\beta} - 1 \right) \\ &= \frac{1}{\beta(\beta - 1)} \frac{1}{n} \sum_{i=1}^n \left[\left(\frac{X_i}{\bar{X}} \right)^\beta - 1 \right] \end{aligned} \tag{5.109}$$

if $\beta \neq 0, 1$. While, for $\beta = 0$ we have

$$G_0(\underline{X}) = -\frac{1}{n} \sum_{i=1}^n \log \frac{X_i}{\bar{X}}, \tag{5.110}$$

and for $\beta = 1$,

$$G_1(\underline{X}) = \frac{1}{n} \sum_{i=1}^n \frac{X_i}{\bar{X}} \log \frac{X_i}{\bar{X}}. \tag{5.111}$$

The asymptotic distribution of (5.109) as $n \to \infty$ and provided that the needed moments exist is given by Thistle (1990)

$$\sqrt{n} \left[G_\beta(\underline{X}) - G_\beta \right] \xrightarrow{d} N(0, v_G^2),$$

where

$$v_G^2 = \frac{k^2}{\mu^{2(\beta+1)}} \left\{ \mu^2 \sigma_\beta^2 - 2\beta \mu \mu_\beta \gamma_{\beta,1} + \beta^2 \mu_\beta^2 \sigma^2 \right\}.$$

As in the case of the Atkinson indices, we can consider estimators of G_β for several different values of β, recognizing that again these estimators are correlated. We denote $G_{\underline{\beta}} = (G_{\beta_1}, \ldots, G_{\beta_m})^\top$, and the vector of the corresponding estimators by $G_{\underline{\beta}}(\underline{X})$. The asymptotic distribution of $G_{\underline{\beta}}(\underline{X})$, provided the appropriate moments exist is

$$\sqrt{n}\left[G_{\underline{\beta}}(\underline{X}) - G_{\underline{\beta}}\right] \xrightarrow{d} N^{(m)}(0, \Omega),$$

where

$$w_{ij} = \frac{k_i k_j}{\mu^{\beta_i+\beta_j+2}}\{\mu^2 \gamma_{\beta_i,\beta_j} - \beta_j \mu_{\beta_j} \mu \gamma_{\beta_i,1}$$
$$-\beta_i \mu_{\beta_i} \mu \gamma_{\beta_j,1} + \beta_i \mu_{\beta_i} \beta_j \mu_{\beta_j} \sigma^2\},\ i, j = 1, \ldots, m$$

where $k_i = 1/\beta_i(\beta_1 - 1)$.

The asymptotic distributions for the two special cases (5.110) and (5.111) can be derived using standard techniques (a multivariate central limit theorem and the delta method).

5.8.12 *The Kolm Inequality Indices*

The Kolm (1976) inequality indices are based on a social welfare approach (see Dalal and Fortini 1982; Atkinson 1970; Kolm 1976), which assumes a social evaluation function for the vector of incomes from which an inequality index is derived. These indices are defined by

$$T_K(\underline{X}) = \frac{1}{\kappa} \log\left(\frac{1}{n}\sum_{i=1}^{n} e^{\kappa(\bar{X}-X_i)}\right), \tag{5.112}$$

where $\kappa > 0$ is a parameter that may be assigned any positive value. The index $T_K(\underline{X}) = 0$ if $\kappa = 0$ and if $\kappa \to \infty$, $T_K(\underline{X})$ tends to $\bar{X} - X_{1:n}$, that is, the gap between the average and the minimum income.

Each member of this Kolm family has the property that if we add the same absolute amount to every X_i, the inequality remains unaltered. Multivariate versions of this class of measures have been provided by Tsui (1995, 1999). The asymptotic distribution of (5.112) can be obtained in routine fashion.

5.8.13 *Additional Sample Inequality Indices*

In this last sub-section we include some additional inequality indices listed in Marshall et al. (2011) and Arnold (2015b), which have not yet received our attention. In general, these measures have not been particularly popular, but there may exist special situations in which they will prove to be "the right tool for the job."

- Emlen's (1973) index

$$T(\underline{X}) = \sum_{i=1}^{n} \frac{X_i}{\bar{X}} \exp(-X_i/n\bar{X})$$

- The minimal majority (Alker 1965)

$$T(\underline{X}) = L_n^{-1}(1/2),$$

where L_n is the sample Lorenz curve.
- Top 100α percent (Alker and Russet 1966),

$$T(\underline{X}) = L_n(\alpha).$$

- Quantile ratios:

$$T(\alpha, \beta) = \frac{X_{[(1-\alpha)n:n]}}{X_{[\beta n]:n}},$$

where $0 < \alpha, \beta < 0.5$.
- Sample income share ratios,

$$T(\alpha, \beta) = \frac{1 - L_n(1-\alpha)}{L_n(\beta)} \tag{5.113}$$

The sample Palma index is a special case of (5.113) when $\alpha = 0.1$ and $\beta = 0.4$.

5.9 A New Class of Inequality Measures

In this section we define a new class of inequality measures defined on the set of n-dimensional vectors of positive real numbers. Let $\underline{x} = (x_1, x_2, \ldots, x_n) \in \mathbb{R}_+^n$. We wish to define the degree of inequality exhibited by $\underline{x}$. To this end define the corresponding share vector

$$\underline{s} = (s_1, s_2, \ldots, s_n),$$

where $s_i = \frac{x_i}{\sum_{j=1}^n x_j}$.

Now let F_0 denote a particular continuous increasing distribution function with $F_0(0) = 0$. For each such distribution function F_0 we will define a measure of inequality $I_{F_0}(\underline{x})$ defined on $\mathbb{R}_+^n$.

Let $X_1, X_2, \ldots, X_n$ be i.i.d. positive random variables with common distribution function F_0. Define the corresponding share vector by

$$\underline{S} = (S_1, S_2, \ldots, S_n),$$

where $S_i = \frac{X_i}{\sum_{j=1}^{n} X_j}$. Then we define

$$I_{F_0}(\underline{x}) = P(\underline{S} \prec \underline{s}).$$

Thus we have a new family of inequality indices, indexed by the class of all distributions of positive random variables.

More generally we could let $\underline{S}$ be an arbitrary n-dimensional random variable with positive coordinates that sum to 1 with probability 1, but we will focus on the exchangeable case defined above.

In particular we consider the case in which the X_i's are i.i.d. $\Gamma(\alpha, 1)$ random variables. In this case we have

$$\underline{S} \sim \text{Dirichlet}(\alpha, \alpha, \ldots, \alpha).$$

We define the ordered shares by $s_{1:n}, s_{2:n}, \ldots, s_{n:n}$ ordered from smallest, $s_{1:n}$, to largest, $s_{n:n}$, and the cumulative ordered shares by

$$t_{1:n}, t_{2:n}, \ldots, t_{n:n},$$

where $t_{i:n} = \sum_{j=1}^{i} s_{i:n}$, with parallel notation for the $S_{i:n}$'s and the $T_{i:n}$'s. With this notation we see that

$$I_{F_0}(\underline{x}) = P(\underline{S} \prec \underline{s}) = P(T_{1:n} > t_{1:n}, T_{2:n} > t_{2:n}, \ldots, T_{n-1:n} > t_{n-1:n}).$$

We have

$$\begin{aligned} & f_{S_{1:n}, S_{2:n}, \ldots, S_{n-1:n}}(s_{1:n}, s_{2:n}, \ldots, s_{n-1:n}) \\ & = \frac{n!\Gamma(n\alpha)}{(\Gamma(\alpha))^n} \left[\prod_{i=1}^{n-1} s_{i:n}\right]^{\alpha-1} \left[1 - \sum_{i=1}^{n-1} s_{i:n}\right]^{\alpha-1}, \\ & 0 < s_{1:n} < s_{2:n} < \cdots < s_{n-1:n} < 1 - \sum_{i=1}^{n-1} s_{i:n} \end{aligned}$$

an ordered Dirichlet distribution. Consequently

$$\begin{aligned} & f_{T_{1:n}, T_{2:n}, \ldots, T_{n-1:n}}(t_{1:n}, t_{2:n}, \ldots, t_{n-1:n}) \\ & = \frac{n!\Gamma(n\alpha)}{(\Gamma(\alpha))^n} \left[t_{i:n} \prod_{i=2}^{n-1} (t_{i:n} - t_{i-1:n})\right]^{\alpha-1} [1 - t_{n-1:n}]^{\alpha-1}, \end{aligned} \tag{5.114}$$

where $0 < t_{1:n} < t_{2:n} - t_{1:n} < t_{3:n} - t_{2:n} < \ldots t_{n-1:n} - t_{n-2:n} < 1 - t_{n-1:n}$.

If we set $\alpha = 1$ in (5.114), we get a uniform distribution, thus

$$f_{T_{1:n},T_{2:n},\ldots,T_{n-1:n}}(t_{1:n}, t_{2:n}, \ldots, t_{n-1:n}) = n!(n-1)! I((t_{1:n}, t_{2:n}, \ldots, t_{n-1:n}) \in B_{n-1}),$$

where $B_{n-1} = \{t_{i:n} : 0 < t_{1:n} < t_{2:n} - t_{1:n} < t_{3:n} - t_{2:n} < \ldots t_{n-1:n} - t_{n-2:n} < 1 - t_{n-1:n}\}$.

In the special, almost trivial, case when $\alpha = 1$ and $n = 3$, we can do the required calculation and we find that explicitly

$$\begin{aligned} I_{F_0}(\underline{x}) &= P(\underline{S} \prec \underline{s}) \\ &= P(T_{1:3} > t_{1:3}, T_{2:3} > t_{2:3}) \\ &= 1 - 3(2s_{1:3}s_{3:3} + s_{2:3}^2) \\ &= 1 - 3(s_{1:3}s_{3:3} + s_{2:3}s_{2:3} + s_{3:3}s_{1:3}) \end{aligned}$$

a nice symmetrical result. Wouldn't it be nice if there were a corresponding symmetric result for $\mathbb{R}_+^n$. This class of inequality measures will be studied in future research.

Note that we can extend the measure $I_{F_0}(\underline{x})$ to deal with the class of positive random variables with finite expectations as follows. For a random variable Y in this class, consider N independent copies $Y_1, Y_2, \ldots, Y_N$. Next define $L_n(F_0)$ to be the sample Lorenz curve of $X_1, X_2, \ldots, X_n$ which are i.i.d with common distribution F_0 and define $L_N(Y)$ to be the sample Lorenz curve of $Y_1, Y_2, \ldots, Y_N$. The F_0-inequality measure of Y is then defined to be

$$I_{F_0}(Y) = \lim_{N\to\infty} P(L_n \geq L_N(Y)),$$

which can be evaluated approximately via simulation.

5.10 Exercises

1. We consider a set of independent random variables $X_1, X_2, \ldots, X_n$ belonging to the class $\mathcal{L}$, and let w_i, $i = 1, 2, \ldots, n$ be non-negative real numbers. Consider the aggregate random variable $X = \prod_{i=1}^n X_i^{w_i}$. Prove that

$$I_7(X) = \sum_{i=1}^n w_i^2 I_7(X_i),$$

where $I_7(X)$ represents the variance of the logarithm of X.

2. Let X be a random variable in $\mathcal{L}$ with $E(X) = \mu > 0$. We define the log-variance as

$$v(X) = E(\log X - \log \mu)^2 = E\left(\log \frac{X}{\mu}\right)^2.$$

(a) If $X \sim LN(\mu, \sigma^2)$, prove $v(X) = \sigma^2 + \frac{\sigma^4}{4}$
(b) If X has a classical Pareto distribution with parameters $\alpha > 1$ and σ, prove $(\alpha > 1)$

$$v(X) = \log \frac{\alpha - 1}{\alpha} + \frac{1}{\alpha} + \frac{1}{\alpha^2}.$$

(c) How is $v(X)$ related to the variance of the logarithm of X?

3. Let X have an exponential distribution with pdf,

$$f(x; a, \tau) = \frac{1}{\tau} \exp(-(x - a)/\tau), \quad x \geq a \geq 0,$$

and $f(x; a, \tau) = 0$ if $x < a$.

(a) Prove that the Shannon entropy is $H(X) = 1 + \log \tau$.
(b) Obtain the Rényi entropy measure $H_\lambda(X)$.

4. The parameterized family of hyperbolic Lorenz curves is defined as

$$L(u; \alpha) = \frac{(\alpha - 1)u}{\alpha - u}, \quad 0 \leq u \leq 1,$$

with $\alpha > 1$ (Rohde 2009; Sarabia et al. 2010b) . Prove that

(a) The Gini index is

$$G = 1 + 2(\alpha - 1)\left[1 + \alpha \log\left(\frac{\alpha - 1}{\alpha}\right)\right].$$

(b) The Pietra index is

$$P = 2(\alpha - \sqrt{\alpha(\alpha - 1)}) - 1$$

(Sarabia et al. 2010b).

5. Let X have a Singh and Maddala (1976) distribution, also called Pareto IV distribution by Arnold (2015b), with cdf

$$F(x) = 1 - \frac{1}{(1 + (x/\sigma)^\gamma)^\alpha}, \quad x \geq 0,$$

where $\alpha, \gamma, \sigma > 0$.

(a) Prove that the sequence of absolute Gini indices are of the form

$$m_n = \frac{\sigma\Gamma(n\alpha - 1/\gamma)\Gamma(1+1/\gamma)}{\Gamma(n\alpha)}, \quad n = 1, 2, \ldots$$

(b) Obtain the sequences of relative Gini indices G_n and $\tilde{G}_n$, $n = 1, 2, \ldots$

6. The Pareto III or Fisk distribution (Fisk 1961a) is defined in terms of its cdf by

$$F(x) = 1 - \frac{1}{1 + \left(\frac{x}{\sigma}\right)^{\gamma}}, \quad x \geq 0,$$

where $\sigma > 0$ and $0 < \gamma < 1$.

(a) Check that the Gini index is given by γ.
(b) Prove that the mean of the Fisk distribution is $\mu = \frac{\sigma\pi\gamma}{\sin(\pi\gamma)}$.
(c) Express the parameter σ in terms of the mean and the Gini index.

7. Consider an exponential distribution with cdf $F(x) = 1 - e^{-x/\lambda}$ for $x \geq 0$ and $F(x) = 0$ if $x < 0$ and $\lambda > 0$. Check that the sequence of absolute Gini indices are $m_n = \frac{1}{n}$ and obtain the sequences G_n and $\tilde{G}_n$. Compute the absolute and relative Gini indices in the case of the translated exponential distribution with cdf $F(x) = 1 - e^{-(x-a)/\lambda}$ if $x \geq a > 0$ and $F(x) = 0$ if $x < a$.

8. Let X be a random variable with cdf $F(x) = x^{\alpha}$ if $0 \leq x \leq 1$, with $\alpha > 0$. Compute the bound for the Gini index (5.66). Verify that equality in (5.66) is achieved when $\alpha = 1$.

9. If $X \sim \Gamma(\alpha, 1)$ is a classical gamma distribution with shape parameter $\alpha > 0$ and unit scale parameter, prove that the Amato index can be written as

$$A(X) = \sum_{n=0}^{\infty} \binom{1/2}{n} \frac{\Gamma(\alpha + 2n)}{\alpha^{2n}\Gamma(\alpha)}.$$

10. For a non-negative random variable X with pdf f and $\mu_r = E(X^r) < \infty$ for $r \in \mathbb{R}$, consider the pdf of the r th moment distribution, which will be denoted by f_r and defined as

$$f_r(x) = \frac{x^r f(x)}{\mu_r}.$$

Let

$$I(f, g) = \int_{-\infty}^{\infty} f(x) \log\left(\frac{f(x)}{g(x)}\right) dx$$

denote the Kullback–Leibler (K-L) divergence between f and g.

(a) Prove that the K-L divergences between f and f_r are given by

$$I(f, f_r) = \log E(X^r) - E(\log X^r) = \log\left(\frac{A(X^r)}{G(X^r)}\right)$$

and

$$I(f_r, f) = E_r(\log X^r) - \log E(X^r) = \log\left(\frac{G_r(X^r)}{A(X^r)}\right),$$

where $E_r(\log X^r)$ is the expected value of $\log X^r$ under the weighted distribution f_r, $A(X^r) = E(X^r)$, $G(X^r) = \exp(E(\log X^r))$ and $H(X^r) = \{E(X^{-r})\}^{-1}$ are the arithmetic, geometric, and harmonic means of the random variable X^r, respectively.

(b) If $X \sim LN(\mu, \sigma^2)$ i.e., a lognormal distribution, verify that $I(f, f_r) = I(f_r, f)$. Conversely, if $I(f, f_r) = I(f_r, f)$, $\forall r \in I$ for some interval I containing 0, then X has a lognormal distribution (Tzavelas and Economou 2012).

11. Let X be a random variable in $\mathcal{L}$ and let $T_1(X) = E[(X/\mu)\log(X/\mu)]$ be the corresponding Theil 1 index. As in the previous exercise we denote by f_r the pdf of the r-th moment distribution.

    (a) Prove that $T_1(X^r) = I(f_r, f)$ for all $r \in \mathbb{R}$.
    (b) Prove that X has a lognormal distribution if

$$T_1(X^r) = \log\left(\frac{A(X^r)}{G(X^r)}\right),$$

for all $r \in I$, for some interval I containing 0, where $A(X^r) = E(X^r)$ and $G(X^r) = \exp(E(\log X^r))$ are the arithmetic and geometric means of the random variable X^r (Tzavelas and Economou 2012).

12. Obtain the asymptotic distribution of the variant Dalton index defined in (5.101).

13. If X has a lognormal distribution with parameters μ and σ^2, obtain the asymptotic distributions of the sample standard deviation and of the sample coefficient of variation.

14. For the exponential distribution with mean 1, prove Eq. (5.92), that is, the expectation of the sample Gini index G_n' is $\frac{n-1}{2n}$.

15. Prove Theorem 5.4.3. *Hint*: Use (5.60) and take into account that $\mu = \sum_{j=1}^{k} p_j \mu_j$, where $E(X) = \mu$ and $E(X_j) = \mu_j$.

Chapter 6
Families of Lorenz Curves

The Lorenz curve is an important instrument for analyzing the size distributions of income, wealth, and inequality. The problem of finding an appropriate functional form for a given data set or class of data sets is an important practical and theoretical problem. In this chapter we study parametric models for Lorenz curves and some of their applications.

We begin studying the basic properties that a function should satisfy in order to be a genuine Lorenz curve. We follow this by identifying the Lorenz curves of some common distributions. Then, we study different ways of generating new and more flexible families of Lorenz curves from one or more baseline Lorenz curves. After that we turn to a more careful study of specific families of Lorenz curves that have been proposed in the literature. In the last section, we analyze some alternative inequality curves. First, we consider the generalized and absolute Lorenz curves. We continue with the Leimkuhler, Bonferroni, and Zenga curves and with two new inequality curves constructed for studying the lower and middle income groups. We also study some reliability curves and the economic concept of relative deprivation, and its connections with the Lorenz curve.

6.1 Basic Results

In this section we study some basic issues regarding Lorenz curves. After considering a characterization theorem, we obtain the Lorenz curves of four common distributions. Then, we study the Lorenz curves of translated and truncated random variables.

B. C. Arnold, J. M. Sarabia, *Majorization and the Lorenz Order with Applications in Applied Mathematics and Economics*, Statistics for Social and Behavioral Sciences,
https://doi.org/10.1007/978-3-319-93773-1_6

6.1.1 A Characterization of the Lorenz Curve

The following characterization Theorem of the Lorenz curve is attributed to Gaffney and Anstin by Pakes (1981).

Theorem 6.1.1 *Assume that* $L(u)$ *is defined and continuous in the interval* $[0, 1]$ *with second derivative* $L''(u)$. *The function* $L(u)$ *is a Lorenz curve iff*

$$L(0) = 0, \quad L(1) = 1, \quad L'(0^+) \geq 0, \quad L''(u) \geq 0, \quad for\ u \in (0, 1). \tag{6.1}$$

The conditions given in (6.1) are readily checked for the curves catalogued in this chapter. It is these conditions that must be checked when we are building new curves from old ones.

Note that if $L_X(u)$ is a Lorenz curve corresponding to a random variable X with cdf $F_X(x)$ and mean μ_X we have

$$F_X^{-1}(x) = \mu_X L'_X(x). \tag{6.2}$$

A Lorenz curve determines the distribution of X up to a scale factor. To obtain the pdf of X from its Lorenz curve we can use Eq. (3.6) in Chap. 3.

Example 6.1.1 Let us consider the Lorenz curve,

$$L(u; \alpha) = \frac{1}{2}(u + u^{\alpha}), \quad 0 \leq u \leq 1,$$

where $\alpha \geq 1$. Using formula (6.2), the cdf corresponding to this Lorenz curve is

$$F_X(x; \alpha, \mu) = \frac{1}{\alpha^{1/(\alpha-1)}}\left(\frac{2x}{\mu} - 1\right)^{1/(\alpha-1)}, \quad \frac{\mu}{2} \leq x \leq \frac{\mu(\alpha+1)}{2},$$

and $F_X(x; \alpha, \mu) = 0$ if $x \leq \frac{\mu}{2}$ and $F_X(x; \alpha, \mu) = 1$ if $x \geq \frac{\mu(\alpha+1)}{2}$. Note the cdf depends on two parameters: α and a scale parameter μ, which represents the mean of the distribution.

6.1.2 Lorenz Curves of Some Common Distributions

In this section we obtain the Lorenz curves of four common probability distributions: the uniform, exponential, classical Pareto and lognormal distributions. These Lorenz curves will be useful for generating more sophisticated families of Lorenz curves.

Let us consider the uniform distribution $U[a, b]$ on the interval $[a, b]$ with cdf $F(x) = \frac{x-a}{b-a}$, with $0 < a \leq x \leq b$. Then,

$$\begin{aligned} L(u) &= \frac{1}{\mu}\int_0^u F^{-1}(t)dt \\ &= \frac{2}{a+b}\int_0^u [(b-a)t+a]dt \\ &= \frac{2}{a+b}\left(\frac{(b-a)u^2}{2}+au\right). \end{aligned}$$

Next, we consider the exponential distribution with cdf $F(x) = 1 - e^{-x/\beta}$ if $x \geq 0$, where $\mu_X = \beta$. It is straightforward to show that

$$L(u) = u + (1-u)\log(1-u), \quad 0 \leq u \leq 1. \tag{6.3}$$

Note that (6.3) does not depend on β because it is a scale parameter.

Now, we consider two relevant families of income distributions: the classical Pareto and the lognormal distributions. The classical Pareto distribution was introduced in Chap. 3, and it is defined in terms of the cdf by

$$F(x) = 1 - \left(\frac{x}{\sigma}\right)^{-\alpha}, \quad x \geq \sigma > 0,$$

and $F(x) = 0$ if $x < \sigma$. The Lorenz curve was obtained in Chap. 3 and is given by,

$$L(u;\alpha) = 1 - (1-u)^{1-1/\alpha}, \quad 0 \leq u \leq 1, \tag{6.4}$$

where $\alpha > 1$.

Finally, we consider the lognormal distribution $X \sim LN(\mu, \sigma^2)$, whose Lorenz curve is

$$L(u) = \Phi(\Phi^{-1}(u) - \sigma), \quad 0 \leq u \leq 1, \tag{6.5}$$

Figure 6.1 shows the Lorenz curves of the classical Pareto and lognormal distributions for some selected values of α and σ. Note that the lognormal curve is self-symmetric, that is, $L(1 - L(u)) = 1 - u$.

Table 6.1 summarizes the Lorenz curves with the corresponding Gini indices for these four common distributions.

6.1.3 *Translated and Truncated Lorenz Curves*

Many distributions can be obtained from a baseline distribution corresponding to a random variable X by adding location and scale parameters to obtain $Y = \lambda + \tau X$. How are the Lorenz curves of Y and X related? First note that, by the scale invariance property of Lorenz curves, X and τX ($\tau > 0$) have identical Lorenz

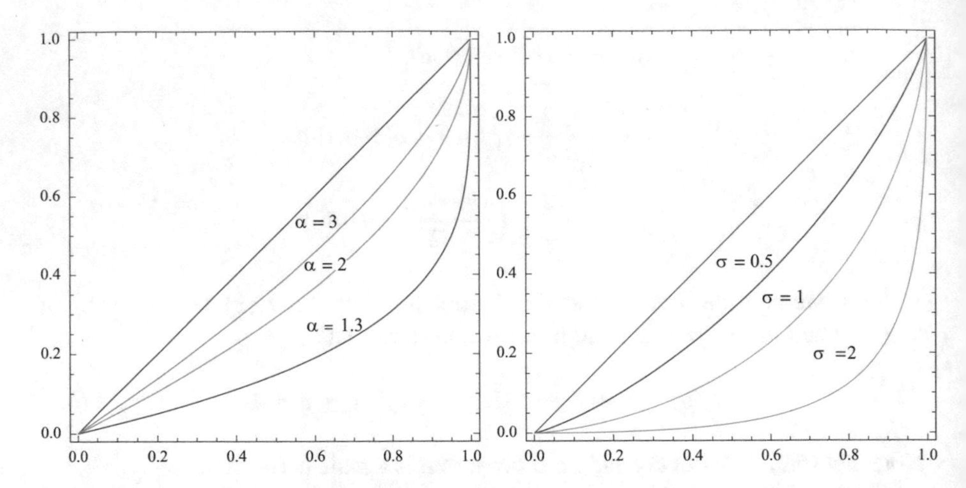

Fig. 6.1 Lorenz curves of the classical Pareto distribution (left) for $\alpha = 1.3$, 2, and 3 (Eq. (6.4)) and Lorenz curve of the lognormal distribution (right) for $\sigma = 0.5$, 1, and 2 (Eq. (6.5))

Table 6.1 Lorenz curves and Gini indices of four common distributions

Distribution	Lorenz curve	Gini index
Uniform $U[a, b]$	$L(u) = \dfrac{2au + (b-a)u^2}{a+b}$	$G = \dfrac{b-a}{3(a+b)}$
Exponential	$L(u) = u + (1-u)\log(1-u)$	$G = \dfrac{1}{2}$
Classical Pareto	$L(u; \alpha) = 1 - (1-u)^{1-1/\alpha}$	$G = \dfrac{1}{2\alpha - 1}, \quad \alpha > 1$
Lognormal	$L(u) = \Phi(\Phi^{-1}(u) - \sigma)$	$G = 2\Phi\left(\dfrac{\sigma}{\sqrt{2}}\right) - 1$

curves. A location change does however affect the Lorenz curve.The following result permits one to obtain the Lorenz curve of $Y = X + \lambda$ from the Lorenz curve of X.

Theorem 6.1.2 *Let X be a random variable in $\mathcal{L}$, with $E(X) = \mu_X$ and Lorenz curve $L_X(u)$. Let us consider the translated random variable $Y = X + \lambda$, with $\lambda \geq 0$. Then, the Lorenz curve of Y is given by*

$$L_Y(u) = \frac{\lambda u + \mu_X L_X(u)}{\lambda + \mu_X} = wu + (1-w)L_X(u), \tag{6.6}$$

where $w = \frac{\lambda}{\lambda + \mu_X}$, that is, $L_Y(u)$ is a convex combination (a finite mixture) of the egalitarian line u and $L_X(u)$ with weights w and $1 - w$.

Example 6.1.2 Consider the three parameter lognormal distribution with pdf,

$$f(y; \mu, \sigma, \lambda) = \frac{1}{(x-\lambda)x\sqrt{2\pi}} \exp\left\{-\frac{[\log(x-\lambda) - \mu]^2}{2\sigma^2}\right\}, \quad x > \lambda \geq 0, \tag{6.7}$$

where $\mu \in \mathbb{R}$ and $\sigma > 0$. Since $Y = X + \lambda$, where $X \sim LN(\mu, \sigma^2)$, using Eqs. (6.5) and (6.6), the Lorenz curve of (6.7) is

$$L_Y(u) = \frac{\lambda u + e^{\mu+\sigma^2/2}\Phi(\Phi^{-1}(u) - \sigma)}{\lambda + e^{\mu+\sigma^2/2}}, \quad 0 \le u \le 1.$$

The following theorem relates the Lorenz curves of truncated random variables (considering both lower and upper truncation) to the Lorenz curve of the untruncated variable.

Theorem 6.1.3 *Let X be a random variable in $\mathcal{L}$ with cdf $F_X(x)$ and Lorenz curve $L_X(u)$.*

(i) Define the lower truncated random variable (the random variable truncated from below at c) by

$$X_{(c)} = \{X|(X \ge c)\},$$

with $c \ge 0$. Then, the Lorenz Curve of $X_{(c)}$ is

$$L_{X_{(c)}}(u) = \frac{L_X[(1 - F_X(c))u + F_X(c)] - L_X(F_X(c))}{1 - L_X(F_X(c))}, \quad 0 \le u \le 1. \tag{6.8}$$

(ii) Define the upper truncated random variable (the random variable truncated from above at c) by

$$X^{(c)} = \{X|(X \le c)\},$$

with $c \ge 0$. Then, the Lorenz Curve of $X^{(c)}$ is

$$L_{X^{(c)}}(u) = \frac{L_X(F_X(c)u)}{L_X(F_X(c))}, \quad 0 \le u \le 1. \tag{6.9}$$

According to Eqs. (6.8) and (6.9) if $L_0(p)$ is a baseline Lorenz curve, the functional forms,

$$L_1(u; w) = \frac{L_0[(1 - w)u + w] - L_0(w)}{1 - L_0(w)}, \quad 0 \le u \le 1,$$

$$L_2(u; w) = \frac{L_0(wu)}{L_0(w)}, \quad 0 \le u \le 1,$$

with $w \in (0, 1)$, are genuine Lorenz curves.

6.1.4 The Modality of the Income Density Function

In this section we study the relation between the Lorenz curve and the modality of its underlying income density. We have the following result (Krause 2014).

Theorem 6.1.4 *If the Lorenz curve $L(u)$ has a third derivative $L'''(u)$ and the cumulative distribution $F(x)$ has a finite positive and differentiable density $f(x)$ in the interval (x_L, x_U), then if and only if $L'''(u)$ has $n \geq 1$ sign changes from $L'''(u) < 0$ to $L'''(u) > 0$ occurring at n points $\tilde{\pi}_i$, with $i = 1, 2, \ldots, n$, then $f'(x)$ has the corresponding sign changes from $f'(x) > 0$ to $f'(x) < 0$ occurring at n points denoted by $\tilde{x}_i$ with $i = 1, 2, \ldots, n$. This means that $f(x)$ is n-modal with modes at $\tilde{x}_i$ with $i = 1, 2, \ldots, n$.*

The proof of this result is based on the Equation,

$$f'(x) = -\frac{f(x)L'''(F(x))}{\mu[L''(F(x))]^2}.$$

For example, if $L'''(u) > 0$ ($L'''(u) < 0$) for all $u \in (0, 1)$, then $f'(x) < 0$ ($f'(x) > 0$) for all $x \in (x_L, x_U)$, and $f(x)$ is zeromodal and downward-sloping (upward-sloping). For the Pareto Lorenz curve $L(u) = 1-(1-u)^\delta$, with $0 \leq u \leq 1$ and $0 < \delta \leq 1$ we have

$$L'''(u) = (2-\delta)(1-\delta)\delta(1-u)^{\delta-2} > 0, \quad u \in (0, 1),$$

confirming that $f(x)$ is zeromodal and downward-sloping. The modality of several other parametric families of Lorenz curves has been studied by Krause (2014).

6.2 The Alchemy of Lorenz Curves

The following set of Theorems permit the generation of new and more flexible families of Lorenz curves from one or more baseline Lorenz curves. These results are readily proved using the Gaffney and Anstis conditions (6.1) given in Theorem 6.1.1. See also Sarabia (2008a).

Theorem 6.2.1 *The expression $L(u) = u^\gamma$ with $\gamma \geq 1$ defines a Lorenz curve. More generally, if $L_0(u)$ is a Lorenz curve, then*

$$L(u) = [L_0(u)]^\gamma, \tag{6.10}$$

with $\gamma \geq 1$ is also a Lorenz curve.

The following result was provided by Sarabia et al. (1999).

Theorem 6.2.2 *Let $L_0(u)$ be a Lorenz curve and consider the function,*

$$L_\alpha(u) = u^\alpha L_0(u), \quad \alpha \geq 0. \tag{6.11}$$

Then, if $\alpha \geq 1$, $L_\alpha(u)$ is a valid Lorenz curve. In addition, if $0 \leq \alpha < 1$ and the third derivative of $L_0(u)$ is non-negative, then $L_\alpha(u)$ is a Lorenz curve as well.

The condition that $L_0'''(u) \geq 0$ in Theorem 6.2.2 can be relaxed as in the following theorem (Wang et al. 2009).

Theorem 6.2.3 *Assume that $L_0(u)$ is a Lorenz curve. Then,*

$$\tilde{L}(u) = u^{\alpha} L_0(u)^{\gamma},$$

is a Lorenz curve for any $\alpha \geq 0$ and $\gamma \geq 1$. Furthermore, if $L_0'''(u) \geq 0$ for all $u \in [0, 1]$, then $\tilde{L}(u)$ is a Lorenz curve if $\alpha \geq 0$, $\gamma \geq \frac{1}{2}$ and $\alpha + \gamma \geq 1$.

The next theorem provides the reflected version of a given Lorenz curve L_0.

Theorem 6.2.4 *Let $L_0(u)$ be a genuine Lorenz curve. Then,*

$$L(u) = 1 - L_0^{-1}(1 - u), \quad 0 \leq u \leq 1,$$

is a genuine Lorenz curve.

Example 6.2.1 If $L_0(u) = 1 - (1 - u)^{\delta}$, with $0 < \delta \leq 1$ is the Pareto's LC (Eq. (6.4)), the corresponding reflected Lorenz curve is the power Lorenz curve $L(u) = u^{1/\delta}$.

Theorem 6.2.5 *If $L_1(u)$ and $L_2(u)$ are genuine Lorenz curves, the following expressions are again Lorenz curves,*

(i) $L(u) = wL_1(u) + (1 - w)L_2(u)$, *with* $0 \leq w \leq 1$,
(ii) $L(u) = L_1(u)^{\alpha_1} L_2(u)^{\alpha_2}$, *with* $\alpha_1, \alpha_2 \geq 1$,
(iii) $L(u) = L_2(L_1(u))$,
(iv) $L(u) = \max\{L_1(u), L_2(u)\}$.

The proof of this theorem is straightforward using the conditions provided by Theorem 6.1.1. Part *(iii)* is included in Sarabia et al. (2017b). See also the discussion of distorted Lorenz curves in Sect. 6.3.3 below.

These results can be extended to the case of more than two Lorenz curves as follows.

Theorem 6.2.6 *If $L_1(u), \ldots, L_m(u)$ are valid Lorenz curves, the following expressions are also Lorenz curves,*

(i) $L(u) = \sum_{i=1}^{m} w_i L_i(u)$, *with* $w_i \geq 0$, $i = 1, \ldots, m$ *and* $\sum_{i=1}^{n} w_i = 1$,
(ii) $L(u) = \prod_{i=1}^{n} L_i(u)^{\alpha_i}$, *with* $\alpha_i \geq 1$, $i = 1, \ldots, m$
(iii) $L(u) = L_m(L_{m-1}(\cdots(L_1(u))\cdots)))$,
(iv) $L(u) = \max\{L_1(u), \ldots, L_m(u)\}$.

Note 1 As noted in Arnold (2015b), it is possible for a linear combination of Lorenz curves of the form $\sum_{k=1}^{m} c_k L_k(u)$ with some c_k's negative to be a valid Lorenz curve.

Note 2 If L_1 and L_2 are Lorenz curves, let us consider the functional form,

$$L(u; \alpha, \nu) = L_1(u)^{\alpha} L_2(u)^{\nu}, \quad 0 \leq u \leq 1, \tag{6.12}$$

with $\alpha, \nu \geq 0$. As an extension of part (ii) in Theorem 6.2.5, Wang et al. (2011) have studied different sets of constraints about the parameters α, ν and conditions on the curves L_1 and L_2 in order to have a genuine Lorenz curve.

As an extension of the finite mixtures given by part (iv) in Theorems 6.2.5 and 6.2.6 we have the following result (Sarabia et al. 2005).

Theorem 6.2.7 *If $L(u; \delta)$ is an indexed collection of valid Lorenz curves where $\delta \in \Delta$, and if $\pi(\delta; \theta)$, $\theta \in \Theta$ denotes a parametric family of densities on the set Δ, it follows that*

$$L(u; \theta) = \int_{\Delta} L(u; \delta)\pi(\delta; \theta)d\delta, \quad \theta \in \Theta, \tag{6.13}$$

is a parametric family of genuine Lorenz curves.

Example 6.2.2 Consider a power Lorenz curve $L(u; \delta) = u^{\delta+1}$, where $\delta > 0$. Assume that the parameter δ is not constant and is modelled according to a Gamma distribution (i.e., $\Gamma(\alpha, \beta)$) with pdf,

$$\pi(\delta; \alpha, \beta) = \frac{\delta^{\alpha-1}e^{-\delta/\beta}}{\beta^{\alpha}\Gamma(\alpha)}, \quad \delta > 0,$$

with $\alpha, \beta > 0$. The mixture Lorenz curve (which might be called a power-gamma curve) is (using Eq. (6.13))

$$L(u; \alpha, \beta) = \int_0^{\infty} u^{\delta+1}\pi(\delta; \alpha, \beta)d\delta = \frac{u}{(1 - \beta \log u)^{\alpha}}, \quad 0 \leq u \leq 1.$$

An extension of part (ii) in Theorem 6.2.6 has been proposed by Sarabia (2013).

Theorem 6.2.8 *If $L(u; \delta)$ is an indexed collection of valid Lorenz curves where $\delta \in \Delta$, and if $\pi(\delta; \theta)$, $\theta \in \Theta$ denotes a parametric family of densities on the set Δ, then*

$$L(u; \theta) = \exp\left\{\int_{\Delta} [\log L(u; \delta)]\pi(\delta; \theta)d\delta\right\}, \quad \theta \in \Theta, \tag{6.14}$$

is a parametric family of genuine Lorenz curves.

6.3 Parametric Families of Lorenz Curves

Using the results in Sects. 6.1 and 6.2 we can generate an extensive array of parametric families of Lorenz curves. In this section we will present some specific parametric families of Lorenz curves which have been proposed in the literature.

6.3.1 Some Hierarchical Families

If we begin with the Lorenz curve of the classical Pareto distribution

$$L_0(u;k) = 1 - (1-u)^k, \ \ 0 < k \le 1,$$

using Theorems 6.2.1 and 6.2.2 we can consider the following hierarchy of Lorenz curves,

$$L_1(u;k,\alpha) = u^\alpha[1-(1-u)^k], \ \ \alpha \ge 0, \tag{6.15}$$

$$L_2(u;k,\gamma) = [1-(1-u)^k]^\gamma, \ \ \gamma \ge 1, \tag{6.16}$$

$$L_3(u;k,\alpha,\gamma) = u^\alpha[1-(1-u)^k]^\gamma, \ \ \alpha \ge 0, \ \gamma \ge 1, \tag{6.17}$$

which will be called the Pareto hierarchy of Lorenz curves, since they originate from the classical Pareto distribution. The family (6.15) coincides with the family proposed by Ortega et al. (1991) while (6.16) can be recognized as the family proposed by Kakwani (1980b) and Rasche et al. (1980). A detailed study of the family (6.17) can be found in Sarabia et al. (1999).

This method has also been used to generate other hierarchies of Lorenz curves, by considering different baseline curves. If we begin with the exponential Lorenz curve, introduced by Chotikapanich (1993), defined by,

$$L_0(u;c) = \frac{e^{cu}-1}{e^c-1}, \ \ 0 \le u \le 1, \tag{6.18}$$

where $c \ge 0$, we obtain a new family of Lorenz curves called the exponential family of Lorenz curves, by Sarabia et al. (2001). Sarabia and Pascual (2002) have considered the following baseline curve,

$$L_0(u;b,c) = \frac{e^{cu}-bu-1}{e^c-b-1}, \ \ 0 \le u \le 1, \tag{6.19}$$

as an extension of (6.18) and have studied the corresponding hierarchical family.

A related general family of Lorenz curves has been considered by Basmann et al. (1990). They constructed a hierarchy of Lorenz curves by beginning with the initial curve,

$$L_0(u) = ue^{-b(u-1)}, \ \ 0 \le u \le 1, \tag{6.20}$$

which was initially proposed by Kakwani and Podder (1973) (see also Rao and Tam 1987). The general family of Lorenz curves obtained in this manner is defined by

$$L(u;\alpha,\beta,\gamma,\delta) = u^{\alpha u+\beta}e^{\gamma(u-1)+\delta(u^2-1)}, \ \ 0 \le u \le 1. \tag{6.21}$$

If we set $\beta = 1$ and $\alpha = \delta = 0$ in (6.21), we obtain the Lorenz curve (6.20).

Sarabia (1997) proposed an alternative methodology for the construction of Lorenz curves which involved specifying an appropriate parametric family of quantile functions and using them to generate families of Lorenz curves. To illustrate this, consider the generalized Tukey's lambda distribution defined in terms of its quantile function as follows:

$$F^{-1}(u) = \lambda_2[\lambda_1 + u^{\lambda_3} + (1-u)^{\lambda_4}], \quad 0 \leq u \leq 1.$$

First, conditions must be imposed on the parameters, the λ_i's, in order to ensure that $F^{-1}(u)$ is non-negative, integrable, and with support in some subset of $[0, \infty)$. Sarabia (1997) has identified the corresponding nested family of Lorenz curves which in the most general case is of the form

$$L(u) = \pi_1 u + \pi_2 u^{\delta_1} + \pi_3[1 - (1-u)^{\delta_2}], \quad 0 \leq u \leq 1, \tag{6.22}$$

where $\pi_1 + \pi_2 + \pi_3 = 1$. In the case in which $\pi_i \geq 0$ and $\delta_1 > 1$ and $\delta_2 \in (0, 1]$ we have a mixture of three Lorenz curves. However, other members of the family (6.22) can also be genuine Lorenz curves, for different choices of the parameters π_i and δ_i (see Sarabia 1997).

6.3.2 General Quadratic Lorenz Curves

In this section we study a simple, but quite flexible, class of Lorenz curves introduced by Villaseñor and Arnold (1984a,b, 1989), called general quadratic Lorenz curves.

To begin, consider a general quadratic form as follows:

$$ax^2 + bxy + cy^2 + dx + ey + f = 0 \tag{6.23}$$

If we impose a constraint that (6.23) pass through the points $(0, 0)$ and $(1, 1)$, Eq. (6.23) becomes in,

$$ax^2 + bxy + cy^2 + dx + ey = 0, \tag{6.24}$$

where $e = -(a + b + c + d)$. If we substitute $(u, L(u))$ for (x, y) in (6.24), the equation can be recognized as one defining implicitly a broad class of what we can call quadratic Lorenz curves.There are three classes of such curves: parabolic, hyperbolic, and elliptical.

Parabolic Lorenz Curves

If we set $b = c = 0$ in (6.24) then, after reparameterization we obtain,

$$L(u; w) = wu + (1-w)u^2, \quad 0 \leq u \leq 1 \tag{6.25}$$

where $w \in [0, 1]$. The Gini index corresponding to this Lorenz curve is given by

$$G(w) = \frac{1-w}{3}.$$

Note that $0 \leq G(w) \leq \frac{1}{3}$, which can be a shortcoming for practical use of this family of curves.

But, in fact, the class of parabolic Lorenz curves is really restricted. Further investigation will reveal that if $w = 1$ the Lorenz curve (6.25) corresponds to a degenerate distribution, while if $w \in [0, 1)$, the corresponding distribution is uniform over a finite interval in $[0, \infty)$.

Hyperbolic Lorenz Curves

If we set $b \neq 0$ and $c = 0$ in (6.24) we have a hyperbola, and the corresponding Lorenz curve can be written of the form,

$$L(u; \delta, \eta) = \frac{u(1+(\eta-1)u)}{1+(\eta-1)u+\delta(1-u)}, \quad 0 \leq u \leq 1, \tag{6.26}$$

where $\delta, \eta > 0$ and $\delta - \eta + 1 > 0$. This functional form has been studied in detail by Arnold (1986b). The corresponding Gini index is

$$G(\delta, \eta) = \frac{\delta}{\delta-\eta+1} + \frac{2\eta\delta}{(\delta-\eta+1)^2}\left[1 + \frac{\delta+1}{\delta-\eta+1}\log\left(\frac{\eta}{\delta+1}\right)\right], \tag{6.27}$$

provided $\delta - \eta + 1 \neq 0$. In the case $\delta - \eta + 1 = 0$ we have

$$G(\delta) = \frac{\delta}{3(1+\delta)}.$$

The probability density function corresponding to (6.26) is

$$f(x; \delta, \eta, \mu) = \frac{K}{[\delta x + (\eta-1)(\mu - x)]^{3/2}}, \quad \mu(1+\delta)^{-1} \leq x \leq \mu(1+\delta\eta^{-1}). \tag{6.28}$$

The pdf (6.28) is decreasing on its support if $\delta > \eta - 1$ and in this situation we can consider the reparameterization,

$$f(x; \alpha, \beta, \gamma) = \frac{K}{[1 + \frac{x-\alpha}{\beta}]^{3/2}}, \quad \alpha \leq x \leq \gamma,$$

where $\beta > 0$ and $0 < \alpha < \gamma < \infty$. On the other hand, if $\delta = \eta - 1$ the resulting pdf is uniform. Finally if $\delta < \eta - 1$, the pdf is increasing on its support, and then the model is not convenient to work with for income data.

A different reparameterization of (6.26) has been considered by Wang and Smyth (2015) and Sarabia et al. (2015). Several subfamilies of (6.26) have also been proposed in the literature. Aggarwal (1984) and Aggarwal and Singh (1984) have considered the subfamily corresponding to the choice $\eta = 1$ and $\delta = (\frac{1+\theta}{1-\theta})^2 - 1$. The resulting family of Lorenz curves is of the form

$$L(u; \theta) = \frac{(1-\theta)^2 u}{(1+\theta)^2 - 4\theta u}, \quad 0 \leq u \leq 1, \tag{6.29}$$

where $\theta \in (0, 1)$. A different reparameterization of (6.29) has also been considered by Rohde (2009) and Sarabia et al. (2010b).

Elliptical Lorenz Curves

The general class (6.24) contains elliptical Lorenz curves when $c \neq 0$ (taking $c = 1$ without loss of generality) and $b^2 - 4a < 0$. The class of elliptical Lorenz curves is given by Villaseñor and Arnold (1989)

$$L(u; a, b, d) = \frac{1}{2}\left[-(bu+e) - \sqrt{\alpha u^2 + \beta u + e^2}\right], \quad 0 \leq u \leq 1, \tag{6.30}$$

where

$$\alpha = b^2 - 4a,$$
$$\beta = 2be - 4d,$$
$$e = -(a+b+d+1),$$

and the parameters satisfy the four conditions:

$$\alpha < 0, \quad e < 0, \quad d \geq 0, \quad a + d \geq 1.$$

The pdfs associated to (6.30) are not complicated and can be written as (Villaseñor and Arnold 1989),

$$f(x; \nu, \tau, \eta) = \frac{K}{[1 + (\frac{x-\nu}{\tau})^2]^{3/2}}, \quad \tau\eta_1 + \nu \leq x \leq \tau\eta_2 + \nu, \tag{6.31}$$

where $0 < \eta_1 < \eta_2 < \infty$, $\tau > 0$, $\tau\eta_1 + \nu \geq 0$ and K is the normalizing constant.

Circular Lorenz Curves

A special subclass of the elliptical Lorenz curves are the circular Lorenz curves (Ogwang and Rao 1996), which are defined by

$$L(u; a) = 1 + a - \sqrt{(1+a)^2 - 2au - u^2}, \quad 0 \leq u \leq 1,$$

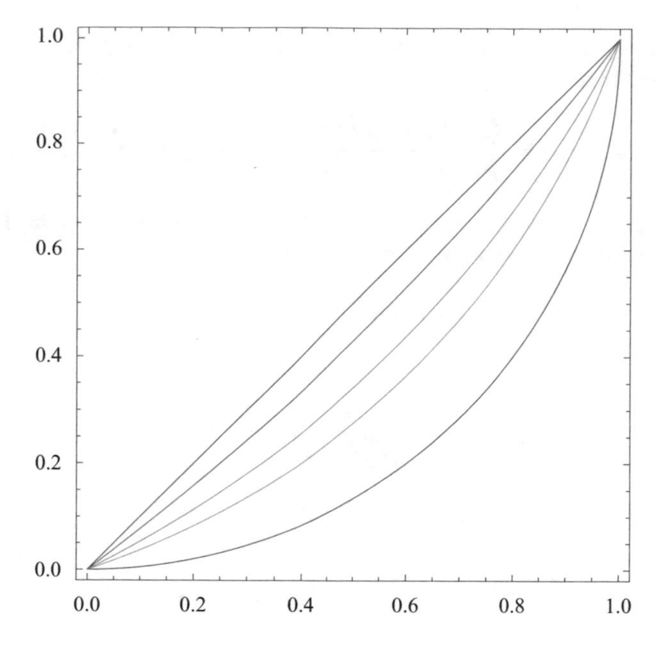

Fig. 6.2 Circular Lorenz curves for $a = 0, 0.5, 1$, and 3

where $a \geq 0$. The circular Lorenz curves are a special case of (6.30) with $a = c = 1$, $b = 0$, $d = 2a$, and $e = -2(1 + a)$ (see also Eq. (6.24)). This family corresponds to arcs of circles, passing through (0, 0) and (1, 1). The center of the circle is the point $(-a, 1 + a)$. Members of this class of curves are self-symmetric. Note that $\lim_{a\to\infty} L(u; a) = u$.

Since $\frac{\partial L(u;a)}{\partial a} \geq 0$, this family of curves is ordered with respect to a, thus,

$$a_1 \geq a_2 \Rightarrow L(u; a_1) \geq L(u; a_2).$$

Figure 6.2 shows circular Lorenz curves for some selected values of a
The Gini indices corresponding to circular Lorenz curves are given by

$$G(a) = -1 - 2a + (1 + 2a + 2a^2) \arcsin\left(\frac{1 + 2a}{1 + 2a + 2a^2}\right). \tag{6.32}$$

The cumulative distribution functions corresponding to the circular Lorenz curves are $(a, \mu > 0)$,

$$F(x; a, \mu) = \frac{-a(\mu^2 + x^2) + x\sqrt{(1 + 2a + 2a^2)(\mu^2 + x^2)}}{\mu^2 + x^2},$$

$$\frac{a\mu}{1+a} \leq x \leq \frac{(1 + a)\mu}{a},$$

and $F(x; a, \mu) = 0$ if $x \leq \frac{a\mu}{1+a}$ and $F(x; a, \mu) = 1$ if $x \geq \frac{(1+a)\mu}{a}$. The pdf's of the circular Lorenz curves are

$$f(x; a, \mu) = \frac{(1 + 2a + 2a^2)^{1/2}}{\mu[1 + (x/\mu)^2]^{3/2}}, \quad \frac{a\mu}{1+a} \leq x \leq \frac{(1+a)\mu}{a},$$

and $f(x; a, \mu) = 0$ otherwise, which are of the form (6.31). If we take $a = 0$ the extreme value of a, we obtain the Lorenz curve

$$L(u) = 1 - \sqrt{1 - u^2}, \quad 0 \leq u \leq 1, \tag{6.33}$$

with cdf

$$F(x; \mu) = \frac{x}{\sqrt{\mu^2 + x^2}}, \quad 0 \leq x < \infty$$

and $F(x; \mu) = 0$ if $x < 0$ and pdf

$$f(x; \mu) = \frac{1}{\mu[1 + (x/\mu)^2]^{3/2}}, \quad 0 \leq x < \infty$$

and $f(x; \mu) = 0$ otherwise. The Gini index of (6.33) is $(\pi/2) - 1$. Consequently, the Gini index (6.32) is bounded,

$$0 \leq G(a) \leq \frac{\pi}{2} - 1 = 0.57.$$

6.3.3 *Other Parametric Families*

In this section we discuss some other relevant models of Lorenz curves proposed in the literature.

Lorenz Curves Generated from Strongly Unimodal Distributions

Arnold et al. (1987) have proposed the class of Lorenz curves of the form,

$$L_F(u; \sigma) = F(F^{-1}(u) - \sigma), \quad 0 \leq u \leq 1, \tag{6.34}$$

where $F(\cdot)$ is any strongly unimodal distribution with unbounded support and $\sigma \geq 0$. For example, if $F = \Phi$, we obtain the Lorenz curve of the lognormal distribution. The Lorenz curve of the classical Pareto distribution given by (6.4) can be written of the form (6.34), with $F(x) = 1 - \exp(-e^x)$, one of the extreme value distributions. Other strongly unimodal choices for F do not seem to have received careful attention as possible income distribution models.

Maximum Entropy Lorenz Curves

Holm (1993) has proposed a class of Lorenz curves obtained from densities which have maximal entropy subject to side conditions on the Gini index and on the distance between the mean and the minimum income. The distributions that are identified in this manner have quantile functions satisfying,

$$\frac{dF^{-1}(p)}{dp} = \frac{c}{1 + \lambda_1(1-p) + \lambda_2 p(1-p)}, \quad 0 \le p \le 1, \tag{6.35}$$

where λ_1 and λ_2 are the corresponding Lagrange multipliers arising from the side conditions. If $\lambda_1 \neq 0$ and $\lambda_2 = 0$, the corresponding Lorenz curve is of the form

$$L(u;r) = u + G\frac{(r-u)\log(1-u/r) - (r-1)\log(u-1/r)u}{r(r-1)\log(1-1/r) + r - 0.5}, \quad 0 \le u \le 1, \tag{6.36}$$

where G is the Gini index and $r \in (-\infty, 0) \cup (1, \infty)$. The limit as $r \downarrow 1$ is

$$L(u) = u + 2G(1-u)\log(1-u), \quad 0 \le u \le 1,$$

which corresponds to a shifted exponential distribution (see Problem 2). In the case $\lambda_1 = 0$ and $\lambda_2 \neq 0$ we obtain a more complicated Lorenz curve, which includes as a limiting case the Lorenz curve,

$$L(u) = u + G[u\log u + (1-u)\log(1-u)], \quad 0 \le u \le 1.$$

Finally, in the case in which $\lambda_i \neq 0$, $i = 1, 2$ a more complicated two parameter family of Lorenz curves is encountered. It includes, as a limiting case, the truncated Pareto distribution.

Lorenz Curves Based on Generating Functions

Sarabia et al. (2010a) have proposed a methodology for constructing Lorenz curves, using the generating functions of positive integer valued random variables. Let X be a discrete random variable with probability generating function

$$P_X(s) = E(s^X) = \sum_{j=1}^{\infty} P(X=j)s^j.$$

and let $L_0(u)$ be an arbitrary baseline Lorenz curve. It follows that the expression

$$L(u) = P_X(L_0(u)), \quad 0 \le u \le 1, \tag{6.37}$$

defines a genuine Lorenz curve. Note that (6.37) can be written as

$$L(u) = P_X(L_0(u)) = \sum_{j=1}^{\infty} P(X = j)\{L_0(u)\}^j,$$

which is a countable mixture of Lorenz curves and, as a consequence, a valid Lorenz curve.

We consider two relevant examples. If X has a geometric distribution with probability mass function $P(X = k) = \pi(1 - \pi)^{k-1}$, $k = 1, 2, \ldots$, then using (6.37) we have

$$L(u) = P_X(L_0(u)) = \frac{\pi L_0(u)}{1 - (1 - \pi)L_0(u)}, \quad 0 < u \leq 1.$$

If instead we consider the random variable $X = Y + 1$ where Y has a Poisson distribution with mean λ, then using (6.37) we obtain

$$L(u) = P_X(L_0(u)) = L_0(u) \exp\{\lambda(L_0(u) - 1)\}, \; 0 < u \leq 1. \tag{6.38}$$

If $L_0(u) = u$ is used in (6.38), we obtain the Lorenz curve proposed by Gupta (1984).

On the other hand, if we consider a random variable X which takes non-negative (rather than strictly positive) values with generating function $P_X(s)$, we can consider the classes of Lorenz curves,

$$L_1(u) = \{L_0(u)\}^{\alpha} P_X(L_0(u))$$

and

$$L_2(u) = \frac{P_X(L_0(u)) - p_0}{1 - p_0},$$

where $L_0(u)$ is any baseline Lorenz curve and $p_0 = P(X = 0) > 0$.

Distorted Lorenz Curves

The class of distorted Lorenz curves has been proposed by Sordo et al. (2013) using the concept of a distortion function. Several models of Lorenz curves can be obtained by distorting a Lorenz curve L by a function h, giving rise to a distorted Lorenz curve $\tilde{L} = h \circ L$. A distortion function is an increasing function $h : [0, 1] \to [0, 1]$ such that $h(0) = 0$ and $h(1) = 1$. If, in addition, $h''(t) \geq 0$ for all $t \in (0, 1)$, then $h(u)$ is itself a Lorenz curve and by Result (iii) of Theorem 6.2.5, $\tilde{L}$ defines a Lorenz curve which can be viewed as a distortion of L. Sordo et al. (2013) have explored the role of these curves in the context of the axiomatic structure of Aaberge (2001) used for ordering on the set of Lorenz curves.

Note that the generating function of a positive integer valued random variable is a distortion function, so that the class of distorted Lorenz curves includes curves of the form (6.37), but the distorted Lorenz class is more extensive than that represented by (6.37).

We have seen that, for a Lorenz curve $L(u)$, if h is also a Lorenz curve then $h(L(u))$ is itself a Lorenz curve. But $h(L(u))$ can be a Lorenz curve even when h is not convex (i.e., not a Lorenz curve). For example, take $L(u) = u^2$ and $h(u) = u^{3/4}$ in which case $h(L(u)) = u^{3/2}$, a valid Lorenz curve.

For a particular Lorenz curve $L(u)$, the class of functions h for which $h(L(u))$ is a valid Lorenz curve will depend on the specific nature of $L(u)$. For a trivial example, if $L(u) = u$ then in order for $h(L(u))$ to be a Lorenz curve, it is necessary and sufficient that h be a Lorenz curve. As a consequence of this observation, a necessary and sufficient condition for $h(L(u))$ to be a valid Lorenz curve for every Lorenz curve $L(u)$ is that h be itself a Lorenz curve.

In parallel fashion one may consider a variant distortion mechanism as follows. For a given Lorenz curve $L(u)$, we consider a composition of the form $\tilde{\tilde{L}}(u) = L(g(u))$ and seek to determine for which choices of the function g will $\tilde{\tilde{L}}(u)$ be a valid Lorenz curve. A simple sufficient condition for this is, of course, that $g(u)$ itself should be a Lorenz curve, for then one can apply condition (iii) of Theorem 6.2.5. But $L(g(u))$ can be a Lorenz curve even when g is not convex (i.e., not a Lorenz curve). For example, again take $L(u) = u^2$ and $g(u) = u^{3/4}$ in which case $L(g(u)) = u^{3/2}$, a valid Lorenz curve.

For a given choice of $L(u)$, necessary and sufficient conditions for $L(g(u))$ to be a valid Lorenz curve will depend on the specific nature of $L(u)$. By again considering the case in which $L(u) = u$, we may however conclude that a necessary and sufficient condition on g for $L(g(u))$ to be a valid Lorenz curve for every choice of $L(u)$ is that g be a Lorenz curve.

The Family of Arctan Lorenz Curves

This family of Lorenz curves makes use of the arctan function acting on a baseline Lorenz curve. If $L_0(u)$ is a Lorenz curve, Gómez-Déniz (2016) defines the following family of Lorenz curves,

$$L(u;\alpha) = 1 - \frac{\arctan(\alpha(1 - L_0(u)))}{\arctan(\alpha)}, \quad 0 \le u \le 1, \tag{6.39}$$

where $\alpha \in \mathbb{R}$ and $\alpha \neq 0$. If $\alpha \to 0$ in (6.39), then $L(u;\alpha) \to L_0(u)$. Note that $L(u;\alpha) = L(u;-\alpha)$, so that we may quite reasonably restrict attention to the case in which $\alpha > 0$. If we substitute the egalitarian Lorenz curve $L_0(u) = u$ in (6.39) we have the family,

$$L(u;\alpha) = 1 - \frac{\arctan(\alpha(1 - u))}{\arctan(\alpha)}, \quad 0 \le u \le 1, \tag{6.40}$$

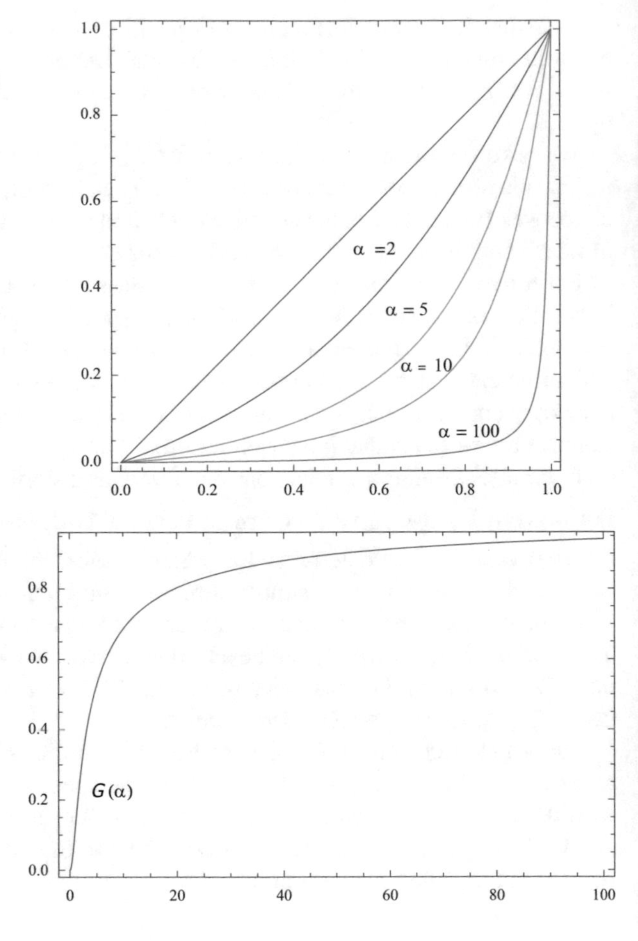

Fig. 6.3 Arctan Lorenz curves (6.40) for some selected values of α (top) and the Gini index (6.41) (bottom) as a function of α

with Gini index,

$$G(\alpha) = 1 - \frac{\log(1+\alpha^2)}{\alpha \arctan(\alpha)}. \tag{6.41}$$

Figure 6.3 includes plots the Lorenz curves (6.40) together the corresponding Gini indices (6.41) for various values of α. A detailed study of this family including Lorenz ordering, inequality measures and applications can be found in Gómez-Déniz (2016).

A more general family of Lorenz curves of the Gómez-Déniz type is obtained by considering H to be an arbitrary increasing and concave function with $H(0) = 0$ and defining

$$L(u;\alpha) = 1 - \frac{H(\alpha(1-u))}{H(\alpha)}, \quad 0 \le u \le 1.$$

6.4 Some Alternative Inequality Curves

Some modified Lorenz curves have proven to be useful in social science settings. Two of them are described below in Sect. 6.4.1. Four alternative inequality curves, intimately related to the Lorenz curve, have been proposed and are described in Sect. 6.4.2. Two additional inequality curves for the lower and middle income groups are presented in Sect. 6.4.3. The section will close with discussion of two other inequality curves, one arising in reliability settings, the other in the context of relative deprivation.

6.4.1 Generalized and Absolute Lorenz Curves

The generalized Lorenz curve (GLC) introduced by Shorrocks (1983) is one of the most important variations of the Lorenz curve. The Lorenz curve is scale invariant and is thus only an indicator of relative inequality. However, it does not provide a complete basis for making social welfare comparisons. Shorrocks' proposal is the following.

Definition 6.4.1 The generalized Lorenz curve is defined by

$$GL_X(u) = \mu_X L_X(u) = \int_0^u F_X^{-1}(t)dt, \quad 0 \leq u \leq 1. \tag{6.42}$$

We have $GL_X(0) = 0$ and $GL_X(1) = \mu_X$. A distribution with a dominating GLC provides greater welfare according to all concave increasing social welfare functions defined on individual incomes (see Kakwani 1980a; Davies et al. 1998). It is evident that the GLC is not scale-free and, as a consequence, it completely determines any distribution with finite mean. The order induced by (6.42) is a new partial ordering, and sometimes it allows a larger percentage of curves to be ordered than does the Lorenz order. Normative interpretations of the restrictions required on the class of social welfare function in order to correspond to GLC dominance have been studied by Shorrocks and Foster (1987) and Davies and Hoy (1994), among others.

An alternative variation of the Lorenz curve has been proposed by Moyes (1987) as follows.

Definition 6.4.2 The absolute Lorenz curve is defined as

$$AL_X(u) = \mu_X[L_X(u) - u] = \int_0^u [F_X^{-1}(t) - \mu_X]dt, \quad 0 \leq u \leq 1.$$

This definition replaces scale invariance by location invariance.

6.4.2 Leimkuhler, Bonferroni and Zenga Curves

The Leimkuhler curve is used in the field of informetrics instead of the Lorenz curve. Sarabia (2008b) provides the following useful representation of this curve which clarifies its relationship with the Lorenz curve. We have the following definition.

Definition 6.4.3 Let X be a random variable in $\mathcal{L}$ with cdf F_X and mean μ_X. The Leimkuhler curve $K_X(u)$ of X is defined by

$$K_X(u) = \frac{1}{\mu_X} \int_{1-u}^{1} F_X^{-1}(t)dt, \quad 0 \le u \le 1.$$

The Leimkuhler curve is a continuous non-decreasing concave function with $K_X(0) = 0$ and $K_X(1) = 1$. A simple representation of the close relationship between the Lorenz and the Leimkuhler curve is provided by the equation,

$$K_X(u) = 1 - L_X(1-u), \quad 0 \le u \le 1. \tag{6.43}$$

The Leimkuhler partial order $\le_K$ can be defined as

$$X \le_K Y \Leftrightarrow K_X(u) \le K_Y(u), \quad u \in [0, 1].$$

Then, from (6.43) it is clear that the Leimkuhler partial order is the same as the Lorenz order. Sarabia and Sarabia (2008) have proposed different parametric families for the Leimkuhler curve.

Next, we define the Bonferroni (1930) curve.

Definition 6.4.4 The Bonferroni curve is defined in terms of the Lorenz curve by

$$B_X(u) = \frac{L_X(u)}{u}, \quad 0 < u \le 1.$$

It is clear that $L(u) \le B(u)$. The Bonferroni partial order $\le_B$ is defined as

$$X \le_B Y \Leftrightarrow B_X(u) \ge B_Y(u), \quad u \in (0, 1].$$

It is evident that the Bonferroni partial order is equivalent to the Lorenz and Leimkuhler orders. An alternative expression for computing the Bonferroni curve is

$$B_X(u) = \int_0^u \frac{[F_X^{(1)}]^{-1}(t)}{F_X^{-1}(t)} dt, \quad 0 < u \le 1,$$

where, as usual, $F^{(1)}(\cdot)$ is the first moment distribution of F_X.

We will continue by discussing two curves proposed by Zenga in 1984 and 2007. We will label these curves as Zenga-I and Zenga-II.

Definition 6.4.5 The Zenga-I curve (Zenga 1984) curve is defined by

$$Z_X^I(u) = 1 - \frac{F_X^{-1}(u)}{[F_X^{(1)}]^{-1}(u)}, \quad 0 \leq u \leq 1.$$

The Zenga-II curve (Zenga 2007) is defined by

$$Z_X^{II}(u) = 1 - \frac{(1-u)L_X(u)}{u(1-L_X(u))}, \quad 0 \leq u \leq 1. \tag{6.44}$$

The Zenga-I curve is scale invariant since it involves the quotient of two related quantile functions. The corresponding partial order is defined by

$$X \leq_{Z^I} Y \Leftrightarrow Z_X^I(u) \geq Z_Y^I(u), \quad 0 < u < 1.$$

The Zenga-I order is not equivalent to the Lorenz order. On the other hand, the Zenga-I curve does not determine the distribution function, that is, it is possible to have two essentially different distributions (not related by a scale change) with the same Zenga-I curve. A detailed discussion regarding characterizations of Zenga's curves has been provided by Arnold (2015a).

The Zenga-II curve is again scale invariant. From (6.44) it is straightforward to verify that

$$Z_X^{II}(u) = \frac{1-B_X(u)}{1-L_X(u)}, \quad 0 \leq u \leq 1. \tag{6.45}$$

We may recover the Lorenz curve from the Zenga-II curve using the expression,

$$L_X(u) = \frac{u(1-Z_X^{II}(u))}{1-uZ_X^{II}(u)}, \quad 0 \leq u \leq 1,$$

and consequently

$$L_X(u) \leq L_Y(u) \Leftrightarrow Z_X^{II}(u) \geq Z_Y^{II}(u), \quad 0 \leq u \leq 1,$$

so that the Zenga-II order and the Lorenz order coincide.

Example 6.4.1 We consider the lognormal distribution $X \sim LN(\mu, \sigma^2)$. The first moment distribution in this case is again of the lognormal form, i.e., $X_1 \sim LN(\mu + \sigma^2, \sigma^2)$. Then the corresponding Zenga-I curve is

$$Z_X^I(u) = 1 - \frac{F_X^{-1}(u)}{[F_X^{(1)}]^{-1}(u)} = 1 - \frac{\exp(\mu + \sigma\Phi^{-1}(u))}{\exp(\mu + \sigma^2 + \sigma\Phi^{-1}(u))} = 1 - e^{-\sigma^2},$$

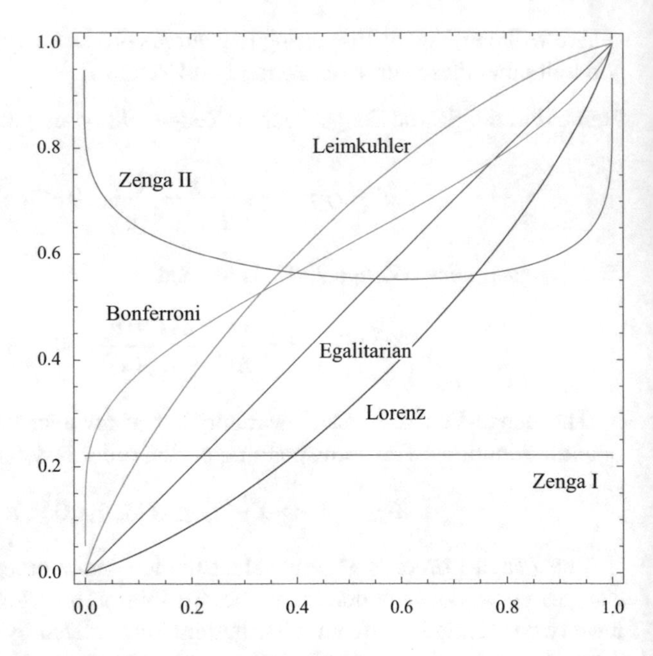

Fig. 6.4 The Lorenz, Leimkuhler, Bonferroni, Zenga I and Zenga II curves for a lognormal distribution with parameter $\sigma = 0.5$

which (perhaps surprisingly) does not depend on u. Using Eq. (6.45) the Zenga-II curve is

$$Z_X^{II}(u) = \frac{1 - B_X(u)}{1 - L_X(u)} = \frac{u - \Phi(\Phi^{-1}(u) - \sigma)}{u\Phi(\sigma - \Phi^{-1}(u))},$$

for $0 \leq u < 1$.

Figure 6.4 displays the Lorenz, Leimkuhler, Bonferroni, Zenga I and Zenga II curves for a lognormal distribution with parameter $\sigma = 0.5$. This figure sheds light on the different nature of these curves.

6.4.3 *Inequality Curves for the Lower and Middle Income Groups*

Here we review two new inequality curves defined by Gastwirth (2016) in terms of the Lorenz curve, for studying the status of the lower and middle income groups.

The first curve is defined by

$$J(u) = \frac{L(u)}{1 - L(1 - u)}, \quad 0 \leq u \leq 1, \tag{6.46}$$

where $L(\cdot)$ is a genuine Lorenz curve. For each u, the curve $J(u)$ is the ratio of the total income of the poorest uth fraction of the population to the total income of the highest uth fraction. This implies that as u increases, $J(u)$ increases as its numerator will increase, while the denominator will decrease.

The second curve $J_m(u)$ considers the status of the middle class. It is defined as,

$$J_m(u) = \frac{L(0.5 + \frac{u}{2}) - L(0.5 - \frac{u}{2})}{1 - L(1-u)}, \quad 0 \leq u \leq 1. \tag{6.47}$$

This curve is the ratio of the total income received by the middle uth fraction of the distribution to that of the upper uth fraction. From their definitions it is clear that $J_m(u) \geq J(u)$ for all u.

According to Gastwirth (2016), because the same proportion of the population is considered in both the numerator and denominator of (6.46) and (6.47), the measures can also be interpreted as the ratio of the average income received by the poorest $100u\%$ or middle $100u\%$ of the population to the average income of the upper $100u\%$.

For the case of the classical Pareto distribution $X \sim P(I)(\sigma, \alpha)$, the curves J and J_m are given by ($\alpha > 1$),

$$J(u) = \frac{1 - (1-u)^{1-1/\alpha}}{u^{1-1/\alpha}}, \quad 0 \leq u \leq 1, \tag{6.48}$$

and

$$J_m(u) = \frac{(0.5 + \frac{u}{2})^{1-1/\alpha} - (0.5 - \frac{u}{2})^{1-1/\alpha}}{u^{1-1/\alpha}}, \quad 0 \leq u \leq 1, \tag{6.49}$$

respectively.

If we consider a lognormal distribution $X \sim LN(\mu, \sigma^2)$, these curves are

$$J(u) = \frac{\Phi(\Phi^{-1}(u) - \sigma)}{\Phi(\sigma - \Phi^{-1}(1-u))}, \quad 0 \leq u \leq 1, \tag{6.50}$$

and

$$J_m(u) = \frac{\Phi(\Phi^{-1}(0.5 + \frac{u}{2}) - \sigma) - \Phi(\Phi^{-1}(0.5 - \frac{u}{2}) - \sigma)}{\Phi(\sigma - \Phi^{-1}(1-u))}, \quad 0 \leq u \leq 1, \tag{6.51}$$

respectively, with $\sigma > 0$ where, as usual, $\Phi(\cdot)$ is the cdf of the standard normal distribution.

Figure 6.5 shows the curves (6.48) and (6.49) in the Pareto case for some selected values of the parameter α. The curves (6.50) and (6.51) for the lognormal distribution are shown in Fig. 6.6.

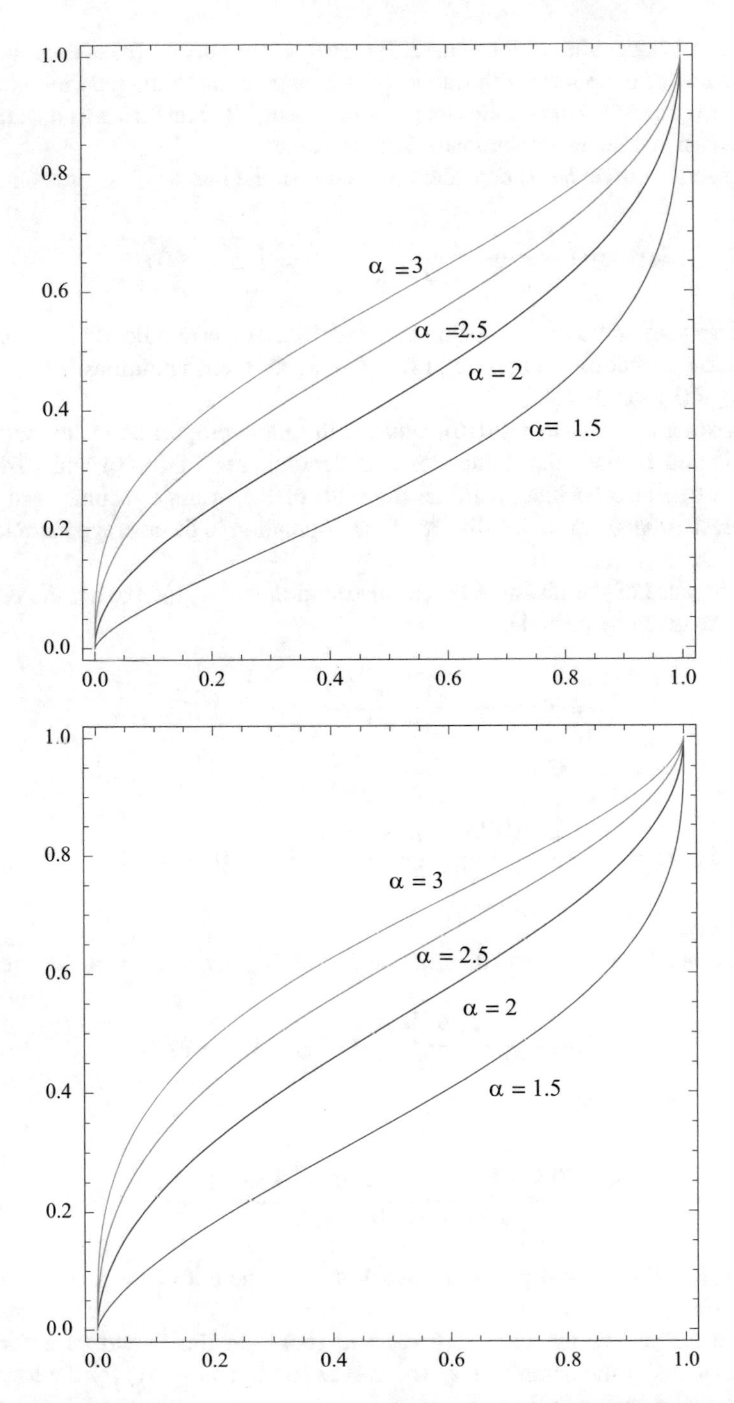

Fig. 6.5 The J (top) and J_m (bottom) curves for the classical Pareto distribution, for some selected values of α

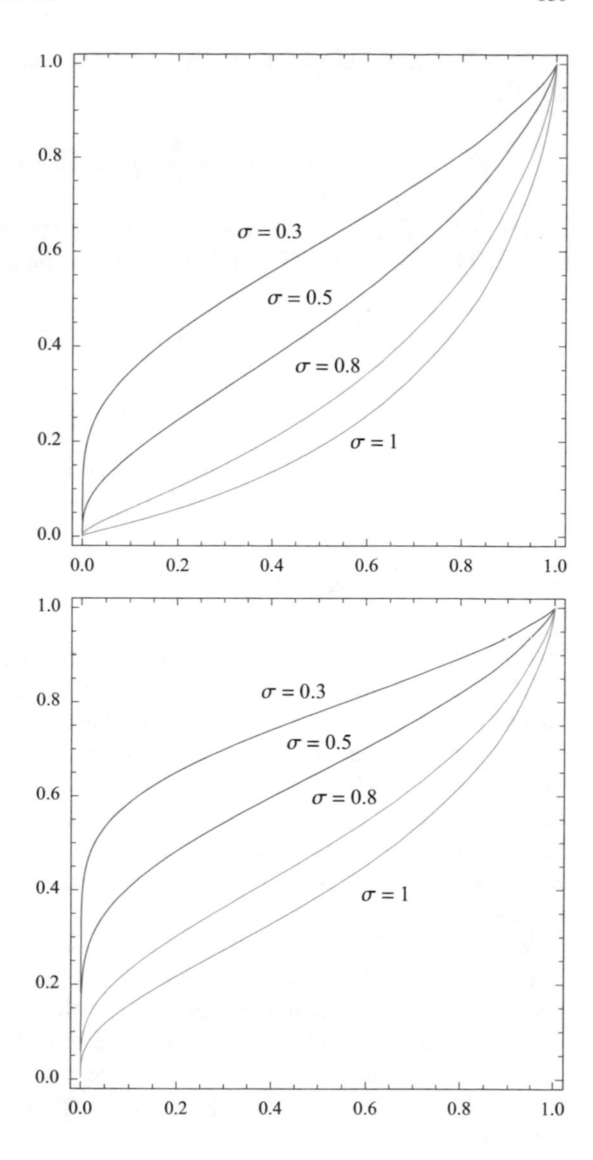

Fig. 6.6 The J (top) and J_m (bottom) curves for the lognormal distribution, for some selected values of σ

6.4.4 Reliability Curves

We next define a concept, arising in reliability contexts, which has some implications in wealth analysis.

Definition 6.4.6 The scaled total time on test transformation (STTTT) is defined as

$$\tilde{T}_X(u) = \frac{1}{\mu} \int_0^{F_X^{-1}(u)} [1 - F_X(v)] dv, \quad 0 \leq u \leq 1. \tag{6.52}$$

If F_X is strictly increasing on its support, we have (see Chandra and Singpurwalla 1981),

$$L_X(u) = \tilde{T}_X(u) - \frac{1}{\mu} F_X^{-1}(u)(1-u), \quad 0 < u < 1.$$

The $STTTT$ function is directly related to the excess wealth transform $W_X(u)$ proposed by Shaked and Shanthikumar (1998) and defined by

$$W_X(u) = \int_{F_X^{-1}(u)}^{\infty} [1 - F_X(v)]dv, \quad 0 < u < 1. \tag{6.53}$$

The functions (6.52) and (6.53) are related by the formula,

$$W_X(u) = \mu[1 - \tilde{T}_X(u)], \quad 0 < u < 1. \tag{6.54}$$

6.4.5 Relative Deprivation

In this section we introduce the concept of relative deprivation, which was proposed by Yitzhaki (1979). It is a concept that is consistent with the theory of attitudes to social inequality, the so-called theory of relative deprivation (Runciman 1966).

Following Yitzhaki (1983), we consider income as the object of relative deprivation: income should be considered as an index of the individual's ability to consume commodities. The range of possible deprivation of a person is $(0, \tilde{y})$, where $\tilde{y}$ is the highest income in the society. For each person with income y_i, we can consider two intervals: first $(y_i, \tilde{y})$, the range of deprived income and $(0, y_i)$, the range of satisfied income.

The total deprivation assigned to a person is the sum of the deprivation inherent in all units of income he is deprived of. Runciman (1966) defines the degree of deprivation inherent in not having y (the ith unit of income) as an increasing function of the proportion of persons in the society who have y. According to previous definition, the degree of relative deprivation of the range $(y, y + dy)$ can be quantified by $1 - F(y)$, which represents the relative frequency of persons with incomes above y.

Definition 6.4.7 For a person with income "y" we define the following two functions:

(i) The relative deprivation function:

$$D(y) = \int_{y}^{\tilde{y}} [1 - F(z)]dz, \tag{6.55}$$

(ii) The relative satisfaction function:

$$S(y) = \int_0^y [1 - F(z)]dz. \tag{6.56}$$

The following results were provided by Yitzhaki (1979). The first result relates the relative satisfaction function with the Lorenz curve of income.

Theorem 6.4.1 *The relative satisfaction function defined in (6.56) can be written as*

$$S(y) = y[1 - F(y)] + \mu L(F(y)), \tag{6.57}$$

where $L(u)$ represents the Lorenz curve of the income distribution.

Proof Integrating (6.56) by parts we have

$$S(y) = y[1 - F(y)] + \int_0^y zf(z)dz,$$

and we obtain directly (6.57). ■

The following Theorem relates the relative satisfaction in a society (that is, in all the population) with the Gini index of the income.

Theorem 6.4.2 *If we define the degree of relative satisfaction in the society, denoted by S, as the average of the relative satisfaction function defined in (6.56), we have that*

$$S = \mu(1 - G), \tag{6.58}$$

where G denotes the Gini index of the income distribution.

Proof We have

$$S = E[S(Y)] = \int_0^{\tilde{y}} S(z)f(z)dz,$$

and using (6.57), we obtain (6.58). ■

6.5 Exercises

1. We consider the polynomial functional form

$$L_k(u; \boldsymbol{\pi}) = \pi_0 + \pi_1 u + \cdots + \pi_k u^k, \quad 0 \le u \le 1,$$

with $k = 1, 2, \ldots$ and $\pi_i \in \mathbb{R}$.

(a) For $k = 1, 2$ and 3, identify the constraints on the parameters, the π_i's, in order to have genuine Lorenz curves.

(b) Compute the Gini indices $G_k(\boldsymbol{\pi}) = 2\int_0^1 [u - L_k(u; \boldsymbol{\pi})]du$, for $k = 1, 2, 3$.

2. Prove Theorem 6.1.3. *Hint*: first identify the cdf's of the random variables $X_{(c)}$ and $X^{(c)}$.

3. Assume both L_1 and L_2 are Lorenz curves and consider the functional form,

$$L(u) = L_1(u)^{\alpha} L_2(u))^{\nu}, \quad 0 \leq u \leq 1.$$

Prove that $L(u)$ is a genuine Lorenz curve if $\alpha \geq 0$, $\nu \geq 1$ and $L_1''(u)/L_1'(u)$ is increasing in [0, 1] (Wang et al. 2011).

4. Consider the two parameter exponential distribution with cdf,

$$F(x) = 1 - e^{-(x-\nu)/\tau}, \quad x \geq \nu,$$

and $F(x) = 0$ if $x < \nu$, with $\nu, \tau > 0$. Using Theorem 6.1.2 and Table 6.1, prove that the corresponding Lorenz curve is

$$L(u) = u + \left(1 + \frac{\nu}{\tau}\right)^{-1} (1 - u)\log(1 - u), \quad 0 \leq u \leq 1.$$

Verify also that the Gini index is given by $G = \frac{\tau}{2(\nu+\tau)}$.

5. The beta-Lorenz curve is defined as

$$L(u; a, b) = \frac{1}{B(a, b)} \int_0^u t^{a-1}(1 - t)^{b-1} dt, \quad 0 \leq u \leq 1.$$

(a) Prove that $L(u; a, b)$ is a genuine Lorenz curve if $a \geq 1$ and $0 < b \leq 1$.
(b) Show that the Gini index is $\frac{a-b}{a+b}$.
(c) Obtain the cdf of the corresponding random variable in the case in which $a = 1$ and $b = 1$.

6. Consider the (Chotikapanich 1993) Lorenz curve given by Eq. (6.18). Prove that the cdf corresponding to this LC is

$$F(x; k, \alpha) = \frac{1}{k} \log\left(\frac{x}{c_k \mu}\right), \quad c_k \mu \leq x \leq c_k \mu e^k,$$

where $c_k = k/(e^k - 1)$ and $F(x; k, \alpha) = 0$ if $x \leq c_k \mu$ and $F(x; k, \alpha) = 1$ if $x \geq c_k \mu e^k$.

7. Consider the two parameter functional form (Wang et al. 2009),

$$L(u; \beta, \gamma) = 1 - (1 - u)^{\beta} e^{-\gamma u}, \quad 0 \leq u \leq 1.$$

Using Theorem 6.1.1 prove that $L(u)$ is a genuine Lorenz curve if $\beta > 0$ and $0 \leq \beta + \gamma \leq \sqrt{\beta}$.

8. The hyperbolic Lorenz curve is of the form,

$$L(u; \sigma) = \frac{u}{e^{\sigma} + (1 - e^{\sigma})u}, \quad 0 \leq u \leq 1,$$

with $\sigma \geq 0$.

(a) Prove that $L(u; \sigma)$ belongs to the class of Lorenz curves (6.34), where $F(x) = \frac{1}{1+e^{-x}}$ is the logistic distribution.

(b) Prove that the cdf associated with the Lorenz curve $L(u; \sigma)$ is given by

$$G(x; \mu, \sigma) = \frac{e^{\sigma} - \sqrt{\mu e^{\sigma}/x}}{e^{\sigma} - 1}, \quad e^{-\sigma}\mu \leq x \leq e^{\sigma}\mu,$$

where $\mu > 0$ is the mean of the distribution (Arnold et al. 1987).

9. Verify Eq. (6.54), which relates the STTTT (Eq. (6.52)) with the excess wealth transform (6.53).

Chapter 7
Multivariate Majorization and Multivariate Lorenz Ordering

7.1 Multivariate Majorization

The temptation to seek multivariate generalizations of majorization and the Lorenz order is strong, and has not been resisted. In an income setting it is reasonable to consider income from several sources or income in incommensurable units. In fact, the idea that income can be measured undimensionally is perhaps the radical point of view, and interest should center on multivariate measures of income. Let us first consider various possible multivariate generalizations of majorization.

Let $\mathbb{R}^+_{n\times m}$ be the set of all $n \times m$ matrices with non-negative real elements. We want to define m-dimensional majorization to be a partial order on $\mathbb{R}^+_{n\times m}$ in such a way as to reduce to ordinary majorization when $m = 1$. One possible definition is to require column by column majorization. For any matrix $X \in \mathbb{R}^+_{n\times m}$ we let $X^{(j)}$ $(j = 1, 2, \ldots, m)$ denote the j'th column. With this notation we have the following definition.

Definition 7.1.1 Let $X, Y \in \mathbb{R}^+_{n\times m}$. We say that X is marginally majorized by Y if, for every $j = 1, 2, \ldots, m$, $X^{(j)} \leq_M Y^{(j)}$, and we write $X \leq_{MM} Y$.

Now from the Hardy, Littlewood, and Polya theorem (Theorem 2.1.1) we know that if $X \leq_{MM} Y$ then there exist doubly stochastic matrices $P_1, P_2, \ldots, P_m$ such that $X^{(j)} = P_j Y^{(j)}$. Of course the P_j's may well not be all the same. If the same choice of doubly stochastic matrix works for every j, we can define a stronger and perhaps more interesting partial order. It is this partial order that is dubbed majorization in Marshall et al. (2011). To capture the spirit of the definition, we will call the relation uniform majorization. We thus have

Definition 7.1.2 Let $X, Y \in \mathbb{R}^+_{n\times m}$. We say that X is uniformly majorized by Y if there exists a doubly stochastic matrix P such that $X = PY$, and we write $X \leq_{UM} Y$.

B. C. Arnold, J. M. Sarabia, *Majorization and the Lorenz Order with Applications in Applied Mathematics and Economics*, Statistics for Social and Behavioral Sciences,
https://doi.org/10.1007/978-3-319-93773-1_7

How are marginal and uniform majorization related? Obviously, we have the implication $X \leq_{UM} Y \Rightarrow X \leq_{MM} Y$, for one can set $P_j = P$ for each j. It is not difficult to verify that the converse fails, i.e., in general $X \leq_{MM} Y \not\Rightarrow X \leq_{UM} Y$ (Exercise 1).

Life in higher dimensions is invariably a little more complicated than a one-dimensional existence. This is exemplified by the fact that Robin Hood loses some of his prominence in higher dimensional versions of majorization.

A Robin Hood operation (refer to Chap. 2) involves a transfer of wealth from a relatively richer individual to a relatively poorer individual in the population. It is equivalent to multiplication by a doubly stochastic matrix of the form $A = (a_{ij})$ where for some k_1, k_2 and some $\lambda \in [0, 1]$

$$\begin{aligned} a_{k_1,k_1} &= 1-\lambda, \quad & a_{k_1,k_2} &= \lambda, \\ a_{k_2,k_1} &= \lambda, & a_{k_2,k_2} &= 1-\lambda, \\ a_{i,j} &= \delta_{ij}, & &\text{otherwise.} \end{aligned} \tag{7.1}$$

Matrices satisfying (7.1) will be called Robin Hood matrices. We may then formulate

Definition 7.1.3 Let $X, Y \in \mathbb{R}^+_{n\times m}$. We say that X is majorized in the Robin Hood sense by Y, if there exists a finite set of Robin Hood matrices (of the form (7.1)) $A_1, A_2, \ldots, A_m$ such that $X = A_1 A_2 \cdots A_m Y$, and we write $X \leq_{RH} Y$.

Now from Chap. 2, we know that majorization can be characterized in terms of Robin Hood operations and relabelings. In our Definition 7.1 we allowed λ to be greater than 1/2, so that the class includes elementary permutation matrices. However, notice that in Definition 7.1.3, when a permutation matrix appears among the A_j's, it permutes all the columns of Y, i.e., the same permutation is applied to all the columns. We have no guarantee that this will allow us to duplicate every doubly stochastic matrix as a finite product of Robin Hood matrices. Thus, we claim that uniform and Robin Hood majorization are different concepts, since provided $n \geq 3$, there exist doubly stochastic matrices which are not finite products of Robin Hood matrices (in the sense of (7.1)). This may seem paradoxical, since in one dimension we know that Robin Hood majorization and majorization do coincide. The following example included in Marshall et al. (2011) shows that the concepts must be distinguished in higher dimensions.

Example 7.1.1 Suppose

$$X = \begin{pmatrix} 1 & 3 \\ 1/2 & 4 \\ 1/2 & 5 \end{pmatrix} \quad \text{and} \quad Y = \begin{pmatrix} 1 & 2 \\ 1 & 4 \\ 0 & 6 \end{pmatrix}$$

then $X = PY$ where

$$P = (1/2)\begin{pmatrix} 1 & 1 & 0 \\ 1 & 0 & 1 \\ 0 & 1 & 1 \end{pmatrix} \tag{7.2}$$

and for no other choice of P is $X = PY$. However, this particular doubly stochastic matrix is not expressible as a finite product of Robin Hood matrices (see Exercise 10). Thus $X \leq_{UM} Y$, but $X \not\leq_{RH} Y$.

There is another attractive possible generalization of majorization to higher dimensions. Recall that we were able to characterize majorization in terms of continuous convex functions on $\mathbb{R}^+$. The definition of convexity is readily extended to functions on $\mathbb{R}_m^+$, and we may formulate.

Definition 7.1.4 Let $X, Y \in \mathbb{R}_{n\times m}^+$. We say that X is convexly majorized by Y if for every $h : \mathbb{R}_m^+ \to \mathbb{R}^+$ that is continuous and convex we have $\sum_{i=1}^n h(X_{(i)}) \leq \sum_{i=1}^n h(Y_{(i)})$ (here $X_{(i)}$ is the i'th row of X), and we write $X \leq_{CM} Y$.

It is a simple consequence of Jensen's inequality that $X \leq_{UM} Y \Rightarrow X \leq_{CM} Y$ (Exercise 2). In dimensions higher than 1 (i.e., $m \geq 2$) it is not obvious whether the converse is true. A proof of the equivalence of uniform majorization and convex majorization was provided by Karlin and Rinott (1988) using a result of Meyer (1966) on dilation in abstract settings. Thus $UM \equiv CM$.

None of the three suggested versions of multivariate majorization thus far introduced (marginal, uniform=convex and Robin Hood) seem to be compelling. Uniform majorization viewed as convex majorization has the advantage of extending readily to cover the case of general non-negative m-dimensional random variables which we will discuss below under the heading multivariate Lorenz ordering. Before considering this generalization, we mention some other possible versions of multivariate majorization. All involve efforts to define the concept in terms of the better understood concept of univariate majorization. First, consider a definition equivalent to one proposed in Marshall et al. (2011).

Definition 7.1.5 Let $X, Y \in \mathbb{R}_{n\times m}^+$. We say that X is positive combinations majorized by Y, if $X\underline{a} \leq_M Y\underline{a}$ for every vector $\underline{a} \in \mathbb{R}_m^+$ with $a_i > 0, i = 1, 2, \ldots, m$ and we write $X \leq_{PCM} Y$.

An economic interpretation of positive combinations majorization is possible. Suppose the rows of X represent m-dimensional income vectors (one for each of n individuals in the population). In particular x_{ij} represents income in currency j accruing to individual i. Suppose now that all the incomes are converted into dollars at rates $a_1, a_2, \ldots, a_m$. The resulting vector of incomes in dollars is then $X\underline{a}$. It is then evident that $X \leq_{PCM} Y$ if the X incomes are majorized by the Y incomes under all possible exchange rates (i.e., if $X\underline{a} \leq_M Y\underline{a}$, for every $\underline{a} \geq \underline{0}$, $\underline{a} \neq \underline{0}$).

A partial order that is similar to, but distinct from, positive combinations majorization is called linear combinations majorization. The only difference is that negative values of the a_i's are now permitted.

Definition 7.1.6 Let $X, Y \in \mathbb{R}^+_{n\times m}$. We say that X is linear combinations majorized by Y, if $X\underline{a} \leq_M Y\underline{a}$ for every vector $\underline{a} \in \mathbb{R}_m$. and we write $X \leq_{LCM} Y$.

Linear-combinations majorization is sometimes called *directional majorization.*

Positive-combinations majorization or positive directional majorization is sometimes called *price majorization* (Mosler 2002). In an economic context, to be discussed further below, it is a more appealing concept than is linear-combinations majorization. However, as will be explained in the section on multivariate Lorenz ordering, linear-combinations majorization admits an attractive interpretation involving the natural extension of Lorenz curves to higher dimensions. The fact that the two orderings are different is illustrated by the following simple example.

Example 7.1.2 (Joe and Verducci 1992) If

$$X = \begin{pmatrix} 1 & 4 \\ 3 & 2 \end{pmatrix}, \quad Y = \begin{pmatrix} 1 & 2 \\ 3 & 4 \end{pmatrix},$$

then $X \leq_{PCM} Y$ but $X \not\leq_{LCM} Y$ [shown by letting $\underline{a} = (1, -1)$].

A different approach to the problem of reducing dimension is provided by

Definition 7.1.7 Let $X, Y \in \mathbb{R}^+_{n\times m}$ and let $g : \mathbb{R}^+_m \to \mathbb{R}^+$. We say that X is g-majorized by Y, if $(g(X_{(1)}), \ldots, g(X_{(n)})) \leq_M (g(Y_{(1)}), \ldots, g(Y_{(m)}))$, and we write $X \leq_{gM} Y$.

Plausible choices for g in this definition include $g(\underline{x}) = \sum_{i=1}^m x_i$, $g(\underline{x}) = \sqrt{\sum_{i=1}^m x_i^2}$ and $g(\underline{x}) = \max_i\{x_i\}$. In an income setting where the x_i's represent incomes of m different types, $g(\underline{x})$ can be interpreted as the utility of the income vector $(x_1, \ldots, x_m)$. If we can agree on a suitable choice for g, then we are enabled to replace an m-dimensional income vector by a suitable one-dimensional utility. g-majorization, together with positive combinations majorization and linear combinations majorization, extends readily to the case of general non-negative m-dimensional random variables to which we now turn.

7.2 Multivariate Lorenz Orderings

Let $\mathcal{L}^{(m)}$ denote the class of m-dimensional random variables whose coordinate random variables are members of $\mathcal{L}$, i.e., are non-negative random variables with positive finite expectations. We consider several generalizations of the Lorenz order to this m-dimensional setting. In the following X_i (respectively Y_i) denotes the i'th coordinate random variable of $\underline{X}$ (respectively $\underline{Y}$), $i = 1, 2, \ldots, m$.

The first extension requires little imagination.

Definition 7.2.1 Let $\underline{X}, \underline{Y} \in \mathcal{L}^{(m)}$. We will say that $\underline{X}$ is marginally Lorenz dominated by $\underline{Y}$, if for each $i = 1, 2, \ldots, m$, $X_i \leq_L Y_i$, and we write $\underline{X} \leq_{ML} \underline{Y}$.

The second extension involves expectations of convex functions:

Definition 7.2.2 Let $\underline{X}, \underline{Y} \in \mathcal{L}^{(m)}$. We will say that $\underline{X}$ is convex-Lorenz dominated by $\underline{Y}$ if for every continuous convex function $h : \mathbb{R}_m^+ \rightarrow \mathbb{R}^+$ we have

$$E\left(h\left(\frac{X_1}{E(X_1)}, \ldots, \frac{X_m}{E(X_m)}\right)\right) \leq E\left(h\left(\frac{Y_1}{E(Y_1)}, \ldots, \frac{Y_m}{E(Y_m)}\right)\right),$$

and we write $\underline{X} \leq_{CL} \underline{Y}$.

A less restrictive concept of Lorenz domination is the following:

Definition 7.2.3 Let $\underline{X}, \underline{Y} \in \mathcal{L}^{(m)}$ and let $g : \mathbb{R}_m^+ \rightarrow \mathbb{R}^+$. We say that $\underline{X}$ is g-Lorenz dominated by $\underline{Y}$ if $g(\underline{X}) \leq_L g(\underline{Y})$, and we write $\underline{X} \leq_{gL} \underline{Y}$.

In Definition 7.2.3, g can be interpreted as a utility function (as in the case of g-majorization). In order to have $E(g(\underline{X})) < \infty$ for every $\underline{X} \in \mathcal{L}^{(m)}$, g should be a bounded function. If g is not bounded one might be tempted to replace it by a new bounded utility, such as $g^* = g/(1+g)$. The partial orders $\leq_{gL}$ and $\leq_{g^*L}$ will however not be equivalent (in the light of the results of Chap. 4 on inequality preserving transformations).

In one dimension, convex Lorenz ordering was interpretable in terms of nested Lorenz curves. The question then arises: in m dimensions, can convex Lorenz ordering be interpreted in terms of m-dimensional Lorenz curves? And, indeed, how should we go about defining a suitable m-dimensional extension of the usual Lorenz curve concept?

Extension of the Lorenz curve concept to higher dimensions was long frustrated by the fact that the usual definitions of the Lorenz curve involved either order statistics or the quantile function of the corresponding distribution, neither of which has a simple multivariate analog.

There is, however, one readily available representation of the univariate Lorenz curve that does not explicitly involve the quantile function, namely, that the Lorenz curve is the set of points in $\mathbb{R}_2^+$ parameterized by x with coordinates

$$\left\{F_X(x), \frac{1}{E(X)} \int_0^x u f_X(u) du\right\}.$$

Analogously, Taguchi (1972a,b) and Lunetta (1972) set out to define, for a bivariate distribution $F_{X,Y}$ with density $f_{X,Y}$, a Lorenz surface parameterized by (x, y) to be the set of points in $\mathbb{R}_3^+$ with coordinates

$$\left\{F_{X,Y}(x,y), \frac{1}{E(X)} \int_0^x \int_0^y u f(u,v) du dv, \frac{1}{E(Y)} \int_0^x \int_0^y v f(u,v) du dv\right\}. \tag{7.3}$$

However the definition used for the explicit description of this Lorenz Surface was that it was a function $L^{(2)} : [0,1]^2 \longrightarrow [0,1]$ defined by

$$(s, t, L^{(2)}(s, t)) =$$

$$\left\{F_{X,Y}(x, y), \tfrac{1}{E(X)} \int_0^x \int_0^y uf(u, v)dudv, \tfrac{1}{E(Y)} \int_0^x \int_0^y vf(u, v)dudv\right\}. \quad (7.4)$$

See Taguchi (1972a,b) for detailed discussion of this surface. Note that it does not treat the coordinate random variables X and Y in a symmetric fashion and an appropriate extension to higher dimensions of (7.4) is problematic (even though an extension of (7.3) is easily envisioned).

Arnold (1987) proposed an alternative parametric definition of a Lorenz surface for bivariate distributions with marginal distributions F_X and F_Y, again indexed by (x, y). The points on this surface are

$$\left\{F_X(x), F_Y(y), \frac{1}{E(XY)} \int_0^x \int_0^y uvf(u, v)dudv\right\}. \quad (7.5)$$

It is easy to visualize how to generalize the latter definition to the case of m dimensional random variables ($m > 2$). One nice feature enjoyed by the surface defined by (7.5) is that, in the case of independence, the bivariate Lorenz surface reduces to the product of the marginal Lorenz curves (Exercise 7). One could, of course, define a "Lorenz surface" ordering on $\mathcal{L}^{(m)}$ by saying that $\underline{X}$ is more unequal in the Lorenz surface sense than $\underline{Y}$ if $L_{\underline{X}}^{(m)}(\underline{u}) \leq L_{\underline{Y}}^{(m)}(\underline{u})$ for all $\underline{u}$, where the surfaces $L_{\underline{X}}^{(m)}(\underline{u})$, $L_{\underline{Y}}^{(m)}(\underline{u})$ are defined by an m-dimensional version of (7.5). The relation of this partial order to the other m-dimensional Lorenz orderings introduced in Definitions 7.2.1–7.2.3 has not been explored. For further discussion of the Arnold surfaces, see Sarabia and Jordá (2013). See also the material in Sects. 7.3 and 7.4 below.

A new definition of the Lorenz curve seems to be required if we are to be able to identify a more natural extension to higher dimensions. An attractive candidate is one involving what are called Lorenz zonoids. Early results on this concept were provided by Koshevoy (1995). For more details, see Koshevoy and Mosler (1996, 1997) and Mosler (2002).

We begin by again considering the Lorenz curve associated with n ordered numbers $x_1 \leq x_2 \leq \ldots \leq x_n$. It is a linear interpolation of the points

$$\left(\frac{i}{n}, \sum_{j=1}^{i} x_j \Big/ \left(\sum_{j=1}^{n} x_j\right)\right) \quad (7.6)$$

For each i, an income interpretation is available for the point (7.6). Its first coordinate represents the fraction of the total population (i.e., i/n) accounted for by the poorest i individuals in the population. The second coordinate corresponds to the fraction of the total income of the population accruing to the poorest i individuals in the population. Instead of considering such extreme subsets of the population we can plot, for every subset of j of the individuals in the population, a point whose coordinates are (1) the proportion of the population accounted for by the j individuals (i.e., j/n), and (2) the proportion of the total income accounted for by the j individuals. No ordering of the x_i's is required to plot these points. The upper envelope of these points can be identified as the "reverse" Lorenz curve in which are plotted (and interpolated) the income share of the richest i individuals against $i/n,\ i = 1, 2, \ldots, n$ (this could be viewed as a discrete Leimkuhler curve).

The region between the Lorenz curve and the reverse Lorenz curve is the convex hull of these 2^n points. If one Lorenz curve is uniformly below a second Lorenz curve, their corresponding reverse Lorenz curves are ordered in the reverse order. It then becomes evident that Lorenz ordering can be defined in terms of the nesting of the convex hulls of the income shares of all subsets of the populations. This avoids ordering and permits a straightforward extension to higher dimensions.

The set of points between the Lorenz curve and the reverse Lorenz curve is called the *Lorenz zonoid*. Before attempting an extension of this concept to higher dimensions, let us first consider an extension of the definition to associate a Lorenz zonoid with every $X \in \mathcal{L}$. To this end, in the finite population setting, we envision computing income shares for subsets of the population that include fractional individuals. Thus, for a given vector $\alpha \in [0, 1]^n$, consider the income share comprising α_1 times the income of individual 1, plus α_2 times the income of individual 2, etc. The size of this subset is $\Sigma_1^n \alpha_i/n$ and its corresponding income share is $\Sigma_{i=1}^n \alpha_i x_i / \Sigma_{i=1}^n x_i$. It is then evident that the Lorenz zonoid corresponding to the population can be envisioned as the set of all points $(\frac{1}{n}\Sigma\alpha_i, \Sigma_{i=1}^n \alpha_i x_i / \Sigma_{i=1}^n x_i)$ in which α ranges over $[0, 1]^n$.

The extension to $\mathcal{L}$ is then straightforward.

Definition 7.2.4 Let Ψ denote the class of all measurable mappings from $\mathbb{R}^+$ to $[0, 1]$. The *Lorenz zonoid* $L(X)$ of the random variable X with distribution function F_X is defined to be the set of points

$$L(X) = \left\{ \left(\int_0^\infty \psi(x) dF_X(x), \frac{\int_0^\infty x\psi(x) dF_X(x)}{E(X)} \right) : \psi \in \Psi \right\}$$

$$= \left\{ \left(E(\psi(X)), \frac{E(X\psi(X))}{E(X)} \right) : \quad \psi \in \Psi \right\}. \tag{7.7}$$

It can be verified that the set of points defined in (7.7) does indeed, in the finite population setting, coincide with the set of points between the Lorenz curve and the reverse Lorenz curve. Again, it is important to emphasize the fact that in this definition, no ordering of the x_i's and no reference to a quantile function are required. Thus the definition has potential for extension to higher dimensions without requiring a suitable definition for higher-dimensional quantiles. Note also that the definition of the Lorenz order on $\mathcal{L}$ is expressible as

$$X \leq_L Y \Longleftrightarrow L(X) \subseteq L(Y), \tag{7.8}$$

where the Lorenz zonoid is defined in (7.7). Nesting of Lorenz zonoids describes the ordering precisely.

An extension to m dimensions can now be developed. Denote by $\mathcal{L}^{(m)}$ the set of all m-dimensional nonnegative (though this can be relaxed) random vectors $\underline{X}$ with finite positive marginal expectations (i.e., such that $E(X_1), \ldots, E(X_m) \in (0, \infty)$). In addition, let $\Psi^{(m)}$ denote the class of all measurable functions from $\mathbb{R}_m^+$ to $[0, 1]$.

Definition 7.2.5 Let $\underline{X} \in \mathcal{L}^{(m)}$. The *Lorenz zonoid* $L(\underline{X})$ of the random vector $\underline{X} = (X_1, \ldots, X_m)$ with distribution $F_{\underline{X}}$ is

$$L(\underline{X}) = \left\{ \left(\int \psi(\underline{x}) dF_{\underline{X}}(\underline{x}), \int x_1 \frac{\psi(\underline{x})}{E(X_1)} dF_{\underline{X}}(\underline{x}), \int x_2 \frac{\psi(\underline{x})}{E(X_2)} dF_{\underline{X}}(\underline{x}), \ldots \right.\right.$$

$$\left.\left. \ldots, \int x_m \frac{\psi(\underline{x})}{E(X_m)} dF_{\underline{X}}(\underline{x}) \right) : \psi \in \Psi^{(m)} \right\}$$

$$= \left\{ \left(E(\psi(\underline{X})), \frac{E(X_1 \psi(\underline{X}))}{E(X_1)}, \ldots, \frac{E(X_m \psi(\underline{X}))}{E(X_m)} \right) : \psi \in \Psi^{(m)} \right\}. \tag{7.9}$$

The Lorenz zonoid is thus a convex "football"-shaped (American football) subset of the $(m + 1)$-dimensional unit cube that includes the points $(0, 0, \ldots, 0)$ and $(1, \ldots, 1)$.

The zonoid Lorenz order is defined in terms of nested Lorenz zonoids as in the one-dimensional case. Thus for, $\underline{X}, \underline{Y} \in \mathcal{L}^{(m)}$

$$\underline{X} \leq_{ZL} \underline{Y} \Longleftrightarrow L(\underline{X}) \subseteq L(\underline{Y}), \tag{7.10}$$

where $L(\underline{X})$ is as defined in (7.9).

An example of such a zonoid, one corresponding to a particular bivariate Pareto distribution, is shown in Fig. 7.1.

There is a link between the Lunetta–Taguchi Lorenz surface (7.3), and its natural m-dimensional extension, and the Lorenz zonoid (7.9). To see this link, suppose that $\underline{X} = (X_1, X_2, \ldots, X_m)$ is an m-dimensional random vector with $0 < E(X_i) < \infty$ for $i = 1, 2, \ldots, k$ and with distribution function $F_{\underline{X}}(\underline{x})$. Then, for each $\underline{x} \in (0, \infty)^m$, define the measurable function

$$\psi_{\underline{x}}(\underline{u}) = I(\underline{u} \leq \underline{x}),$$

and let $\Psi_{L-T}^{(m)}$ denote the class of all such functions. A comparison of the relevant definitions will confirm the fact that the definition of the m-dimensional extension of the bivariate Lunetta–Taguchi surface (7.3) is the same as the definition of the Lorenz zonoid (7.9) except that $\Psi^{(m)}$ has been replaced by $\Psi_{L-T}^{(m)}$. Thus the Lunetta–Taguchi surface corresponds to a particular subset of the Lorenz zonoid. In one dimension the Lunetta–Taguchi surface coincides with the lower bound of the Lorenz zonoid, i.e., with the usual Lorenz curve. In m-dimensions, the Lunetta–Taguchi surface has a similar role as a subset of the Lorenz zonoid.

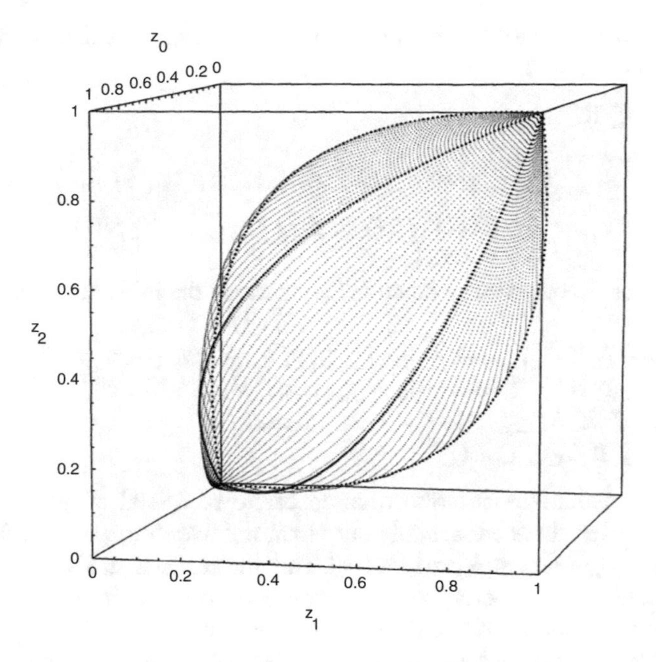

Fig. 7.1 The Lorenz zonoid for a bivariate Pareto (II) distribution with $\alpha = 9$ and parameters $(\mu_1, \mu_2, \sigma_1, \sigma_2) = (0, 0, 1, 1)$

It will be recalled that a definition of Lorenz order in one dimension was possible in terms of expectations of convex functions. Thus in one dimension, $X \leq_L Y$ if and only if $E(g(X/EX)) \leq E(g(Y/EY))$ for every continuous convex function g. The convex Lorenz order in $\mathcal{L}^{(m)}$ is the natural k-dimensional version of this ordering. In the section on multivariate majorization, attention was directed to linear combinations and positive combinations versions of multivariate majorization, both of which permitted use of one-dimensional majorization in ordering vectors. Obvious parallel definitions can be formulated for Lorenz ordering. Thus

Definition 7.2.6 Let $\underline{X}, \underline{Y} \in \mathcal{L}^{(m)}$. We say that $\underline{X}$ is positive combinations Lorenz ordered with respect to $\underline{Y}$, if $\underline{X}\,\underline{a}^T \leq_L \underline{Y}\,\underline{a}^T$ for every vector $\underline{a} \in \mathbb{R}_m^+$ with $a_i > 0$ $i = 1, 2, \ldots, m$ and we write $\underline{X} \leq_{PCL} \underline{Y}$.

Definition 7.2.7 Let $\underline{X}, \underline{Y} \in \mathcal{L}^{(m)}$. We say that $\underline{X}$ is linear combinations Lorenz ordered with respect to $\underline{Y}$, if $\underline{X}\,\underline{a}^T \leq_{GL} \underline{Y}\,\underline{a}^T$ for every vector $\underline{a} \in \mathbb{R}_m$, and we write $\underline{X} \leq_{LCL} \underline{Y}$.

In the definition of $\leq_{LCL}$ above it should be noted that use has been made of a generalized one-dimensional Lorenz order. This is required since, for some choices of the vector $\underline{a}$ the random variable $\sum_{i=1}^m a_i X_i$ can take on negative values and the usual Lorenz order will not be well-defined. Recall that the generalized Lorenz order was defined in Chap. 6 as follows. For arbitrary one-dimensional random variables X and Y, not necessarily non-negative, we write $X \leq_{GL} Y$ if $\int_0^u F_X^{-1}(u)\,du \geq \int_0^u F_Y^{-1}(u),\ \ \forall\, u \in (0, 1)$.

We have thus apparently identified five competing versions of multivariate Lorenz orders, namely:

(i) $\underline{X} \leq_{CL} \underline{Y}$ if

$$E(g(X_1/E(X_1), X_2/E(X_2), \ldots, X_m/E(X_m)) \\ \leq E(g(Y_1/E(Y_1), Y_2/E(Y_2), \ldots, Y_m/E(Y_m))$$

for all continuous convex functions g for which the indicated expectations are finite.

(ii) $\underline{X} \leq_{LCL} \underline{Y}$ if $\sum_{i=1}^m a_i X_i \leq_{GL} \sum_{i=1}^m a_i Y_i, \ \forall \underline{a} \in (-\infty, \infty)^m$.

(iii) $\underline{X} \leq_{PCL} \underline{Y}$ if $\sum_{i=1}^m c_i X_i \leq_L \sum_{i=1}^m c_i Y_i, \ \forall \underline{c} \in [0, \infty)^m$.

(iv) $\underline{X} \leq_{ML} \underline{Y}$ if $X_i \leq_L Y_i, \ i = 1, 2, \ldots, m$.

(v) $\underline{X} \leq_{ZL} \underline{Y}$ if $L(\underline{X}) \subset L(\underline{Y})$

The above definitions describe what appear to be a total of five partial orders on $\mathcal{L}^{(m)}$. However, there are actually only four, not five. It may be verified that the partial orders $\leq_{LCL}$ and $\leq_{ZL}$ are identical. Thus the nested zonoid order on $\mathcal{L}^{(m)}$ can be reinterpreted as an ordering which corresponds to generalized Lorenz ordering of all linear combinations of the coordinate random variables.

The first four partial orders in the list are distinct and they are actually listed in the order of decreasing strength. Thus we have

$$\underline{X} \leq_{CL} \underline{Y} \Rightarrow \underline{X} \leq_{LCL} \underline{Y} \Rightarrow \underline{X} \leq_{PCL} \underline{Y} \Rightarrow \underline{X} \leq_{ML} \underline{Y}. \tag{7.11}$$

Names can be associated with these four m-dimensional Lorenz orders. The last order appearing in (7.11), $\leq_{ML}$, the weakest of the group, is naturally called marginal Lorenz ordering. It ignores any dependence relations between the coordinate random variables. The first ordering in (7.11), $\leq_{CL}$, can be called convex ordering. The partial order, $\leq_{LCL} \equiv \leq_{ZL}$, will be called the zonoid ordering or the linear combinations ordering. The partial order $\leq_{PCL}$ is sometimes called the price Lorenz order, or the positive combinations order, or the exchange rate Lorenz order. The genesis of the last of these names is as follows (analogous to the discussion following the definition in Sect. 7.1 of a version of multivariate majorization). Suppose that the coordinates of $\underline{X}$ and $\underline{Y}$ represent holdings (or earnings) in m different currencies. Suppose that we exchange all of the holdings into one currency, perhaps Euros, according to m exchange rates $c_1, c_2, \ldots, c_m$, then it is natural to compare $\sum_{i=1}^m c_i X_1$ and $\sum_{i=1}^m c_i Y_1$ with regard to inequality by the usual univariate Lorenz order. The ordering $\leq_{L_3}$ requires that $\sum_{i=1}^m c_i X_1 \leq_L \sum_{i=1}^m c_i Y_1$ for every vector of exchange rates $\underline{c}$. In this context, it is natural to require that the c_i's be positive. How would you interpret a negative exchange rate?

This last observation highlights a possible lacuna in the arguments in support of the zonoid ordering. Since $\leq_{ZL}$ and $\leq_{PCL}$ are equivalent, the zonoid ordering requires generalized Lorenz ordering of the univariate random variables $\sum_{i=1}^m a_i X_1$ and $\sum_{i=1}^m a_i Y_1$ even when some of the a_i's are negative (corresponding, if you wish, to negative exchange rates!). See Koshevoy and Mosler (1996) for an insightful introduction to the Lorenz zonoid order, including discussion of the role of the univariate generalized Lorenz order in the zonoid order. A more extensive discussion may be found in Mosler (2002).

A positive feature of the convex order, $\leq_{CL}$, is that the averaging theorem survives intact, i.e., Strassen's Theorem 3.2.3 is still true in the m-dimensional setting. Thus

Theorem 7.2.1 *For* $\underline{X}, \underline{Y} \in \mathcal{L}^{(m)}$, $\underline{Y} \leq_{CL} \underline{X}$ *if and only if there exist jointly distributed random variables* $\underline{X}'$ *and* Z' *such that* $\underline{X} \stackrel{d}{=} \underline{X}'$ *and* $\underline{Y} \stackrel{d}{=} E(\underline{X}'|Z')$.

In fact Strassen proved the result in a much more abstract setting than we require (see Meyer (1966), for further related discussion). Whitt (1980) discussed application of these ideas in a reliability context.

Despite the fact that the zonoid ordering will fail to have Strassen's balayage equivalence theorem (Theorem 7.2.1) and will involve comparisons of generalized Lorenz curves with curious exchange rate interpretations, it appears that the zonoid order ($\leq_{ZL} \equiv \leq_{LCL}$) is the most defensible of the competing partial orders. Nevertheless it cannot be ruled out that, in some applications, the other three partial orders might be more appropriate.

Notes

(1) If one considers multivariate majorization instead of multivariate Lorenz ordering, it is not necessary to introduce the concept of a "generalized Lorenz curve" into the discussion. Majorization is well-defined for vectors in $(-\infty, \infty)^m$, rather than just $[0, \infty)^m$ for Lorenz ordering. Consequently there is no concern about linear combinations taking on negative values. Refer to Chapter 15 of Marshall et al. (2011) for more details.

(2) The convex order is stronger than the order $\leq_{ZL}$. Examples of simple two-dimensional cases in which $\underline{X} \leq_{ZL} \underline{Y}$ but $\underline{X} \not\leq_{CL} \underline{Y}$ are provided by Elton and Hill (1992).

7.3 Explicit Expressions for the Arnold Lorenz Surface

One of the difficulties with the Koshevoy and Mosler (1996) definition of the Lorenz zonoid is the lack of explicit analytic expressions for such zonoids.

In this sense, the definition proposed by Arnold is more appealing since it allows one to obtain explicit expressions for the Lorenz surface. This aspect has been explored by Sarabia and Jordá (2013, 2014a) and by Arnold and Sarabia (2018). In this section we present some of these explicit expressions based on Sarabia and Jordá (2014a,b), focussing on the bivariate case.

The different results are based on an explicit expression of the bivariate Lorenz surface defined in (7.5), which admits the following simple representation.

Theorem 7.3.1 *Let (X_1, X_2) be a bivariate random variable belonging to $\mathcal{L}^{(2)}$, with joint cdf and pdf given by F_{12} and f_{12}, respectively, and marginal distributions with cdfs and pdfs given by F_i and f_i, $i = 1, 2$, respectively. Then, the bivariate Arnold Lorenz surface defined in (7.5) can be written in the explicit form,*

$$L_{12}(u_1, u_2) = \frac{1}{E(X_1X_2)} \int_0^{u_1} \int_0^{u_2} A(x_1, x_2)dx_1dx_2, \quad 0 \leq u_1, u_2 \leq 1, \tag{7.12}$$

where

$$A(x_1, x_2) = \frac{F_1^{-1}(x_1)F_2^{-1}(x_2)f_{12}(F_1^{-1}(x_1), F_2^{-1}(x_2))}{f_1(F_1^{-1}(x_1))f_2(F_2^{-1}(x_2))}. \tag{7.13}$$

Proof The proof is direct making the change of variable $(u_1, u_2) = (F_1(x_1), F_2(x_2))$ in (7.5). ■

Now, we consider an example.

Example 7.3.1 Let (X_1, X_2) be a bivariate Farlie-Gumbel-Morgenstern (FGM) distribution with uniform marginals and joint probability density function,

$$f_{12}(x_1, x_2) = 1 + w(1 - 2x_1)(1 - 2x_2), \quad 0 \leq x_1, x_2 \leq 1,$$

where $w \in (-1, 1)$ is a dependence parameter. The marginal distributions are uniform on the interval [0, 1] and the corresponding marginal Lorenz curves are $L_{X_i}(u) = u^2, 0 \leq u \leq 1, i = 1, 2$. The (1, 1)-mixed moment is

$$E(X_1X_2) = \frac{1}{36}(9 + w),$$

and the $A(x_1, x_2)$ function defined in (7.13) is, for this model, of the form

$$A(x_1, x_2) = x_1x_2\{1 + w(1 - 2x_1)(1 - 2x_2)\}.$$

Using (7.12) the bivariate Lorenz surface is

$$L_{12}(u_1, u_2; w) = \frac{u_1^2u_2^2[9 + w(4u_1 - 3)(4u_2 - 3)]}{9 + w}, \quad 0 \leq u_1, u_2 \leq 1. \tag{7.14}$$

If we set $w = 0$ (the independence case) in (7.14), we obtain $L_{12}(u_1, u_2) = u_1^2u_2^2$, which is the product of the marginal Lorenz curves.

Figure 7.2 shows the bivariate Lorenz surface (7.14) for two selected values of the parameter w.

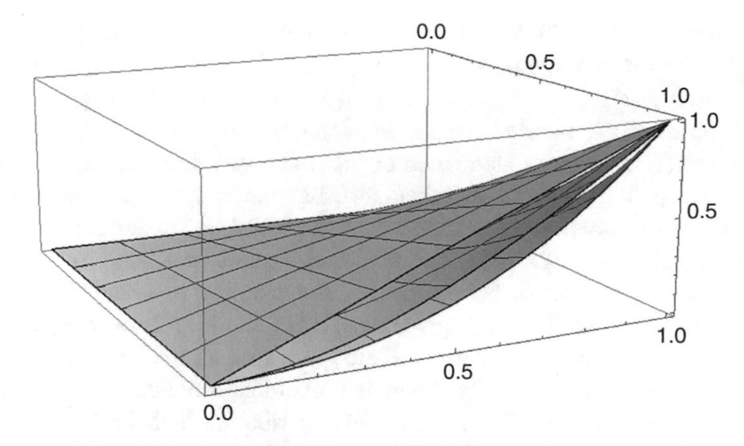

Fig. 7.2 Bivariate Lorenz surface (7.14) of the Farlie-Gumbel-Morgenstern family with uniform marginals, $w = -0.9$ and 0.9 and the egalitarian bivariate surface $L_{12}(u_1, u_2) = u_1 u_2$

7.3.1 *The Bivariate Sarmanov–Lee Lorenz Surface*

In this section, we introduce the Lorenz surface corresponding to the bivariate Sarmanov–Lee distribution.

As an initial step we define the concentration curve introduced by Kakwani (1977).

Definition 7.3.1 Let $X \in \mathcal{L}$ be a random variable and let $g(x)$ be a continuous function of x whose first derivative exists and is such that $g(x) \geq 0$. If the mean $E[g(X)]$ exists, the concentration curve of $g(X)$ in the Kakwani sense is defined as

$$L(x; g) = \frac{1}{E[g(X)]} \int_0^x g(u) dF(u),$$

where $F(x)$ is the cdf of the random variable X.

Now, we define the bivariate Sarmanov–Lee distribution. The random vector (X_1, X_2) is said to have a bivariate Sarmanov–Lee distribution if its joint pdf is given by

$$f(x_1, x_2) = f_1(x_1) f_2(x_2) \{1 + w\phi_1(x_1)\phi_2(x_2)\}, \tag{7.15}$$

where $f_1(x_1)$ and $f_2(x_2)$ are the univariate marginal pdf's, $\phi_i(t)$, $i = 1, 2$ are bounded nonconstant functions such that

$$\int_{-\infty}^{\infty} \phi_i(t) f_i(t) dt = 0, \quad i = 1, 2,$$

and w is a real number which satisfies the condition $1+w\phi_1(x_1)\phi_2(x_2) \geq 0$ for all x_1 and x_2. We use the notation $\mu_i = E(X_i) = \int_{-\infty}^{\infty} tf_i(t)dt, i = 1, 2, \sigma_i^2 = \text{var}(X_i) = \int_{-\infty}^{\infty}(t - \mu_i)^2 f_i(t)dt, i = 1, 2$ and $\nu_i = E[X_i\phi_i(X_i)] = \int_{-\infty}^{\infty} t\phi_i(t)f_i(t)dt, i = 1, 2$. Properties of this family have been explored by Lee (1996).

Note that (7.15) and its associated copula have two components: a first component corresponding to the marginal distributions and the second component which defines the structure of dependence, determined by the parameter w and the functions $\phi_i(u)$, $i = 1, 2$. These two components will be related to the structure of the associated Lorenz surface, and the corresponding bivariate Gini index. In relation with other families with given marginals, the Sarmanov–Lee copula has several advantages: its joint pdf and cdf are quite simple; the covariance structure in general is not limited and its different probabilistic features can be obtained in explicit forms. On the other hand, the SL distribution includes several relevant special cases, including the classical FGM distribution and the variations proposed in Huang and Kotz (1999) and Bairamov and Kotz (2003).

The bivariate SL Lorenz surface is obtained using (7.15) in Eq. (7.12).

Theorem 7.3.2 *Let $(X_1, X_2) \in \mathcal{L}^{(2)}$ have a bivariate Sarmanov–Lee distribution with joint pdf (7.15). Then, its bivariate Lorenz surface is given by*

$$L_{12}(u_1, u_2) = \pi L_1(u_1)L_2(u_2) + (1 - \pi)L_1(u_1; g_1)L_2(u_2; g_2), \tag{7.16}$$

where

$$\pi = \frac{\mu_1\mu_2}{E(X_1X_2)} = \frac{\mu_1\mu_2}{\mu_1\mu_2 + w\nu_1\nu_2},$$

and $L_i(u_i)$, $i = 1, 2$ are the Lorenz curves of the marginal distributions of the X_i's, $i = 1, 2$, respectively, and $L_i(u_i; g_i)$, $i = 1, 2$ represent the concentration curves (in the sense of Definition 7.3.1) of the random variables $g_i(X_i) = X_i\phi_i(X_i)$, $i = 1, 2$, respectively.

Proof Exercise 11. ■

The interpretation of (7.16) is quite direct: the bivariate Lorenz surface can be expressed as a convex linear combination of two components: a first component corresponding to the product of the marginal Lorenz curves (the marginal component) and a second component corresponding to the product of the concentration Lorenz curves (the dependence structure component).

Next, we consider the Pareto Lorenz surface based on the FGM family with Pareto marginals. Note that the FGM bivariate distribution is a special case of (7.15).

Example 7.3.2 Let (X_1, X_2) be a bivariate FGM with classical Pareto marginals and joint pdf,

$$f_{12}(x_1, x_2; \alpha, \sigma) = f_1(x_1)f_2(x_2)\{1 + w[1 - 2F_1(x_1)][1 - 2F_2(x_2)]\}, \tag{7.17}$$

where

$$F_i(x_i) = 1 - \left(\frac{x}{\sigma_i}\right)^{-\alpha_i}, \quad x_i \geq \sigma_i, \quad i = 1, 2,$$

$$f_i(x_i) = \frac{\alpha_i}{\sigma_i}\left(\frac{x}{\sigma_i}\right)^{-\alpha_i - 1}, \quad x_i \geq \sigma_i, \quad i = 1, 2,$$

are the cdf and the pdf of the classical Pareto distributions, respectively, with $\alpha_i > 1$, $\sigma_i > 0$, $i = 1, 2$, $-1 \leq w \leq 1$ and $\phi_i(x_i) = 1 - 2F_i(x_i)$, $i = 1, 2$ in (7.15).

Using (7.16) with $g_i(x_i) = x_i[1 - 2F_i(x_i)]$, $i = 1, 2$ and after some computations, the bivariate Lorenz surface associated to (7.17) is

$$L_{12}(u_1, u_2) = \pi_w L(u_1; \alpha_1) L(u_2; \alpha_2) + (1 - \pi_w) L_1(u_1; g_1) L_2(u_2; g_2), \tag{7.18}$$

where the Lorenz and the concentration curves are given, respectively, by

$$L(u_i; \alpha_i) = 1 - (1 - u_i)^{1 - 1/\alpha_i}, \quad 0 \leq u \leq 1, \quad i = 1, 2,$$

$$L_i(u_i; g_i) = 1 - (1 - u_i)^{1 - 1/\alpha_i}[1 + 2(\alpha_i - 1)u_i], \quad 0 \leq u \leq 1, \quad i = 1, 2,$$

and

$$\pi_w = \frac{(2\alpha_1 - 1)(2\alpha_2 - 1)}{(2\alpha_1 - 1)(2\alpha_2 - 1) + w}.$$

Figure 7.3 shows the Pareto Lorenz surface (7.18) for two selected values of the parameters.

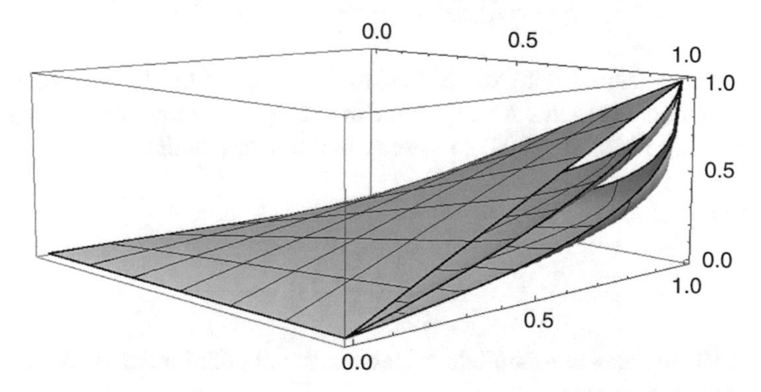

Fig. 7.3 Bivariate Pareto Lorenz surface (7.14) for (α_1, α_2) equal to (2,3), (1.5,1.5) with $w = 0.5$ together with the egalitarian bivariate surface $L_{12}(u_1, u_2) = u_1 u_2$

7.4 Summary Measures of *m*-Dimensional Inequality

A variety of summary measures of inequality have been suggested for m-dimensional distributions. We mention but a few. For the convex order $\leq_{CL}$, any specific choice of a continuous convex function g could be used to measure inequality by the quantity

$$E\left(g\left(\frac{X_1}{E(X_1)}, \ldots, \frac{X_m}{E(X_m)}\right)\right).$$

If we could agree on a suitable definition of the Lorenz surface for m-dimensional random variables, we could measure inequality by the $(m + l)$-dimensional volume between the Lorenz surface of a given distribution and the Lorenz surface of a degenerate distribution (in direct analogy to one interpretation of the Gini index in the case $m = 1$). Alternatively (mimicking Kakwani-Lunetta), we could use the m-dimensional volume of the Lorenz surface as an inequality measure.

If the Lorenz ordering via nested Lorenz zonoids [as in (7.10)] is used, then attractive analogs to univariate measures are available: (1) the $(m + 1)$-dimensional volume of the Lorenz zonoid, (2) the m-dimensional volume of the boundary of the Lorenz zonoid, (3) the maximal distance between two points in the Lorenz zonoid. When $m = 1$, relatively simple expressions for these indices are available. In higher dimensions, this is not true. Koshevoy and Mosler (1997) do provide an analytic expression for the volume of the Lorenz zonoid as follows, though it is not easy to evaluate. For $\underline{X} \in \mathcal{L}^{(m)}$, define a normalized version of $\underline{X}$, denoted by $\underline{\widetilde{X}}$, in which $\widetilde{X}_i = X_i/E(X_i)$, $i = 1, 2, \ldots, m$. Consider $m + 1$ independent, identically distributed m-dimensional random vectors $\underline{\widetilde{X}}^{(1)}, \ldots, \underline{\widetilde{X}}^{(m+1)}$ each with the same distribution as $\underline{\widetilde{X}}$. Let Q be an $(m + 1) \times (m + 1)$ matrix whose ith row is $(1, \underline{\widetilde{X}}^{(i)})$, $i = 1, 2, \ldots, m + 1$. It follows that

$$\text{volume}(L(\underline{X})) = \frac{1}{(m+1)!} E(|\det Q|). \tag{7.19}$$

A drawback associated with the use of the volume of the Lorenz zonoid as a measure of inequality is that it can assume the value of 0 for certain nondegenerate distributions. See Mosler (2002) for discussion of a variant definition avoiding this pitfall.

When $m = 1$,

$$Q = \begin{pmatrix} 1 & \widetilde{X}_1 \\ 1 & \widetilde{X}_2 \end{pmatrix},$$

so that (7.19) reduces to a familiar expression for the Gini index of X in the one-dimensional case, namely,

$$G(X) = \frac{1}{2} E \left| \frac{X^{(1)}}{E(X^{(1)})} - \frac{X^{(2)}}{E(X^{(2)})} \right|, \tag{7.20}$$

where $X^{(1)}$ and $X^{(2)}$ are independent and identically distributed copies of X; i.e., it is one half of the expected distance between independent normalized copies of X.

The expression (7.20) leads to the following extension to m dimensions. For $\underline{X} \in \mathcal{L}^m$, define

$$G(\underline{X}) = \frac{1}{2m} E(||\underline{\widetilde{X}}^{(1)} - \underline{\widetilde{X}}^{(2)}||), \tag{7.21}$$

where $\underline{\widetilde{X}}^{(1)}$ and $\underline{\widetilde{X}}^{(2)}$ are independent, identically distributed normalized copies of $\underline{\widetilde{X}}$ (i.e., rescaled so that the marginal means are all equal to 1) and where $||\cdot||$ denotes the m-dimensional Euclidean norm. A proof that if $\underline{X} \leq_{ZL} \underline{Y}$, then $G(\underline{X}) \leq G(\underline{Y})$ [where G is defined by (7.21)] is given by Mosler (2002). Other norms or other measures of distance (instead of Euclidean distance) can be used in (7.21), perhaps with an advantage of computational simplicity.

7.4.1 *Bivariate Gini Index for the Arnold Lorenz Surface*

For a bivariate Lorenz surface $L_{12}(u_1, u_2)$ defined in (7.5) or (7.12) the bivariate Gini index can be defined as

$$G_{12} = 4 \int_0^1 \int_0^1 [u_1 u_2 - L_{12}(u_1, u_2)] du_1 du_2. \tag{7.22}$$

This bivariate Gini index has a reasonable interpretation in terms of equality (equal to (1-the inequality), i.e. $1 - G$). In the case of independence it does since in that case we have

$$1 - G_{12} = (1 - G_1)(1 - G_2), \tag{7.23}$$

where G_i, $i = 1, 2$ are the marginal Gini indices.

Example 7.4.1 The bivariate Gini index of the FGM Lorenz surface with uniform marginals defined in (7.14) is

$$G_{12}(w) = 4 \int_0^1 \int_0^1 [u_1 u_2 - L_{12}(u_1, u_2; w)] \, du_1 du_2 = \frac{5 + w}{9 + 4}. \tag{7.24}$$

If we set $w = 0$ in (7.24), we get $G_{12} = \frac{5}{9}$, and (7.23) holds.

The following result provides a convenient expression for the two-attribute bivariate Gini defined in (7.22), which permits a simple decomposition of the equality into two factors. The first component represents the equality within variables and the second factor represents the equality between variables (Sarabia and Jordá 2013, 2014a).

Theorem 7.4.1 *Let (X_1, X_2) have a bivariate Sarmanov–Lee distribution with bivariate Lorenz surface $L_{12}(u_1, u_2)$. The two-attribute bivariate Gini index defined in (7.22) admits the following representation,*

$$1 - G_{12} = \pi(1 - G_1)(1 - G_2) + (1 - \pi)(1 - G_{g_1})(1 - G_{g_2}),$$

where G_i, $i = 1, 2$ are the Gini indices of the marginal Lorenz curves, and G_{g_i}, $i = 1, 2$ represent the concentration indices of the concentration curves $L(u_i, g_i)$, $i = 1, 2$ in the sense of Definition 7.3.1.

Proof The proof is direct using expression (7.16) and definition (7.22). ∎

The overall equality (OE) given by $1 - G_{12}$ thus can be decomposed into two factors,

$$\text{OE} = \text{EW} + \text{EB},$$

where

$$\begin{aligned} \text{OE} &= 1 - G_{12}, \\ \text{EW} &= \pi(1 - G_1)(1 - G_2), \\ \text{EB} &= (1 - \pi)(1 - G_{g_1})(1 - G_{g_2}). \end{aligned}$$

Here, EW represents the equality within variables and the second factor, EB represents the equality between variables which involves the structure of the dependence of the underlying bivariate income distribution through the functions g_i, $i = 1, 2$. Note that the decomposition is well defined since $0 \leq \text{OE} \leq 1$ and $0 \leq \text{EW} \leq 1$ and hence $0 \leq \text{EB} \leq 1$.

7.5 Alternative Multivariate Inequality Indices

In this section we will provide multivariate versions of some of the inequality measures previously discussed in Chap. 5.

7.5.1 Multivariate Shannon and Rényi Entropies

If $\underline{X}$ is a multivariate random variable with joint pdf $f_{X_1,\ldots,X_m}(x_1, \ldots, x_m)$, the Shannon entropy of $\underline{X}$ is defined by

$$H(\underline{X}) = \int f_{X_1,\ldots,X_m}(x_1, \ldots, x_m) \log f_{X_1,\ldots,X_m}(x_1, \ldots, x_m) dx_1 \cdots dx_m,$$

and the multivariate Rényi entropy by

$$H_\lambda(\underline{X}) = \frac{1}{1-\lambda} \log\left(\int f^{\lambda}_{X_1,\ldots,X_m}(x_1,\ldots,x_m)dx_1\cdots dx_m\right),$$

with $\lambda > 0$ and $\lambda \neq 1$.

Example 7.5.1 Consider a multivariate Pareto type IV distribution (see Arnold 2015b) with joint survival function,

$$P(X_1 > x_1,\ldots,X_m > x_m) = \left\{1+\sum_{i=1}^{m}\left(\frac{x_i-\lambda_i}{\theta_i}\right)^{1/\gamma_i}\right\}^{-\alpha},$$

with $x_i > \lambda_i$, $\gamma_i, \theta_i > 0$, $i = 1, 2, \ldots, m$ and $\alpha > 0$. The multivariate Shannon entropy is of the form (Darbellay and Vajda 2000a),

$$\begin{aligned} H(\underline{X}) = &-\sum_{i=1}^{m}\log\left(\frac{\alpha+i-1}{\gamma_i\theta_i}\right) + (\alpha+m)\{\Psi(\alpha+m)-\Psi(\alpha)\} \\ &-\{\Psi(1)-\Psi(\alpha)\}\left(m-\sum_{i=1}^{m}\gamma_i\right), \end{aligned}$$

where, in this expression, $\Psi(t) = (d/dt)\log\Gamma(t)$ denotes the digamma function.

The multivariate Rényi entropy is (Zografos and Nadarajah 2005),

$$H_\lambda(\underline{X}) = \frac{\Gamma(\lambda(\alpha+m) - \sum_{i=1}^{m}\beta_i)\prod_{i=1}^{m}\Gamma(\beta_i)}{\Gamma(\lambda(\alpha+m))}\prod_{i=1}^{m}(\alpha+i-1)^{\lambda}(\gamma_i\theta_i)^{1-\lambda},$$

where $\beta_i = \gamma_i(1-\lambda)+\lambda > 0$, $i = 1, 2, \ldots, m$ and $\lambda(\alpha+n) - \sum_{i=1}^{k}\beta_i > 0$.

Example 7.5.2 If $\underline{X} \sim N^{(m)}(\underline{\mu}, \Sigma)$ is an m-dimensional normal distribution, we have

$$H(\underline{X}) = \frac{1}{2}\log\left((2\pi e)^m|\Sigma|\right).$$

In the multivariate case, the Shannon entropy possesses an additive decomposition property. We focus on the bivariate case, but this decomposition is equally valid in the case of two random vectors of dimension $m > 2$. Let (X, Y) be a bivariate random variable with joint density $f_{X,Y}(x, y)$, and associated marginal and conditional densities $f_X(x)$, $f_Y(y)$, $f_{X|Y}(x|y)$ and $f_{Y|X}(y|x)$. If we define $\tilde{H}(\cdot)$ as

$$\tilde{H}(f_{X|Y}) = E[H(f_{X|Y})] = -\int\left\{\int f_{X|Y}(x|y)\log f_{X|Y}(x|y)dx\right\} f_Y(y)dy, \tag{7.25}$$

we have the following Theorem.

Theorem 7.5.1 *The entropy of $f_{X,Y}(x, y)$ can be written as*

$$H(f_{X,Y}) = \tilde{H}(f_{X|Y}) + H(f_Y) = \tilde{H}(f_{Y|X}) + H(f_X), \tag{7.26}$$

where $\tilde{H}(f_{X|Y})$ is defined in (7.25). If X and Y are independent, then

$$H(f_{X,Y}) = H(f_X) + H(f_Y).$$

Proof From (7.25) we have $\tilde{H}(f_{X|Y}) = -\int\int f_{X,Y}(x, y)\log(f_{X|Y}(x|y))dxdy$.
Now, writing

$$H(f_{X,Y}) = -\int\int f_{X,Y}(x, y)\{\log(f_{X|Y}(x|y)) - \log(f_Y(y))\}dxdy,$$

we obtain (7.26). ■

7.5.2 Multivariate Generalized Entropy and Theil Indices

Let $X = (X_1, \ldots, X_m)$ be an m-dimensional random vector with non-negative components. We define the multivariate generalized entropy indices as (Sarabia et al. 2017a)

$$\mathrm{MGE}(\underline{X};\theta) = \frac{1}{\theta(\theta-1)}\left\{E\left[\left(\frac{X_1\cdots X_m}{\mu_{12\cdots m}}\right)^{\theta}\right]-1\right\},\quad \theta\in\mathbb{R}-\{0,1\},$$

where

$$\mu_{12\cdots m} = E(X_1X_2\cdots X_m) < \infty$$

If we take limits as θ converges to 0 and 1, we obtain

$$\lim_{\theta\to 0}\mathrm{MGE}(\underline{X};\theta) = T_0^{(m)}(\underline{X}),$$

and

$$\lim_{\theta\to 1}\mathrm{MGE}(\underline{X};\theta) = T_1^{(m)}(\underline{X}),$$

where

$$T_0^{(m)}(\underline{X}) = -E\left[\log\left(\frac{X_1\cdots X_m}{\mu_{12\ldots m}}\right)\right], \tag{7.27}$$

and

$$T_1^{(m)}(\underline{X}) = -E\left[\frac{X_1 \cdots X_m}{\mu_{12\ldots m}} \log\left(\frac{X_1 \cdots X_m}{\mu_{12\ldots m}}\right)\right], \tag{7.28}$$

are the m-dimensional Theil(0) and Theil(1) indices, respectively.

For the bivariate case $k = 2$, expression (7.27) can be written as

$$T_0^{(2)}(\underline{X}) = T_0(X_1) + T_0(X_2) + \log \frac{\mu_{12}}{\mu_1 \mu_2}, \tag{7.29}$$

and (7.28) can be written as

$$T_1^{(2)}(\underline{X}) = \mu_{12}^{-1}\left[E(X_1 X_2 \log X_1) + E(X_1 X_2 \log X_2) - \mu_{12} \log \mu_{12}\right], \tag{7.30}$$

which can be computed in terms of the derivatives of $u(r_1, r_2) = E(X_1^{r_1} X_2^{r_2})$.

Example 7.5.3 Let $\underline{X} = (X_1, X_2)$ be a bivariate lognormal distribution with joint pdf,

$$f_{X_1,X_2}(x_1, x_2) = \frac{1}{2\pi x_1 x_2 \sqrt{|\Sigma|}} \exp\left\{-\frac{1}{2}(\log \boldsymbol{x} - \boldsymbol{\mu})\Sigma^{-1}(\log \boldsymbol{x} - \boldsymbol{\mu})^T\right\},$$

where $\boldsymbol{x} = (x_1, x_2)$, and $\boldsymbol{\mu} = (\mu_1, \mu_2)$. We have

$$E(X_1^{r_1} X_2^{r_2}) = \exp\left(\boldsymbol{r}\boldsymbol{\mu}^T + \frac{1}{2}\boldsymbol{r}\Sigma\boldsymbol{r}^T\right),$$

where $\boldsymbol{r} = (r_1, r_2)$. Using (7.29) and (7.30) we get

$$T_0^{(2)}(\underline{X}) = T_1^{(2)}(\underline{X}) = \frac{\sigma_1^2}{2} + \frac{\sigma_2^2}{2} + \sigma_{12},$$

where both bivariate indices are the same, as in the univariate case.

7.6 Exercises

1. By considering the two-dimensional case ($m = 2$), verify that $X \leq_{MM} Y \not\Rightarrow X \leq_{UM} Y$.

2. Prove that $X \leq_{UM} Y \Rightarrow X \leq_{CM} Y$.

3. Prove that $X \leq_{MO} Y \Rightarrow X \leq_{MM} Y$ and that $X \leq_{CM} Y \Rightarrow X \leq_{MM} Y$.

4. Suppose $\underline{X}$ and $\underline{Y} \in \mathcal{L}^{(m)}$ have independent coordinate random variables such that $X_i \leq_L Y_i, i = 1, 2, \ldots, n$. Prove that $\underline{X} \leq_{CL} \underline{Y}$.

5. Suppose $\underline{X} \in \mathcal{L}^{(m)}$ and $U \geq 0$ are independent. Assume $E(U) \in (0, \infty)$ and define $\underline{Y} = U\underline{X}$. Prove $\underline{X} \leq_{CL} \underline{Y}$. This generalizes the misreported income model of Chap. 4.

6. Discuss the relationships between the three orders $\leq_{CL}, \leq_{PCL}$ and $\leq_{gL}$ defined on $\mathcal{L}^{(m)}$.

7. Suppose that X_1 and $X_2 \in \mathcal{L}$ with corresponding Lorenz curves $L_1(u)$ and $L_2(u)$. Show that if X_1 and X_2 are independent, the Lorenz surface of the random variable (X_1, X_2) defined by (7.5) is of the form $L(u, v) = L_1(u)L_2(v)$.

8. Suppose

$$f_{X_1,X_2}(x_1, x_2) = \begin{cases} x_1 + x_2, \; 0 < x_1 < 1, \quad 0 < x_2 < 1, \\ \quad 0, \qquad \text{otherwise} \end{cases}$$

and

$$f_{X_1,X_2}(x_1, x_2) = \begin{cases} 1, \; 0 < x_1 < 1, \quad 0 < x_2 < 1, \\ 0, \; \text{otherwise} \end{cases}$$

Evaluate and compare the corresponding Lorenz surfaces defined by (7.5). What happens if we use the Taguchi definition (7.4)?

9. In accordance with custom, inequality measures vary from 0 to 1 with 1 representing the sometimes unachievable case of greatest inequality. In Eq. (7.21) verify that this k-dimensional Gini index does indeed have range [0, 1].

10. Verify that the matrix P given in Eq. (7.2) is not expressible as a finite product of Robin Hood matrices. [Verify that finite products of Robin Hood matrices, in the 3×3 case, cannot have three zero elements.]

11. Prove Theorem 7.16.

12. For the Pareto Lorenz surface based on the FGM family defined in (7.18), prove that the bivariate Gini index is given by

$$G_{12}(\alpha_1, \alpha_2) = \frac{(3\alpha_1 - 1)(3\alpha_2 - 1)(2\alpha_1 + 2\alpha_2 - 3) + [h(\alpha_1, \alpha_2)]w}{(3\alpha_1 - 1)(3\alpha_2 - 1)[(1 - 2\alpha_1)(1 - 2\alpha_2) + w]},$$

where

$$h(\alpha_1, \alpha_2) = -3 - 4\alpha_1^2(\alpha_2 - 1)^2 + (5 - 4\alpha_2)\alpha_2 + \alpha_1(5 + \alpha_2(8\alpha_2 - 7)).$$

Chapter 8
Stochastic Majorization

We return to the original setting in which Hardy, Littlewood, and Polya discussed majorization, i.e., vectors in $\mathbb{R}^n$. However, we now consider random variables which take on values in $\mathbb{R}^n$. If $\underline{X}$ and $\underline{Y}$ are such n-dimensional random variables, then for given realizations of $\underline{X}$ and $\underline{Y}$, say $\underline{x}$ and $\underline{y}$, we may or may not have $\underline{x} \leq_M \underline{y}$. If we have such a relation for every realization of $(\underline{X}, \underline{Y})$, then we have a very strong version of stochastic majorization holding between $\underline{X}$ and $\underline{Y}$. However, we may be interested in weaker versions. Certainly, for most purposes $P(\underline{X} \leq_M \underline{Y}) = 1$ would be more than adequate. We will content ourselves with the requirement that there be versions of $\underline{X}$ and $\underline{Y}$ for which $\underline{X}$ is almost surely majorized by $\underline{Y}$. However, this will not be transparent from the Definition (8.1.2 below). We will focus attention on one particular form of stochastic majorization, the one proposed by Nevius et al. (1977). We will mention in passing other possible definitions and refer the interested reader to the rather complicated diagram on page 426 of Marshall et al. (2011), which summarizes known facts about the interrelationships between the various brands of stochastic majorization.

8.1 Definition and Main Results

To motivate our definition, recall that by the definition of Schur convexity (Definition 2.2.1) we know that $\underline{x} \leq_M \underline{y}$ if and only if $g(\underline{x}) \leq g(\underline{y})$ for every Schur convex function g. Now, if $\underline{X}$ and $\underline{Y}$ are random variables, then we need to decide in what sense will $g(\underline{X})$, a random variable, provided g is Borel measurable, be required to be less than the random variable $g(\underline{Y})$. The comparison used by Nevius, Proschan, and Sethuraman was based on stochastic ordering (which we will study more thoroughly in Chap. 9). The relevant concepts are defined by

B. C. Arnold, J. M. Sarabia, *Majorization and the Lorenz Order with Applications in Applied Mathematics and Economics*, Statistics for Social and Behavioral Sciences,
https://doi.org/10.1007/978-3-319-93773-1_8

Definition 8.1.1 Let X and Y be one-dimensional random variables. We will write $X \leq_{\text{st}} Y$ (X is not stochastically larger than Y), if

$$P(X \leq x) \geq P(Y \leq x), \quad \forall x \in \mathbb{R}.$$

Definition 8.1.2 Let $\underline{X}$ and $\underline{Y}$ be n-dimensional random variables. We will say that $\underline{X}$ is stochastically majorized by $\underline{Y}$ and write $\underline{X} \leq_{\text{st}.M} \underline{Y}$ if $g(\underline{X}) \leq_{\text{st}} g(\underline{Y})$ for every Borel measurable Schur convex function $g : \mathbb{R}^n \to \mathbb{R}$.

It will be recalled from Chap. 2 that in the definition of non-stochastic majorization one does not have to check every Schur convex g, in order to determine whether majorization obtains. Certain subclasses suffice. For example, we could consider g's which are continuous, symmetric, and convex or, instead, g's that are separable convex or in the extreme, by the HLP theorem (Theorem 2.2.6), g's which are separable angle functions. Now in the definition of stochastic majorization (Definition 8.1.2), if we replace "for all Borel measurable Schur convex functions g" by for all g in some specified subclass of well-behaved Schur convex functions such as those just mentioned, we find that different partial orders may be defined. Some putative definitions for stochastic majorization derived in this manner may be described as follows. First, define

$$G_1 = \{g \ : \ g \ \text{ is Borel measurable and Schur convex on } \ \mathbb{R}^n\},$$

$$G_2 = \{g \ : \ g \ \text{ is continuous, symmetric and convex on } \ \mathbb{R}^n\},$$

$$G_3 = \left\{ g \ : \ g(\underline{x}) = \sum_{i=1}^{n} h(x_i) \ \text{ for some continuous convex function h on } \mathbb{R} \right\},$$

and

$$G_4 = \left\{ g \ : \ g(\underline{x}) = \sum_{i=1}^{n} (x_i - c)^+ \ \text{ for some } \ c \in \mathbb{R} \ \text{ or } \ g(\underline{x}) = \sum_{i=1}^{n} x_i \right\}.$$

Now, for $i = 1, 2, 3, 4$ we define the partial order $\leq_{\text{st}.M(i)}$ by $\underline{X} \leq_{\text{st}.M(i)} \underline{Y}$ iff $g(\underline{X}) \leq_{\text{st}} g(\underline{Y})$, $\forall g \in G_i$. Of course, $\leq_{\text{st}.M(1)}$ is just the same as stochastic majorization as described in Definition 8.1.2.

Since $G_4 \subset G_3 \subset G_2 \subset G_1$, it is obvious that

$$\leq_{\text{st}.M(1)} \Rightarrow \leq_{\text{st}.M(2)} \Rightarrow \leq_{\text{st}.M(3)} \Rightarrow \leq_{\text{st}.M(4)} \cdot$$

Marshall et al. (2011) provide an example to show that $\leq_{\text{st}.M(2)} \not\Rightarrow \leq_{\text{st}.M(1)}$. It is not known whether $\leq_{\text{st}.M(2)}$ and $\leq_{\text{st}.M(3)}$ are equivalent. The relation between $\leq_{\text{st}.M(3)}$ and $\leq_{\text{st}.M(4)}$ is considered in Exercise 1.

Faced with such a surfeit of possible definitions (Marshall et al. (2011) even consider several more), we will stick with the original choice exemplified by Definition 8.1.2. Two equivalent versions of that definition can be obtained from the following

Theorem 8.1.1 *The following conditions are equivalent:*

(1) $\underline{X} \leq_{st.M} \underline{Y}$
(2) $E[g(\underline{X})] \leq E[g(\underline{Y})]$ *for every Schur convex g for which both expectations exist.*
(3) $P(\underline{X} \in A) \leq P(\underline{Y} \in A)$ *for every measurable Schur convex set A (i.e., for every A such that* $\underline{x} \in A$ *and* $\underline{x} \leq_M \underline{y} \Rightarrow \underline{y} \in A$*).*

Proof (1) $\Rightarrow$ (2), since $U \leq_{\text{st}} V$ implies $E(U) \leq E(V)$ when expectations exist. (2) $\Rightarrow$ (3), since the indicator function of a Schur convex set is Schur convex. (3) $\Rightarrow$ (1), since one may let $A = \{\underline{x} : g(\underline{x}) > c\}$. ■

Stochastic majorization is preserved under mixing, normalization, a strong mode of convergence and, to a certain extent, convolution. Specifically, we have

Theorem 8.1.2 *Let* $\{\underline{X}_\lambda : \lambda \in \Lambda\}$ *and* $\{\underline{Y}_\lambda : \lambda \in \Lambda\}$ *be two indexed collections of random variables where* Λ *is a subset of* $\mathbb{R}$*. Let G be the distribution function of a random variable whose range is in* Λ*, and let* $\underline{X}_G$ *and* $\underline{Y}_G$ *be the corresponding G-mixtures of* $\{\underline{X}_\lambda\}$ *and* $\{\underline{Y}_\lambda\}$*, respectively, i.e.,*

$$F_{\underline{X}_G}(\underline{x}) = \int_\Lambda P(\underline{X}_\lambda \leq \underline{x})\, dG(\lambda) \tag{8.1}$$

and $F_{\underline{Y}_G}$ *is analogously defined. If* $\underline{X}_\lambda \leq_{\text{st}.M} \underline{Y}_\lambda$ *for every* $\lambda \in \Lambda$*, then* $\underline{X}_G \leq_{\text{st}.M} \underline{Y}_G$*.*

Proof For any function g for which $E[g(\underline{X}_G)]$ exists, it follows from (8.1) that

$$E[g(\underline{X}_G)] = \int_\Lambda E(g(\underline{X}_\lambda))\, dG(\lambda)$$

and that an analogous expression is available for $E[g(\underline{Y}_G)]$. If g is Schur convex, then since $\underline{X}_\lambda \leq_{\text{st}.M} \underline{Y}_\lambda$ we know by condition (2) of Theorem 8.1.1 that $E[g(\underline{X}_\lambda)] \leq E[g(\underline{Y}_\lambda)]$, for every $\lambda \in \Lambda$. Integrating this with respect to the distribution G using Eq. (8.1), we conclude that $E[g(\underline{X}_G)] \leq E[g(\underline{Y}_G)]$. Since this is true for any Schur convex g for which the expectations exist, we conclude that $\underline{X}_G \leq_{\text{st}.M} \underline{Y}_G$ (again applying condition (2) of Theorem 8.1.1). ■

Theorem 8.1.3 *Suppose that* $\underline{X} \leq_{st.M} \underline{Y}$ *and* $\underline{X}' = f\left(\sum_{i=1}^n X_i\right)\underline{X}$*,* $\underline{Y}' = f\left(\sum_{i=1}^n Y_i\right)\underline{Y}$ *where f is a Borel measurable function on* $\mathbb{R}$*. It follows that* $\underline{X}' \leq_{st.M} \underline{Y}'$

Proof Since $\underline{X} \leq_{\text{st}.M} \underline{Y}$ it follows that $\sum_{i=1}^n X_i \stackrel{d}{=} \sum_{i=1}^n Y_i$ (Exercise 2). Now since $\underline{X} \leq_{\text{st}.M} \underline{Y} \Rightarrow c\underline{X} \leq_{\text{st}.M} c\underline{Y}$ for any c, the theorem follows from our mixture theorem, since we have

$$F_{\underline{X}'}(\underline{x}) = \int_{-\infty}^{\infty} P(c\underline{X} \le \underline{x})\, dG(c)$$

and

$$F_{\underline{Y}'}(\underline{x}) = \int_{-\infty}^{\infty} P(c\underline{Y} \le \underline{x})\, dG(c)$$

where G is the common distribution of $f\left(\sum_{i=1}^{n} X_i\right)$ and $f\left(\sum_{i=1}^{n} Y_i\right)$. ∎

Theorem 8.1.4 *Suppose* $\{\underline{X}_m\}$ *and* $\{\underline{Y}_m\}$ *are two sequences of n-dimensional random variables and that* $\underline{X}_m \to \underline{X}$ *and* $\underline{Y}_m \to \underline{Y}$ *in the sense that* $E(g(\underline{X}_n)) \to E(g(\underline{X}))$ *for every bounded measurable function g. If* $\underline{X}_m \le_{st.M} \underline{Y}_m$ *for each m, then* $\underline{X} \le_{st.M} \underline{Y}$.

Proof Let g be an arbitrary Schur convex Borel measurable function. Let z be a point of continuity of both of the distributions of $g(\underline{X})$ and $g(\underline{Y})$. Observe that $I(g(\underline{x}) \le \underline{z})$ is a bounded Borel measurable function. We may thus argue as follows.

$$\begin{aligned}
P(g(\underline{X}) \le z) &= E(I(g(\underline{X}) \le z)) \\
&= \lim_{n\to\infty} E(I(g(\underline{X}_m) \le z)) \\
&= \lim_{n\to\infty} P(g(\underline{X}_m \le z) \\
&\ge \lim_{n\to\infty} P(g(\underline{Y}_m \le z) \quad \text{since } \underline{X}_m \le_{\text{st}.M} \underline{Y}_m \\
&= \lim_{n\to\infty} E(I(g(\underline{Y}_m) \le z)) \\
&= E(I(g(\underline{Y}) \le z)) \\
&= P(g(\underline{Y}) \le z).
\end{aligned}$$

Since such points z are dense in $\mathbb{R}$, it follows that $g(\underline{X}) \le_{\text{st}} g(\underline{Y})$ and since this holds for every Borel measurable Schur convex g, we conclude $\underline{X} \le_{\text{st}} \underline{Y}$. ∎

The hypothesis in this theorem can be relaxed slightly. It suffices that $\underline{X}_m \xrightarrow{d} \underline{X}$ and $\underline{Y}_m \xrightarrow{d} \underline{Y}$. See Marshall et al. (2011, p. 424).

Closure under convolution will elude us. It is possible to construct an example of random vectors $\underline{X}^{(1)}, \underline{X}^{(2)}, \underline{Y}^{(1)}$ and $\underline{Y}^{(2)}$ (even in the case $n = 2$) such that $\underline{X}^{(1)}, \underline{X}^{(2)}$ are independent, $\underline{Y}^{(1)}, \underline{Y}^{(2)}$ are independent, $\underline{X}^{(1)} \le_{\text{st}.M} \underline{Y}^{(1)}$ and $\underline{X}^{(2)} \le_{\text{st}.M} \underline{Y}^{(2)}$ yet $\underline{X}^{(1)} + \underline{X}^{(2)} \not\le_{\text{st}.M} \underline{Y}^{(1)} + \underline{Y}^{(2)}$. See Marshall et al. (2011, p. 424). If, by chance, the random vectors always have coordinates which are in increasing order, we can get our result. Recall from (2.5) the notation

$$O_n = \{x : x_1 \le x_2 \le \cdots \le x_n\}.$$

Theorem 8.1.5 *If $\underline{X}^{(1)}$, $\underline{X}^{(2)}$ are independent, $\underline{Y}^{(1)}$, $\underline{Y}^{(2)}$ are independent and all four random vectors take on values restricted to the set O_n. If $\underline{X}^{(1)} \leq_{\text{st}.M} \underline{Y}^{(1)}$ and $\underline{X}^{(2)} \leq_{\text{st}.M} \underline{Y}^{(2)}$, then $\underline{X}^{(1)} + \underline{X}^{(2)} \leq_{\text{st}.M} \underline{Y}^{(1)} + \underline{Y}^{(2)}$.*

Proof If $\underline{u}, \underline{v}, \underline{w} \in O_n$, then $\underline{u} \leq_M \underline{v} \Rightarrow \underline{u} + \underline{w} \leq_M \underline{v} + \underline{w}$. Thus, if g is Schur convex on O_n, then for every $\underline{w} \in O_n$, $g_{\underline{w}}(\underline{x}) = g(\underline{x} + \underline{w})$ is Schur convex on O_n. Now, we can write for an arbitrary Schur convex g

$$\begin{aligned} E[g(\underline{X}^{(1)} + \underline{X}^{(2)})] &= \int_{O_n} E[g(\underline{X}^{(1)} + \underline{w})]\, dF_{\underline{X}^{(2)}}(\underline{w}) \\ &\leq \int_{O_n} E[g(\underline{Y}^{(1)} + \underline{w})]\, dF_{\underline{X}^{(2)}}(\underline{w}) \\ &= E[g(\underline{Y}^{(1)} + \underline{X}^{(2)})]. \end{aligned}$$

Here we have assumed, without loss of generality, that all four random vectors $\underline{X}^{(1)}, \underline{X}^{(2)}, \underline{Y}^{(1)}, \underline{Y}^{(2)}$ are independent. By conditioning on $\underline{Y}^{(1)}$ we may then prove $E[g(\underline{Y}^{(1)} + \underline{X}^{(2)})] \leq E[g(\underline{Y}^{(1)} + \underline{Y}^{(2)})]$. Since g was an arbitrary Schur convex function, we conclude $\underline{X}^{(1)} + \underline{X}^{(2)} \leq_{\text{st}.M} \underline{Y}^{(1)} + \underline{Y}^{(2)}$. ■

We have alluded to the fact that if $\underline{X} \leq_{\text{st}.M} \underline{Y}$ then $\sum_{i=1}^n X_i \stackrel{d}{=} \sum_{i=1}^n Y_i$. What happens if we have stochastic majorization in both directions, i.e., $\underline{X} \leq_{\text{st}.M} \underline{Y}$ and $\underline{Y} \leq_{\text{st}.M} \underline{X}$? We cannot conclude that $\underline{X} = \underline{Y}$ or even that $\underline{X}$ and $\underline{Y}$ are identically distributed. However, it is true that they must have identically distributed order statistics. This is not surprising since an analogous result is encountered in the non-stochastic case. If $\underline{x} \leq_M \underline{y}$ and $\underline{y} \leq_M \underline{x}$ it does not follow that $\underline{x} = \underline{y}$ but it does follow that $x_{i:n} = y_{i:n}$, $i = 1, 2, \ldots, n$. In the stochastic case, we argue as follows.

For $j = 1, 2, \ldots, n$ and $z \in \mathbb{R}$ define

$$g_z^{(j)}(\underline{x}) = \mathrm{I}\left(\sum_{i=1}^{j} x_{i:n} \leq z\right). \tag{8.2}$$

It is clear that these $g_z^{(j)}(\cdot)$'s are Schur convex functions. Since products of Schur convex functions are again Schur convex, we conclude that for any vector $z_1, z_2, \ldots, z_n$ satisfying $z_{i+1} - z_i > z_i - z_{i-1}$, $i = 1, 2, \ldots, n-1$ where by convention $z_0 = 0$, we can conclude that

$$E\left[\prod_{j=1}^{n} g_{z_j}^{(j)}(\underline{X})\right] = E\left[\prod_{j=1}^{n} g_{z_j}^{(j)}(\underline{Y})\right] \tag{8.3}$$

since both $\underline{X} \leq_{\text{st}.M} \underline{Y}$ and $\underline{Y} \leq_{\text{st}.M} \underline{X}$. However,

$$E\left[\prod_{j=1}^{n} g_{z_j}^{(j)}(\underline{X})\right] = P\left(\sum_{i=1}^{j} X_{i:n} \le z_j,\ j = 1, 2, \ldots, n\right),$$

so that (8.3) is enough to guarantee that

$$\left(X_{1:n}, \sum_{i=1}^{2} X_{i:n}, \ldots, \sum_{i=1}^{n} X_{i:n}\right) \stackrel{d}{=} \left(Y_{1:n}, \sum_{i=1}^{2} Y_{i:n}, \ldots, \sum_{i=1}^{n} Y_{i:n}\right),$$

from which it follows directly that

$$(X_{1:n}, X_{2:n}, \ldots, X_{n:n}) \stackrel{d}{=} (Y_{1:n}, Y_{2:n}, \ldots, Y_{n:n}).$$

Note that the vector

$$(X_{1:n}, X_{1:n} + X_{2:n}, \ldots, X_{1:n} + X_{2:n} + \cdots + X_{n:n})$$

alluded to above is just the un-normalized Lorenz curve of the vector $(X_1, \ldots, X_n)$. We can define $\le_{\text{st}.M}$ in terms of functions of the un-normalized Lorenz curve. Thus, if we let $\mathbb{L}$ be the class of all possible un-normalized Lorenz curves, i.e.,

$$\mathbb{L} = \{(z_1, z_2, \ldots, z_n) : z_{i+1} - z_i > z_i - z_{i-1},\ i = 1, 2, \ldots, n\} \tag{8.4}$$

(again $z_0 = 0$ by convention), and for any random vector $\underline{X}$ we denote its un-norma lized Lorenz curve by $L^*(\underline{X})$, we can verify:

Theorem 8.1.6 *$\underline{X} \le_{st.M} \underline{Y}$ if and only if $h(L^*(\underline{X})) \le_{st} h(L^*(\underline{Y}))$ for every $h(\underline{z}) : \mathbb{L} \to \mathbb{R}$ which for fixed z_n is a monotone decreasing function of each of the arguments $z_1, z_2, \ldots, z_{n-1}$.*

Proof Exercise 3. ■

When we introduced stochastic majorization, it was remarked that our definition of $\underline{X} \le_{\text{st}.M} \underline{Y}$ was, in fact, equivalent to the existence of versions of $\underline{X}$ and $\underline{Y}$ for which almost sure majorization obtains. This observation, attributed by Marshall and Olkin to T. Snijders, is relatively easy to prove, if we are willing to consider rather complicated probability spaces.

Theorem 8.1.7 *$\underline{X} \le_{st.M} \underline{Y}$ if and only if there exist two random variables $\underline{X}'$, $\underline{Y}'$ defined on the same probability space for which $\underline{X} \stackrel{d}{=} \underline{X}'$, $\underline{Y} \stackrel{d}{=} \underline{Y}'$ and $P(\underline{X}' \le_M \underline{Y}') = 1$.*

Proof Let

$$w = \{(\underline{x}, \underline{y}) ;\ \underline{x} \in \mathbb{R}^n,\ \underline{y} \in \mathbb{R}^n,\ \underline{x} \le_M \underline{y}\} \subset \mathbb{R}^{2n}.$$

Let μ and ν be, respectively, probability measures on $\mathbb{R}^n$ determined by the distributions of $\underline{X}$ and $\underline{Y}$. Since w is a closed subset of $\mathbb{R}^{2n}$, and since Strassen's (1965) condition (30) is satisfied, there exists a probability measure λ on $\mathbb{R}^{2n}$ with marginals μ and ν which is supported by w, i.e., with $\lambda(w) = 1$. Let $(\underline{X}', \underline{Y}')$ be a $2n$-dimensional random variable with distribution λ, then the conditions of the theorem are clearly satisfied. ■

In the remainder of this chapter we will give several examples of stochastic majorization. The first is an example in which stochastic majorization occurs in the almost sure sense.

Theorem 8.1.8 *Let $X_1, X_2, \ldots, X_n$ be positive random variables and let g be a positive star shaped function defined on $[0, \infty)$. Define $Z_i = g(X_i)$, $i = 1, 2, \ldots, n$. It follows that*

$$P\left(\left(\sum_{i=1}^{n} X_i\right)^{-1} \underline{X} \leq_M \left(\sum_{i=1}^{n} Z_i\right)^{-1} \underline{Z}\right) = 1. \tag{8.5}$$

Proof For each point w in the probability space consider a discrete uniform random variable $\hat{X}_w$ with n possible values $X_1(w), X_2(w), \ldots, X_n(w)$. Apply Theorem 4.1.2 (note: g star shaped implies $g(x)/x$ is non-decreasing and g is non-decreasing) and conclude that for each w

$$\left(\sum_{i=1}^{n} X_i(w)\right)^{-1} \underline{X}(w) \leq_M \left(\sum_{i=1}^{n} Z_i(w)\right)^{-1} \underline{Z}(w),$$

but this implies (8.5). ■

An important source of examples of stochastic majorization depends on the following preservation theorem of Nevius, Proschan, and Sethuraman.

Theorem 8.1.9 *Let $\{f(x, \lambda)\}_{\lambda > 0}$ be a family of densities on $(0, \infty)$. Suppose that $f(x, \lambda)$ is totally positive of order two, i.e., for $x_1 < x_2$ and $\lambda_1 < \lambda_2$*

$$\begin{vmatrix} f(x_1, \lambda_1) & f(x_1, \lambda_2) \\ f(x_2, \lambda_1) & f(x_2, \lambda_2) \end{vmatrix} \geq 0.$$

In addition, assume that $f(x, \lambda)$ satisfies the following "semi-group" property in λ

$$f(x, \lambda_1 + \lambda_2) = \int_0^{\infty} f(x - y, \lambda_1) f(y, \lambda_2)\, d\mu(y),$$

where μ denotes either Lebesgue measure on $(0, \infty)$ or counting measure on $\{0, 1, 2, \ldots\}$.

Let $(X_{\lambda_1}, X_{\lambda_2}, \ldots, X_{\lambda_n})$ denote a vector with independent components such that X_{λ_i} has density $f(x, \lambda_i)$, $i = 1, 2, \ldots, n$. If $\underline{\lambda} \leq_M \underline{\lambda}$, then $X_{\underline{\lambda}} \leq_{st.M} X_{\underline{\lambda}'}$.

Proof Without loss of generality, we may assume $n = 2$. The result will then follow, provided we can show that for any Schur convex g the function ϕ defined by

$$\phi(\lambda_1, \lambda_2) = \int_0^\infty \int_0^\infty g(x_1, x_2) f(x_1, \lambda_1) f(x_2, \lambda_2)\, d\mu(x_1)\, d\mu(x_2)$$

is again Schur convex. Suppose $\lambda_1 < \lambda_2$ and ϵ is small, then $(\lambda_1 + \epsilon, \lambda_2 - \epsilon) \leq_M (\lambda_1, \lambda_2)$, and we may write

$$\begin{aligned}
&\phi(\lambda_1, \lambda_2) - \phi(\lambda_1 + \epsilon, \lambda_2 - \epsilon) \\
&\quad = \int_0^\infty \int_0^\infty [f(x_1, \lambda_1) f(x_2, \lambda_2) \\
&\qquad - f(x_1, \lambda_1 + \epsilon) f(x_2, \lambda_2 - \epsilon)] g(x_1, x_2)\, d\mu(x_1)\, d\mu(x_2) \\
&\quad = \int_0^\infty f(y, \epsilon) \int_0^\infty \int_0^\infty f(x_1, \lambda_1) f(x_2 - y, \lambda_2 - \epsilon) \\
&\qquad - f(x_1 - y, \lambda_1 + \epsilon) f(x_2, \lambda_2 - \epsilon)] g(x_1, x_2)\, d\mu(x_1)\, d\mu(x_2)\, d\mu(y).
\end{aligned}$$

Now, change variables and recall g is symmetric. Thus,

$$\begin{aligned}
&\phi(\lambda_1, \lambda_2) - \phi(\lambda_1 + \epsilon, \lambda_2 - \epsilon) \\
&\quad = \int_0^\infty f(y, \epsilon) \int\int_{z_1 \leq z_2} [f(z_2, \lambda_1) f(z_1, \lambda_2 - \epsilon) - f(z_1, \lambda_1) f(z_2, \lambda_2 - \epsilon)] \\
&\qquad \times [g(z_1 + y, z_2) - g(z_1, z_2 + y)]\, d\mu(z_1)\, d\mu(z_2)\, d\mu(y).
\end{aligned}$$

The Schur convexity of g guarantees that $g(z_1 + y, z_2) - g(z_1, z_2 + y) \leq 0$, and the total positivity of order 2 of f guarantees that $f(z_2, \lambda_1) f(z_1, \lambda_2 - \epsilon) - f(z_1, \lambda_1) f(z_2, \lambda_2 - \epsilon) \leq 0$. The Schur convexity of ϕ is thus confirmed. ■

The hypotheses of Theorem 8.1.9 are strong. Four examples are available: binomial, Poisson, negative binomial, and gamma. More examples can be generated by considering mixtures as follows.

Theorem 8.1.10 *Let $X_{\underline{\lambda}}$ be a collection of n-dimensional random variables indexed by $\underline{\lambda} \in \mathbb{R}_n$ such that $\lambda \leq_M \lambda \Rightarrow \underline{X}_{\underline{\lambda}} \leq_{st.M} \underline{X}_{\underline{\lambda}'}$. Let $\underline{Z}$ and $\underline{Z}'$ be two n-dimensional random vectors with corresponding distribution functions $F_{\underline{Z}}(\underline{z})$ and $F_{\underline{Z}'}(\underline{z})$. Define $\underline{U}$ and $\underline{U}'$ by*

$$F_{\underline{U}}(\underline{u}) = \int_{\mathbb{R}^n} P(\underline{X}_{\underline{\lambda}} \leq \underline{u})\, dF_{\underline{Z}}(\underline{\lambda})$$

and

$$F_{\underline{U}'}(\underline{u}) = \int_{\mathbb{R}^n} P(\underline{X}_{\underline{\lambda}} \leq \underline{u})\, dF_{\underline{Z}'}(\underline{\lambda})$$

If $\underline{Z} \leq_{\text{st}.M} \underline{Z}'$*, then* $\underline{U} \leq_{\text{st}.M} \underline{U}'$ *[more briefly we could write* $\underline{Z} \leq_{\text{st}.M} \underline{Z}' \Rightarrow \underline{X}_{\underline{Z}} \leq_{\text{st}.M} \underline{X}_{\underline{Z}'}$*.]*

Proof By Theorem 8.1.7 we can assume that $\underline{Z} \leq_M \underline{Z}'$ a.s. Then, for any bounded Schur convex g we have $E(g(\underline{U})|\underline{Z}) \leq E(g(\underline{U}')|\underline{Z}')$ which may be integrated to yield $E(g(\underline{U})) \leq E(g(\underline{U}'))$. It follows that $\underline{U} \leq_{\text{st}.M} \underline{U}'$. ∎

Example 8.1.1 Let $\underline{X}_{\underline{\lambda}} \sim \text{multinomial}(N, \underline{\lambda})$ where $\lambda_i > 0$, $\sum_{i=1}^n \lambda_i = 1$, i.e.,

$$P(\underline{X}_{\underline{\lambda}} = \underline{k}) = N! \prod_{i=1}^n \frac{\lambda_i^{k_i}}{k_i!},$$

$0 \leq k_i \leq N$, $\sum_{i=1}^n k_i = N$. If $\underline{\lambda} \leq_M \underline{\lambda}'$, then $\underline{X}_{\underline{\lambda}} \leq_{s.t.M} \underline{X}_{\underline{\lambda}'}$.

The simplest way to verify this assertion is to observe that if $Z_{\lambda_1}, \ldots, Z_{\lambda_n}$ are independent random variables with $Z_i \sim \text{Poisson}(\lambda_i)$, then $\underline{\lambda} \leq_M \underline{\lambda}'$ implies $\underline{Z}_{\underline{\lambda}} \leq_{\text{st}.M} \underline{Z}_{\underline{\lambda}'}$, by Theorem 8.1.9. But then if we define

$$X_{\lambda_i} = I\left(\sum_{i=1}^n Z_{\lambda_i} = N\right) Z_{\lambda_i},$$

we have $\underline{X}_{\underline{\lambda}} \leq_{\text{st}.M} \underline{X}_{\underline{\lambda}'}$ by Theorem 8.1.3. But the conditional distribution of such an $\underline{X}_{\underline{\lambda}}$ given $\sum_{i=1}^n Z_{\lambda_i} \neq \underline{0}$ is just multinomial $(N, \underline{\lambda})$. The desired result follows, since $\sum_{i=1}^n Z_{\lambda_i} \overset{d}{=} \sum_{i=1}^n Z_{\lambda_i'}$ (cf. Exercise 2).

Other examples are described in Exercise 7.

8.2 Exercises

1. Determine whether $\leq_{s.t.M(4)} \Rightarrow \leq_{s.t.M(3)}$.
2. Suppose $\underline{X} \leq_{s.t.M} \underline{Y}$. Prove that $\sum_{i=1}^n X_i \overset{d}{=} \sum_{i=1}^n Y_i$.
3. Prove Theorem 8.1.6.
4. Prove that the random variables $\underline{X}'$ and $\underline{Y}'$ alluded to in Theorem 8.1.7 cannot be independent.
5. Suppose $\underline{X} \leq_{\text{st}.M} \underline{Y}$. Prove that for any $\underline{a}$ with $\sum_{i=1}^n a_i = 1$,

$$P\left(\left(\sum_{i=1}^{n} X_i\right)^{-1} \underline{X} \geq_M \underline{a}\right) \leq P\left(\left(\sum_{i=1}^{n} Y_i\right)^{-1} \underline{Y} \geq_M \underline{a}\right).$$

6. Suppose $X_1, \ldots, X_n$ are independent random variables with X_i Poisson λ_i and that $X_1^*, X_2^*, \ldots, X_n^*$ are i.i.d. Poisson $(\bar{\lambda})$ random variables $(\bar{\lambda} = \frac{1}{n}\sum_{i=1}^{n}\lambda_i)$. Define $g(\underline{x}) = \left(\frac{1}{n}\sum_{i=1}^{n} x_i^2\right) / \left(\frac{1}{n}\sum_{i=1}^{n} x_i\right)^2$. Verify that $P(g(\underline{X}) > \delta) \leq P(g(\underline{X}^*) > \delta), \delta > 0$.

7. (a) Show that the family of n-dimensional Dirichlet random variables $\{\underline{X}_{\underline{\lambda}}\}$ satisfies $\underline{\lambda} \leq \underline{\lambda}' \Rightarrow \underline{X}_{\underline{\lambda}} \leq_{\text{st}.M} \underline{X}_{\underline{\lambda}'}$

   (b) Multivariate negative binomial. Consider a sequence of independent experiments each with $(n+1)$ possible outcomes $0, 1, 2, \ldots, n$ with associated probabilities

$$\left(\frac{\lambda_0}{\sum_{i=0}^{n}\lambda_i}, \frac{\lambda_1}{\sum_{i=0}^{n}\lambda_i}, \ldots, \frac{\lambda_n}{\sum_{i=0}^{n}\lambda_i}\right).$$

   Define $\underline{X}_{\underline{\lambda}} = (X_1, \ldots, X_n)$ by $X_i =$ the number of outcomes of type i that precede the N'th outcome of type 0 (where N is a fixed integer). Let $\underline{\lambda} = (\lambda_1, \ldots, \lambda_n)$. Prove that $\underline{\lambda} \leq_M \underline{\lambda}' \Rightarrow \underline{X}_{\underline{\lambda}} \leq_{\text{st}.M} \underline{X}_{\underline{\lambda}'}$

Chapter 9
Some Related Orderings

In Sect. 6.4, we encountered some alternative inequality curves and the corresponding orderings that they can be used to define (they were Bonferroni, Leimkuhler, and Zenga-I and II). In this chapter we consider certain other partial orders defined on $\mathcal{L}$ (the class of non-negative random variables with positive finite expectations) that are closely related to the Lorenz ordering and are sometimes useful to aid in the determination of whether or not random variables are Lorenz ordered. The first, $*$-ordering, is often easier to deal with than Lorenz ordering, and is a prime example of an ordering that can sometimes be used to verify Lorenz ordering which it implies. The other group of partial orderings to be discussed are those known as stochastic dominance of degree k, $k = 1, 2, \ldots$ Degree 1 is just stochastic ordering. Degree 2 is intimately related to the Lorenz order, but distinct. Higher degree stochastic orders are most frequently encountered in economic contexts. The treatment provided here is brief.

9.1 Star-Ordering

Let $X, Y \in \mathcal{L}$ with corresponding distribution functions F_X and F_Y. Star-shaped ordering or, more briefly, star ordering is defined as follows.

Definition 9.1.1 We say that X is star-shaped with respect to Y, write $X \leq_* Y$ if $F_X^{-1}(u)/F_Y^{-1}(u)$ is a non-increasing function of u.

Since $F_{cX}^{-1}(u) = cF_X^{-1}(u)$ for any positive c and any $X \in \mathcal{L}$, it is obvious that $*$-ordering is scale invariant. A simple sufficient condition for $F_Y^{-1}(F_X(x))/x$ to be increasing is that $F_Y^{-1}(F_X(x))$ be convex on the support of F_X (see Exercise 1). If $F_X^{-1}(u)/F_Y^{-1}(u)$ has a simple differentiable form, then we can check for $*$-ordering

B. C. Arnold, J. M. Sarabia, *Majorization and the Lorenz Order with Applications in Applied Mathematics and Economics*, Statistics for Social and Behavioral Sciences,
https://doi.org/10.1007/978-3-319-93773-1_9

by verifying that its derivative is non-positive. We can use $*$-ordering to verify that Lorenz ordering obtains as a consequence of the following theorem.

Theorem 9.1.1 *Suppose $X, Y \in \mathcal{L}$. If $X \leq_* Y$, then $X \leq_L Y$.*

Proof Without loss of generality, since both partial orders are scale invariant, we may assume that $E(X) = E(Y) = 1$. In such a case

$$L_X(u) - L_Y(u) = \int_0^u [F_X^{-1}(v) - F_Y^{-1}(v)]dv \tag{9.1}$$

Since $F_X^{-1}(v)/F_Y^{-1}(v)$ is non-increasing, the integrand is first positive and then negative as v ranges from 0 to 1 [cf. the proof of Theorem 4.1.1]. Thus, the integral assumes its smallest value when $u = 1$. It follows that $L_X(u) - L_Y(u) \geq L_X(1) - L_Y(1) = 1 - 1 = 0$, and consequently $X \leq_L Y$. ■

If we return now to the proof of Theorem 4.1.1, it is clear that conditions (ii) of that theorem actually are sufficient for $g(X) \leq_* X \; \forall X \in \mathcal{L}$, and the theorem could be restated as follows

Theorem 9.1.2 *Let $g : \mathbb{R}^+ \to \mathbb{R}^+$. The following are equivalent.*

(i) $g(X) \leq_L X, \forall X \in \mathcal{L}$
(ii) $g(X) \leq_* X, \forall X \in \mathcal{L}$
(iii) $g(x) > 0, \forall x > 0$, *$g(x)$ is non-decreasing on $[0, \infty)$ and $g(x)/x$ is non-increasing on $(0, \infty)$.*

Theorem 4.1.2 can be analogously restated. If we return again to the proof of Theorem 9.1.1, we can see that the important consequence of $*$-ordering was that $F_X^{-1}(v) - F_Y^{-1}(v)$ had only one sign change $(+, -)$ on the interval $[0, 1]$. This sign-change property is not scale invariant, but it does permit us to formulate the following simple sufficient condition for Lorenz ordering.

Theorem 9.1.3 *Suppose $X, Y \in \mathcal{L}$ and that $[F_X^{-1}(v)/E(X)] - [F_Y^{-1}(v)/E(Y)]$ has at most one sign change (from + to −) as v ranges from 0 to 1. It follows that $X \leq_L Y$.*

Proof The result is obvious from the discussion following (9.1). ■

If inverse functions cross at most once then the same holds for the original functions. Consequently, we can restate the last sufficient condition in the form: A sufficient condition that $X \leq_L Y$ is that $F_X(E(X)x) - F_Y(E(Y)x)$ has at most one sign change (from − to +) as x ranges from 0 to ∞. If $X \leq_* Y$ one can show that $F_X(\lambda x) - F_Y(\nu x)$ has at most one sign change (from − to +) for any choices of λ and ν. It is thus evident that if we use the sign-change property alluded to in Theorem 9.1.3 to define a partial order on $\mathcal{L}$, it will occupy an intermediate position between $\leq_*$ and $\leq_L$.

Definition 9.1.2 We will say that X is sign-change ordered with respect to Y and write $X \leq_{s.c.} Y$, if $[F_X^{-1}(v)/E(X)] - [F_Y^{-1}(v)/E(Y)]$ has at most one sign change (from + to −) as v ranges from 0 to 1.

A simple sufficient condition for sign change ordering can be stated in terms of density crossings (assuming densities exist). Thus,

Theorem 9.1.4 *Let $X, Y \in \mathcal{L}$ have corresponding densities $f_X(x)$ and $f_Y(y)$ (with respect to a convenient dominating measure on $\mathbb{R}^+$, in the most abstract setting). If the function*

$$E(X)f_X(E(X)x) - E(Y)f_Y(E(Y)x) \tag{9.2}$$

has two sign changes (from $-$ to $+$ to $-$) as x ranges from 0 to ∞, then $X \leq_{s.c.} Y$.

Proof (9.2) is merely the density of $X/E(X)$ minus the density of $Y/E(Y)$. The difference between the distribution functions

$$F_X(E(X)x) - F_Y(E(Y)x) = F_{X/E(X)}(x) - F_{Y/E(Y)}(x)$$

has the sign sequence $-,+$ since it is a function which begins at 0 when $x = 0$, ends at 0 (as $x \to \infty$) and whose derivative by (9.2) has sign sequence $-,+,-$. ■

Verification that $X \leq_L Y$ is frequently most easily done by using the density crossing argument (Theorem 9.1.4) or by showing $*$-ordering obtains (if $F_X^{-1}(u)$ and $F_Y^{-1}(u)$ are available in convenient tractable forms).

Example 9.1.1 If $X \leq_* Y$, then $X_{i:n} \leq_* Y_{i:n}$ for every $i \leq n$. This follows since $X_{i:n} \stackrel{d}{=} F_X^{-1}(U_{i:n})$ where $U_{i:n}$ is the i'th order statistic from a sample of size n from a uniform (0, 1) distribution. Similarly, $Y_{i:n} \stackrel{d}{=} F_Y^{-1}(U_{i:n})$. So if we let $G_{i:n}$ be the distribution function of $U_{i:n}$, we can write

$$\frac{F_{X_{i:n}}^{-1}(u)}{F_{Y_{i:n}}^{-1}(u)} = \frac{F_X^{-1}[G_{i:n}(u)]}{F_Y^{-1}[G_{i:n}(u)]}.$$

But this ratio is non-increasing as u increases, since $G_{i:n}^{-1}$ is monotone increasing and $F_X^{-1}(v)/F_Y^{-1}(v)$ is a non-increasing function of v (since $X \leq_* Y$).

Example 9.1.2 Suppose X has a symmetric distribution on the interval $[0, c]$. The sample median (for samples of odd size) decreases in inequality in the Lorenz sense as sample size increases. Specifically, if $m < m'$

$$X_{m'+1:2m'+1} \leq_L X_{m+1:2m+1}.$$

If we consider $m' = m + 1$, and denote the density of $X_{m+1:2m+1}$ by $f_m(x)$, we find that the ratio of densities $f_{m+1}(x)/f_m(x)$ is given by

$$\frac{f_{m+1}(x)}{f_m(x)} = F_X(x)[1 - F_X(x)]\frac{2(2m+3)}{m+1}. \tag{9.3}$$

By symmetry $E(X_{m+1:2m+1}) = E(X_{m+2:2m+3}) = c/2$, so that we do not have to correct for differences in means. Since expression (9.3) is greater than 1 for intermediate values of x and less than 1 for large and small values, it follows that, by Theorem 9.1.4, $X_{m+2:2m+3} \leq_{s.c.} X_{m+1:2m+1}$ and consequently they are similarly Lorenz ordered.

9.2 Stochastic Dominance

The initial work on stochastic dominance was executed in the context of bounded non-negative random variables. Subsequent researchers (especially Fishburn 1980) have extended consideration to unbounded random variables but subtle changes are required in the definitions and interpretations. In the present development we will content ourselves with consideration of the bounded case. It may be and frequently has been remarked on that the restriction to bounded random variables is not a serious restriction when dealing with real world data.

For the remainder of this section all random variables are nonnegative and are assumed to be bounded above by $B > 0$. We will denote the class of such random variables by $\mathcal{L}_B$. Evidently if $X \in \mathcal{L}_B$, $E(X^k) < \infty \; \forall k$. Associated with $X \in \mathcal{L}_B$ is a sequence of "distribution" functions defined by repeated integration. Thus for $x \in [0, B)$,

$$\tilde{F}_X^{(1)}(x) = P(X \leq x) = F_X(x) \tag{9.4}$$

and for $i = 2, 3, \ldots$

$$\tilde{F}_X^{(i)}(x) = \int_0^x \tilde{F}_X^{(i-1)}(y)\, dy \tag{9.5}$$

The word "distribution" is placed in quotation marks because, although $\tilde{F}_X^{(i)}(x)$ $(i = 2, 3, \ldots)$ is continuous and non-decreasing, it has not been normalized to be a true distribution function, i.e. in general we do not have $\tilde{F}_X^{(i)}(B) = 1$. An interpretation of the i'th distribution function is available as follows.

$$\begin{aligned}
\tilde{F}_X^{(i)}(x) &= \int_0^x \tilde{F}_X^{(i-1)}(y_{i-1})\, dy_{i-1} \\
&= \int_0^x dy_{i-1} \int_0^{y_{i-1}} \tilde{F}_X^{(i-2)}(y_{i-2})\, dy_{i-2} \\
&= \ldots \\
&= \int_0^x dy_{i-1} \int_0^{y_{i-1}} dy_{i-2} \cdots \int_0^{y_2} dy_2 \int_0^{y_1} dF_X(y_0)
\end{aligned}$$

The range of integration is

$$\{(y_0, \ldots, y_{i-1}) : \; y_0 \le y_1 \le y_2 \cdots \le y_{i-1} \le x\}.$$

Performing the integration in reverse order we have

$$\begin{aligned} \tilde{F}_X^{(i)}(x) &= \int_0^x dF_X(y_0) \int_{y_0}^x dy_1 \int_{y_1}^x dy_2 \cdots \int_{y_{i_2}}^x dy_{i-1} \\ &= \frac{1}{(i-1)!} \int_0^x (x - y_0)^{i-1} \, dF_X(y_0). \end{aligned} \tag{9.6}$$

It is thus possible to write $\tilde{F}_X^{(i)}(x)$ as a function of the moment distributions of F_X. Recall that the j'th moment distribution of X is defined by

$$F_X^{(j)}(x) = \int_0^x t^j \, dF_X(t)/E(X^j) \tag{9.7}$$

($F_X^{(1)}$ was introduced in Chap. 3 in the discussion of Lorenz curves) and, for $j \ge 2$, the j'th moment distribution was defined in Sect. 5.6. From (9.6) we have

$$\tilde{F}_X^{(i)}(x) = \frac{1}{(i-1)!} \sum_{j=0}^{i-1} x^{i-1-j} (-1)^j E(X^j) F_X^{(j)}(x). \tag{9.8}$$

Observe that setting $x = B$ in (9.8) yields

$$\tilde{F}_X^{(i)}(B) = \frac{1}{(i-1)!} E[(B - X)^{i-1}]. \tag{9.9}$$

Stochastic dominance is defined in terms of the sequence of "distribution" functions (9.5).

Definition 9.2.1 Let X and $Y \in \mathcal{L}_B$. For $n = 1, 2, \ldots$ we say that X is n'th degree stochastically dominated by Y and write $X \le_{s.d.(n)} Y$, if

$$\tilde{F}_X^{(j)}(B) \ge \tilde{F}_Y^{(j)}(B), \quad j = 1, 2, \ldots, n-1. \tag{9.10}$$

and

$$\tilde{F}_X^{(n)}(x) \ge \tilde{F}_Y^{(n)}(x), \quad \forall x \in [0, B]. \tag{9.11}$$

Fishburn's (1976; 1980) variant definition of n'th degree stochastic dominance involves only condition (9.11) (see Exercise 7). The present definition is in the spirit of Whitmore and Findlay (1978). Stochastic dominance of degrees 2 and 3 has been much used in decision making contexts. Observe that first degree stochastic dominance is just the usual stochastic ordering denoted earlier by $\le_{\text{st}}$.

In certain economic, financial and decision theoretical contexts interest centers on expectations of the form $E(u(X))$ where u is a completely monotone function in the following sense.

Definition 9.2.2 $u : \mathbb{R}^+ \to \mathbb{R}^+$ is completely monotone, if it is infinitely differentiable and if

$$(-1)^n \frac{d^n}{dx^n} u(x) \geq 0$$

for every x and every n.

A completely monotone function is thus, succinctly, one whose successive derivatives alternate in sign according to the sequence +,−,+,... A remarkable observation found, for example, in Feller (1971, p. 415), is that such functions are necessarily Laplace transforms. Specifically, if u is completely monotone, then there exist two distribution functions G_1 and G_2 on $[0, \infty)$ such that for some c_1 and c_2

$$u(x) = c_1 \int_0^\infty e^{-xy}\, dG_1(y) - c_2 \int_0^\infty e^{-xy}\, dG_2(y) \tag{9.12}$$

Less restrictive conditions on u would involve requirements that the signs of the successive derivatives of u alternate, but just up to a point. For $n = 1, 2, \ldots$ we may define classes $\{\mathcal{U}_n\}_{n=1}^\infty$ of functions as follows:

$$\mathcal{U}_n = \{u \mid u : \mathbb{R}^+ \to \mathbb{R}^+,\ (-1)^{k+1} D^k u \geq 0,\ k = 1, 2, \ldots, n\} \tag{9.13}$$

where D^k is the k'th derivative operator. Thus, $\mathcal{U}_1$ consists of the differentiable non-decreasing functions. $\mathcal{U}_2$ consists of twice differentiable non-decreasing concave functions. $\mathcal{U}_3$ consists of members of $\mathcal{U}_2$ which have a non-negative third derivative, etc. Evidently,

$$\mathcal{U}_n \subset \mathcal{U}_{n-1},\ n = 2, 3, \ldots. \tag{9.14}$$

Our definitions of stochastic dominance of degrees $1, 2, \ldots$ assume a particularly simple form when related to expectations of functions in the classes $\mathcal{U}_1, \mathcal{U}_2, \ldots$

Theorem 9.2.1 *For* $X, Y \in \mathcal{L}_B$, $X \leq_{s.d.(n)} Y$ *if and only if* $E(u(X)) \leq E(u(Y))$ *for every* $u \in \mathcal{U}_n$ *for which the expectations exist.*

Proof Suppose $X \leq_{s.d.(n)} Y$ and $u \in \mathcal{U}_n$. Repeated integration by parts yields

$$\begin{aligned} E(u(Y)) - E(u(X)) &= \int_0^B u(x)\, d[F_Y(x) - F_X(x)] \\ &= u(x)[\tilde{F}_Y^{(1)}(x) - \tilde{F}_X^{(1)}(x)]|_0^B \end{aligned}$$

$$- u'(x)[\tilde{F}_Y^{(2)}(x) - \tilde{F}_X^{(2)}(x)]|_0^B$$
$$+ \ldots$$
$$+(-1)^{n-1}u^{(n-1)}(x)[\tilde{F}_Y^{(n)}(x) - \tilde{F}_X^{(n)}(x)]|_0^B$$
$$+(-1)^n \int_0^B u^{(n)}(x)[\tilde{F}_Y^{(n)}(x) - \tilde{F}_X^{(n)}(x)]\, dx$$

where $u^{(i)}(x) = D^i u(x)$. However all these terms are non-negative by (9.10) and (9.11).

To prove the converse, we write, as above

$$E(u(Y)) - E(u(X)) = \sum_{i=0}^{n-1}(-1)^i u^{(i)}(B)[\tilde{F}_Y^{(i+1)}(B) - \tilde{F}_X^{(i+1)}(B)] + \int_0^B u^{(n)}(x)[\tilde{F}_Y^{(n)}(x) - \tilde{F}_X^{(n)}(x)]\, dx.$$

If (9.10) fails to hold for some j, then we will have $E(u(Y) - u(X)) < 0$ for any $u \in \mathcal{U}_n$ for which $u^{(j)}(B)$ is much bigger than the other $u^{(i)}(B)$'s and $u^{(n)}(x)$. If (9.11) fails to hold at some point x_0, we will have $E(u(Y) - u(X)) < 0$ for a $u \in \mathcal{U}_n$ for which $u^{(n)}(x_0)$ is large compared with other values of $u^{(n)}(x)$ and for which the $u^{(i)}(B)$'s, $i < n$ are small. It is tedious to construct such members of $\mathcal{U}_n$, but it can be done. The cases $n = 1, 2,$ and 3 have been frequently dealt with in the literature. ∎

In the light of (9.9), condition (9.10) is seen to involve an inequality involving complicated functions of the first $n - 2$ moments of X and Y. For large values of n it is difficult to interpret the implications of such a condition. The Fishburn variant definition (see Exercises 3 and 7) avoids this problem. The resulting partial order, requiring (9.11) only, is interpretable in terms of a slightly smaller class of utility functions say $\tilde{\mathcal{U}}_n$, a proper subset of $\mathcal{U}_n$ (again see Exercise 7). The price we pay is that although utility classes $\{\mathcal{U}_n\}$ are readily motivated in terms involving risk aversion, it is not clear why one should be happy to restrict attention to the more structured classes $\tilde{\mathcal{U}}_n$.

An intimate relationship between second degree stochastic dominance and Lorenz ordering is immediately evident.

Theorem 9.2.2 *Let $X, Y \in \mathcal{L}_B$. $X \leq_L Y$ if and only if $Y/E(Y) \leq_{s.d.(2)} X/E(X)$.*

Proof Without loss of generality, assume $E(X) = E(Y) = 1$. Suppose $X \leq_L Y$ and consider $u \in \mathcal{U}_2$. Since u is concave, $-u$ is convex and by Theorem 3.2.1, $E(u(Y)) \leq E(u(X))$. Since u was arbitrary in $\mathcal{U}_2$, we conclude that $Y \leq_{s.d.(2)} X$. Suppose $Y \leq_{s.d.(2)} X$. Consider angle functions of the form

$$g_c(x) = \begin{cases} -x + c, & x \leq c, \\ 0, & x > c \end{cases}$$

The functions $-g_c$ can be approximated arbitrarily closely by members of $\mathcal{U}_2$ and, consequently, $E[g_c(X)] \leq E[g_c(Y)]$ for every c. But this is enough to ensure that $X \leq_L Y$. ■

A caveat is in order regarding the application of Theorem 9.2.2. It says that Lorenz ordering and stochastic dominance of degree 2 are intimately related. It does *not* say that one is the reverse of the other. To see this, first observe that from Theorem 9.2.1 we know that

$$X \leq_{s.d.(n)} Y \Rightarrow X \leq_{s.d.(n+1)} Y. \tag{9.15}$$

Also note that if $X \leq_{s.d.(n)} Y$ and $Y \leq_{s.d.(n)} X$, we must have $X \stackrel{d}{=} Y$. This already tells us that stochastic dominance of degree 2 is distinctly different from reverse Lorenz ordering, since Lorenz ordering is scale invariant, i.e. if $X \leq_L Y$ and $Y \leq_L X$ we can only conclude $X \stackrel{d}{=} cY$ for some c.

Now we seek a pair of random variables, not identically distributed, for which $X \leq_L Y$ and $X \leq_{s.d.(1)} Y$. For such random variables we will have $X \leq_L Y$ yet $Y \nleq_{s.d.(2)} X$. A simple example is provided by considering $X \sim$ uniform(1, 1.5) and $Y \sim$ uniform(2, 4). Clearly $X \leq_L Y$, $X \leq_{s.d.(1)} Y$, $X \leq_{s.d.(2)} Y$ and $Y \nleq_{s.d.(2)} X$. In this example we also have $X \leq_{s.d.(2)} Y$, yet $Y \nleq_L X$. Thus division by expectations in Theorem 9.2.2 is indispensable.

What can be said about preservation and attenuation of stochastic dominance? The analog of Theorem 4.1.1 is not very interesting.

Theorem 9.2.3 *Let $g : [0, B] \to [0, B]$. The following are equivalent*

(i) $g(X) \leq_{s.d.(n)} X, \forall X \in \mathcal{L}_B$
(ii) $g(x) \leq x, \forall x \in [0, B]$

Neither is the characterization of functions which preserve stochastic dominance.

Theorem 9.2.4 *Let $g : [0, B] \to [0, B]$. A necessary and sufficient condition that $X \leq_{s.d.(n)} Y$ imply $g(X) \leq_{s.d.(n)} g(Y)$ is that g is non-decreasing on $[0, B]$.*

In both cases the necessity part of the proof involves considering degenerate X's and Y's. Among degenerate random variables the stochastic dominance relations of differing degrees coincide, i.e., if $P(X = a) = P(Y = b) = 1$, then $X \leq_{s.d.(n)} Y \iff a \leq b$.

Theorems 9.2.3 and 9.2.4 reinforce the evidence that stochastic dominance of degree 2 is a different breed of cat than Lorenz ordering.

9.3 Exercises

1. Let F_X and F_Y be (for convenience) strictly increasing distribution functions on $(0, \infty)$. We say that X is convex with respect to Y and write $X \leq_c Y$, if $F_Y^{-1}(F_X(x))$ is a convex function on $(0, \infty)$. Prove that if $X \leq_c Y$ then necessarily $X \leq_* Y$.

2. Construct examples to verify that convex ordering, sign change ordering, star ordering, and Lorenz ordering are distinct partial orders.

3. Compare the present definition of stochastic dominance of order 3 with that used in Fishburn (1976). (As in Exercise 7 below with $\alpha = 2$, but restricted to $\mathcal{L}_B$).

4. Prove the converse of Theorem 9.2.1 when $n = 1, 2$.

5. Prove that $X \leq_* Y$ if and only if for every $c > 0$, X and cY have distribution functions which cross at most once.

6. Show by example that the converse to Theorem 9.1.4 does not hold.

7. (Fishburn's stochastic dominance for unbounded random variables). Let X and Y be non-negative random variables. We write $X \leq_{s.d.(\alpha)} Y$, if

$$F_X^{(\alpha)}(x) \geq F_Y^{(\alpha)}(x), \ \forall x \geq 0$$

where

$$F_X^{(\alpha)}(x) = \frac{1}{\Gamma(\alpha)} \int_0^x (x-y)^{\alpha-1)} dF_X(y), \quad \alpha \geq 1.$$

(a) Verify that for $\alpha < \alpha'$

$$F_X^{(\alpha')}(x) = \frac{1}{\Gamma(\alpha' - \alpha)} \int_0^x (x-y)^{(\alpha'-\alpha-1)} F_X^{(\alpha)}(y)\, dy$$

and conclude that α dominance implies α' dominance.

(b) Let $\mathcal{V}_\alpha = \{v : v$ is real valued continuous on $[0, \infty)$, positive on $(0, \infty)$ and satisfies $\int_0^\infty x^{\alpha-1} v(x)\, dx < \infty\}$. Define $\tilde{\mathcal{U}}_\alpha$ by: $u \in \tilde{\mathcal{U}}_\alpha$, if there exists some $v \in \mathcal{V}_\alpha$ and some real c such that

$$u(x) = -\int_x^\infty v(y)(y-x)^{\alpha-1}\, dy + c.$$

Show that $X \leq_{s.d.(\alpha)} Y$ if and only if $E(u(X)) \leq E(u(Y))$ for every $u \in \tilde{\mathcal{U}}_\alpha$.

8. X and Y are said to be ordered in dispersion ($X \leq_{disp} Y$), if $F_X^{-1}(\beta) - F_X^{-1}(\alpha) \leq F_Y^{-1}(\beta) - F_Y^{-1}(\alpha)$ whenever $0 < \alpha < \beta < 1$. Show that $X \leq_{disp} Y$ if and only if for every real c the distribution functions of $X + c$ and Y cross at most once, and if there is a sign change, $F_{X+c} - F_Y$ changes sign from $-$ to $+$ (Shaked 1982). Compare this result with that of Exercise 5.

9. If X has a symmetric density on $(0, 2)$ and $f_X(x)$ is decreasing on $(1, 2)$, show that $\mathrm{var}(X) < 1/3$. (Compare X with a uniform$(0, 2)$ random variable].

Chapter 10
Inequality Analysis in Families of Income Distributions

10.1 Introduction

In this chapter we will study inequality measures for two flexible families of income distributions: the McDonald family (McDonald 1984) and the family of generalized Pareto distributions proposed by Arnold (1983, 2015b). We will include analytic expressions for the Lorenz curve, Gini indices, and other inequality measures for the different distributions of the two families.

As well, we will study stochastic comparisons within certain parametric subfamilies of the MacDonald family.

10.2 The McDonald Family: Definitions and Basic Properties

We begin with the generalized functions for the size distribution of income proposed by McDonald (1984). These will be said to belong to the "McDonald family of income distributions" or simply the "McDonald family." The McDonald family is composed of three subfamilies of distributions: the generalized gamma distributions, the generalized beta distributions of the first kind, and the generalized beta distributions of the second kind, sometimes called Generalized Beta Prime distributions. This family has several advantages for practical use. First, the family includes many of the most popular distributions used in the analysis of income, wealth and risk analysis, as special or limiting cases. A second advantage is that many of the properties and features of McDonald distributions can be obtained in simple analytic forms. Some references which discuss different properties of this

B. C. Arnold, J. M. Sarabia, *Majorization and the Lorenz Order with Applications in Applied Mathematics and Economics*, Statistics for Social and Behavioral Sciences,
https://doi.org/10.1007/978-3-319-93773-1_10

family are: McDonald (1984), Wilfling (1996a,b), Wilfling and Krämer (1993), Kleiber (1996), Kleiber (1999), Kleiber and Kotz (2003), Sarabia and Jordá (2014b), Sarabia et al. (2002), Belzunce et al. (2013), Sarabia et al. (2017a,b) among others.

The three subfamilies that together comprise the McDonald family are, as mentioned above: the generalized gamma (GG) distributions, the generalized beta distributions of the first kind (GB1), and the generalized beta distributions of the second kind (GB2). These three types of distributions, GG, GB1, and GB2 are defined in terms of their probability density functions as follows (each of the three models has positive parameters $a, b, p, q > 0$):

$$f_{\rm GG}(x; a, p, b) = \frac{ax^{ap-1}\exp(-(x/b)^a)}{b^{ap}\Gamma(p)}, \quad x \geq 0, \tag{10.1}$$

$$f_{\rm GB1}(x; a, p, q, b) = \frac{ax^{ap-1}[1-(x/b)^a]^{q-1}}{b^{ap}B(p,q)}, \quad 0 \leq x \leq b, \tag{10.2}$$

and

$$f_{\rm GB2}(x; a, p, q, b) = \frac{ax^{ap-1}}{b^{ap}B(p,q)[1+(x/b)^a]^{p+q}}, \quad x \geq 0, \tag{10.3}$$

respectively, and 0 otherwise. Here $\Gamma(\alpha) = \int_0^\infty t^{\alpha-1}\exp(-t)dt$ represents the gamma function and $B(p,q) = \int_0^1 t^{p-1}(1-t)^{q-1}dt$ denotes the beta function, where $\alpha, p, q > 0$.

For each of the three distributions, the parameter b is a scale parameter while a, p, and q are shape parameters. We will indicate that a random variable X has (10.1), (10.2), or (10.3) as its density by writing $X \sim \text{GG}(a, p, b)$, $X \sim \text{GB1}(a, p, q, b)$ and $X \sim \text{GB2}(a, p, q, b)$, respectively. If $b = 1$, we omit mention of it and write $X \sim \text{GG}(a, p)$, $X \sim \text{GB1}(a, p, q)$ and $X \sim \text{GB2}(a, p, q)$.

These McDonald models include a wide spectrum of candidate income distributions including many that have been proposed in the existing literature, as listed below.

The $\text{GG}(a, p, b)$ distributions include as special cases the following distributions:

- Exponential distributions: $\text{Exp}(\lambda)$ (with intensity or rate parameter λ) $= \text{GG}(1, 1, \lambda^{-1})$
- Classical gamma distribution $\Gamma(\alpha, \beta) = \text{GG}(1, \alpha, \beta)$
- Chi-squared distribution $\chi_n^2 = \text{GG}(1, n/2, 2)$
- Weibull distribution $W(a) = \text{GG}(a, 1, 1)$
- Half normal distribution $\text{GG}(2, 1/2, \sqrt{2})$

The usual two-parameter lognormal distribution with density of the form,

$$f(x; \mu, \sigma) = \frac{1}{\sigma x\sqrt{2\pi}}\exp\left\{-\frac{1}{2}\left(\frac{\log x - \mu}{\sigma}\right)^2\right\}, \quad x > 0, \tag{10.4}$$

where $\mu \in \mathbb{R}$ and $\sigma > 0$ (denoted by $X \sim LN(\mu, \sigma^2)$) is a limiting case of the GG distribution (McDonald 1984).

The GB1(a, p, q, b) distribution includes as special cases the following distributions.

- Classical beta distribution: $B(\alpha, \beta) = \text{GB1}(1, \alpha, \beta)$
- Power distribution GB1$(a, p, 1)$
- Kumaraswamy distribution $K(a, q) = \text{GB1}(a, 1, q)$
- Arcsin distribution GB1$(1, \frac{1}{2}, \frac{1}{2})$

The GB2(a, p, q, b) distribution includes the following special cases.

- Beta distribution of the second kind $B^{(2)}(\alpha, \beta) = \text{GB2}(1, \alpha, \beta)$
- Singh–Maddala distribution (Singh and Maddala 1976) $\text{SM}(a, q) = \text{GB2}(a, 1, q)$
- Dagum distribution (Dagum 1977) distribution $D(a, p) = \text{GB2}(a, p, 1)$
- Lomax or Pareto II distribution $L(q) = \text{GB2}(1, 1, q)$
- Fisk or log-logistic distribution $F(a) = \text{GB2}(a, 1, 1)$

The GB2 distribution was referred to as a Feller–Pareto distribution by Arnold (1983, 2015b), who included an additional location parameter.

On the other hand, the GG distribution can be obtained as a limit of the GB1 distribution, when $q \to \infty$ and $b = \tilde{b}(p+q)^{1/a}$, and as a limit of the GB2 distribution when $q \to \infty$ and $b = \tilde{b}q^{1/a}$.

Simple stochastic representations of the McDonald distributions are available in terms of independent gamma variables. Thus, with independent random variables $V \sim \Gamma(p, 1)$ and $W \sim \Gamma(q, 1)$ we have

$$bV^a \sim \text{GG}(a, p, b), \tag{10.5}$$

$$b\left(\frac{V}{V+W}\right)^a \sim \text{GB1}(a, p, q, b), \tag{10.6}$$

$$b\left(\frac{V}{W}\right)^a \sim \text{GB2}(a, p, q, b). \tag{10.7}$$

These stochastic representations will prove useful for simulation of McDonald family variables.

Next, we review some properties of the McDonald family, which will be useful for the computation of various inequality measures.

First, to obtain the cdf of the GG distribution, we consider the incomplete gamma function ratio defined by

$$G(x; \nu) = \frac{1}{\Gamma(\nu)} \int_0^x t^{\nu-1} \exp(-t) dt, \quad x > 0, \tag{10.8}$$

with $\nu > 0$. Note that (10.8) corresponds to the cdf of the classical gamma distribution with shape parameter $\nu > 0$ and scale parameter $b = 1$. Using (10.8), the cdf of (10.1) is given by

$$F_{\text{GG}}(x; a, p, b) = G((x/b)^a; p), \quad x \geq 0. \tag{10.9}$$

Next, we consider the incomplete beta function ratio defined by

$$B(x; p, q) = \frac{1}{B(p, q)} \int_0^x t^{p-1}(1-t)^{q-1} dt, \quad 0 \leq x \leq 1 \tag{10.10}$$

with $p, q > 0$. The function (10.10) corresponds to the cdf of the classical beta distribution of the first kind $B(p, q)$. The cdf of the GB1 distribution is

$$F_{\text{GB1}}(x; a, p, q, b) = B((x/b)^a; p, q), \quad 0 \leq x \leq 1, \tag{10.11}$$

where $B(\cdot; \cdot, \cdot)$ is defined in (10.10).

Despite the somewhat complicated aspect of (10.3), the corresponding cdf can be easily defined in terms of the incomplete beta function ratio (10.10). Thus, the cdf of the GB2 distribution is given by

$$F_{\text{GB2}}(x; a, p, q, b) = B\left(\frac{(x/b)^a}{1+(x/b)^a}; p, q\right), \quad x \geq 0. \tag{10.12}$$

The cdf's, the quantile functions, the k-th moment distribution, and the corresponding moments for the McDonald family are summarized in Table 10.1. For the lognormal distribution, $X \sim LN(\mu, \sigma^2)$, the cdf is

$$F(x; \mu, \sigma) = \Phi\left(\frac{\log x - \mu}{\sigma}\right), \quad x \geq 0,$$

and the quantile function is

$$F^{-1}(u; \mu, \sigma) = \exp\{\mu + \sigma\Phi^{-1}(u)\}, \quad 0 \leq u \leq 1. \tag{10.13}$$

The k-th moment for the lognormal distribution is

$$E(X^k) = \exp\left(k\mu + k^2\sigma^2/2\right),$$

and the k-th moment distribution is

$$X^{(k)} \sim LN(\mu + k\sigma^2, \sigma^2). \tag{10.14}$$

Table 10.1 Main features of the McDonald family: cdf, quantile function, k-th moment distribution and the k-th raw moment

Distribution	GG(a, p, b)	GB1(a, p, q, b)	GB2(a, p, q, b)
CDF	$G\left((x/b)^a; p\right)$	$B\left((x/b)^a; p, q\right)$	$B\left(\frac{(x/b)^a}{1+(x/b)^a}; p, q\right)$
Quantile	$b\left\{F_G^{-1}(u; p)\right\}^{1/a}$	$b\left\{F_B^{-1}(u; p, q)\right\}^{1/a}$	$b\left\{\frac{F_B^{-1}(u; p, q)}{1-F_B^{-1}(u; p, q)}\right\}^{1/a}$
$F^{(k)}$	GG$(a, p+\frac{k}{a})$	GB1$(a, p+\frac{k}{a}, q)$	GB2$(a, p+\frac{k}{a}, q-\frac{k}{a})$
$E(X^k)$	$\frac{b^k\Gamma(p+\frac{k}{a})}{\Gamma(p)}$	$\frac{b^k B(p+\frac{k}{a}, q)}{B(p, q)}$	$\frac{b^k B(p+\frac{k}{a}, q-\frac{k}{a})}{B(p, q)}$

Note: We denote by $G(x; \nu)$ the incomplete gamma function ratio defined in (10.8) and $B(x; p, q)$ the incomplete beta function ratio defined in (10.10). As well, $F_G^{-1}(u, p)$ with $0 \leq u \leq 1$ denotes the quantile function of the classical gamma distribution and $F_B^{-1}(u, p, q)$ with $0 \leq u \leq 1$ the quantile function of the classical beta distribution

10.2.1 Lorenz Curves and Gini Indices

In this section, we include the Lorenz curves and the Gini indices for the three classes of distributions in the McDonald family. In his 1984 paper, McDonald reported the Gini indices for the three families. However, he did not include expressions for the corresponding Lorenz curves. To obtain expressions for these Lorenz curves,we will make use of the generalized hypergeometric function ${}_pF_q(\boldsymbol{a}; \boldsymbol{b}; x)$ defined by

$$
{}_pF_q(a_1, \ldots, a_p; b_1, \ldots, b_q; x) = \sum_{k=0}^{\infty} \frac{(a_1)_k \cdots (a_p)_k}{(b_1)_k \cdots (b_q)_k} \frac{x^k}{k!}, \tag{10.15}
$$

where $(a)_k$ represent the Pochhammer symbol defined by $(a)_k = a(a+1)\cdots(a+k-1)$.

If a random variable X has a GG(a, p) distribution, then its Lorenz curve is given by

$$
L_{\text{GG}}(u; a, p) = G\left(G^{-1}(u; p); p + \frac{1}{a}\right), \quad 0 \leq u \leq 1, \tag{10.16}
$$

where $G(x; \nu)$ is defined in (10.8). The proof is direct, using the formula $L_X(u) = F_{X^{(1)}}(F_X^{-1}(u))$, and the results included in Table 10.1. Some special cases of the Lorenz curve (10.16) are:

- Lorenz curve of the classical gamma distribution ($a = 1$):

$$
L_G(u; p) = G(G^{-1}(u; p); p + 1), \quad 0 \leq u \leq 1.
$$

Table 10.2 Special cases of the Gini index for the GG family

Distribution	Gini index
Classical Gamma $G(p) = \text{GG}(1, p)$ $(= \Gamma(p, 1))$	$\dfrac{\Gamma(1+\frac{1}{2})}{\sqrt{\pi}\Gamma(p+1)}$
Weibull $W(a) = \text{GG}(a, 1)$	$1 - \dfrac{1}{2^{1/a}}$
Standard exponential $exp(1) = \text{GG}(1, 1)$	$\dfrac{1}{2}$

- Lorenz curve of the Weibull distribution ($p = 1$):

$$L_W(u; a) = G\left(-\log(1-u); \frac{1}{a} + 1\right), \quad 0 \le u \le 1.$$

- Lorenz curve of the standard exponential distribution ($a = p = 1$):

$$L_E(u) = u + (1-u)\log(1-u), \quad 0 \le u \le 1.$$

The Gini index of the GG family is, as derived by (McDonald 1984),

$$G_{\text{GG}}(a, p, q) = \frac{1}{2^{2p+1/a}B(p, p+1/a)}\left(\frac{1}{p^2}G^{(1)} - \frac{1}{p+1/a}G^{(2)}\right), \qquad (10.17)$$

where

$$G^{(1)} = {}_2F_1\left(1, 2p + \frac{1}{a}; p+1; \frac{1}{2}\right),$$

$$G^{(1)} = {}_2F_1\left(1, 2p + \frac{1}{a}; p + \frac{1}{a} + 1; \frac{1}{2}\right),$$

where ${}_2F_1(a, b; c; x)$ represents the Gauss hypergeometric function, which is a special case of the generalized hypergeometric function defined in (10.15). Some special cases of the Gini index for the GG family appear in Table 10.2.

We continue with the generalized beta distributions of the first kind. If a random variable X has a $\text{GB1}(a, p, q)$ distribution, then its Lorenz curve is given by

$$L_{\text{GB1}}(u; a, p, q) = B\left(B^{-1}(u; p, q); p + \frac{1}{a}, q\right), \quad 0 \le u \le 1, \qquad (10.18)$$

where the function $B(x; p, q)$ is defined in (10.10). Special cases of the Lorenz curve (10.18) of the GB1 family are:

- Lorenz curve of the power distribution ($q = 1$):

$$L_P(u; a, p) = u^{1/\theta+1}, \quad 0 \le u \le 1,$$

where $\theta = ap$.

Table 10.3 Special cases of the Gini index for the GB1 family

Distribution	Gini index
Classical Beta $B(p,q) = \text{GB1}(1, p, q)$	$\dfrac{2B(2p,2q)}{pB^2(p,q)}$
Kumaraswamy $K(a,q) = \text{GB1}(a, 1, q)$	$1 - 2\dfrac{\Gamma(2q)\Gamma(q+1/a+1)}{\Gamma(q)\Gamma(2q+1/a+1)}$
Power GB1$(a, p, 1)$	$\dfrac{1}{1+2\theta}, \quad \theta = ap$
Complementary power GB1$(1, 1, q)$	$\dfrac{1}{1+2q}$
Arcsine GB1$(1, \frac{1}{2}, \frac{1}{2})$	$\dfrac{4}{\pi^2}$

- Lorenz curve of the complementary power distribution ($a = p = 1$):

$$L_{\text{CP}}(u;q) = 1 - (q+1)(1-u) + q(1-u)^{1+1/q}, \quad 0 \le u \le 1$$

- Lorenz curve of the arcsine distribution ($a = 1$ and $p = q = 1/2$):

$$L_A(u) = \frac{\pi u - \sin(\pi u)}{\pi}, \quad 0 \le u \le 1.$$

If a random variable X has a GB1(a, p, q) distribution, then its Gini index is given by (McDonald 1984),

$$G_{\text{GB1}} = K^{(1)}{}_4F_3\left(2p+\frac{1}{a}, p, p+\frac{1}{a}, 1-q; 2p+q+\frac{1}{a}, p+1, p+\frac{1}{a}+1; 1\right),$$

in which

$$K^{(1)} = \frac{B(2p+1/a, q)}{B(p,q)B(p+1/a,q)p(ap+1)}$$

and ${}_4F_3(\boldsymbol{a}; \boldsymbol{b}; x)$ represents the generalized hypergeometric function defined in (10.15), with $p = 4$ and $q = 3$. Special cases of the Gini index of the GB1 family are given in Table 10.3.

We turn now to consider the generalized beta distributions of the second kind. If a random variable X has a GB2(a, p, q) distribution, then its Lorenz curve is given by

$$L_{\text{GB2}}(u; a, p, q) = B\left(B^{-1}(u; p, q); p + \frac{1}{a}, q - \frac{1}{a}\right), \quad 0 \le u \le 1, \tag{10.19}$$

where $q > 1/a$ and $B(x; p, q)$ is defined in (10.10). Some relevant special cases of the Lorenz curve of the GB2 distribution (10.19) are the following,

- Lorenz curve of the Singh–Maddala distribution ($p = 1$):

$$L_{SM}(u; a, q) = B\left(1 - (1-u)^{1/q}; 1 + \frac{1}{a}, q - \frac{1}{a}\right), \quad 0 \le u \le 1,$$

valid if $q > 1/a$.

- Lorenz curve of the Dagum distribution ($q = 1$):

$$L_D(u; a, p) = B\left(u^{1/p}; p + \frac{1}{a}, 1 - \frac{1}{a}\right), \quad 0 \le u \le 1,$$

valid if $a > 1$.

- Lorenz curve of the beta distribution of the second kind ($a = 1$):

$$L_{B2}(u; p, q) = B\left(B^{-1}(u; p, q); p + 1, q - 1\right), \quad 0 \le u \le 1,$$

valid if $q > 1$.

The Gini index for the GB2 family is given by (McDonald 1984),

$$G_{GB2} = \frac{B(2q - 1/a, 2p + 1/a)}{B(p, q)B(p + 1/a, q - 1/a)}\left(\frac{1}{p}H^{(1)} - \frac{1}{p + 1/a}H^{(2)}\right),$$

where

$$H^{(1)} = {}_3F_2\left(1, p + q, 2p + \frac{1}{a}; p + 1, 2(p + q); 1\right),$$

$$H^{(2)} = {}_3F_2\left(1, p + q, 2p + \frac{1}{a}; p + \frac{1}{a} + 1, 2(p + q); 1\right),$$

if $q > 1/a$, where ${}_3F_2(a_1, a_2, a_3; b_1, b_2; x)$ is a special case of the generalized hypergeometric function defined in (10.15). Some relevant special cases of the Gini index for the GB2 family appear in Table 10.4.

We now consider the lognormal distribution. It has been remarked that the lognormal distribution can be viewed as a limiting case of McDonald distributions. However, direct computation of the lognormal Lorenz curve and Gini index is preferable, as follows. Let $X \sim LN(\mu, \sigma^2)$ be a lognormal random variable with pdf given by (10.4), then its Lorenz curve is given by

$$L_X(u; \sigma) = \Phi(\Phi^{-1}(u) - \sigma), \quad 0 \le u \le 1, \tag{10.20}$$

where $\Phi(\cdot)$ represents the cdf of the standard normal distribution. From (10.14) the first moment distribution of X is again of the lognormal form, i.e., $X^{(1)} \sim LN(\mu + \sigma^2, \sigma^2)$, and using the quantile function given in (10.13) we obtain (10.20). The Gini index of the lognormal distribution is

$$G = 2\Phi\left(\frac{\sigma}{\sqrt{2}}\right) - 1, \tag{10.21}$$

Table 10.4 Special cases of the Gini index for the GB2 family

Distribution	Gini index
Beta of the second kind $B^{(2)}(p,q) = \text{GB2}(1,p,q)$	$\dfrac{2B(2p,2q-1)}{pB^2(p,q)}, \quad q>1$
Singh–Maddala $\text{SM}(a,q) = \text{GB2}(a,1,q)$	$1-\dfrac{\Gamma(q)\Gamma(2q-\frac{1}{a})}{\Gamma(q-\frac{1}{a})\Gamma(2q)}, \quad q>1/a$
Dagum $D(a,p) = \text{GB2}(a,p,1)$	$\dfrac{\Gamma(p)\Gamma(2p+\frac{1}{a})}{\Gamma(2p)\Gamma(p+\frac{1}{a})}-1, \quad a>1$
Fisk or log-logistic $F(a) = \text{GB2}(a,1,1)$	$\dfrac{1}{a}, \quad a>1$
Lomax or Pareto II $Pa(q) = \text{GB2}(1,1,q)$	$\dfrac{q}{2q-1}, \quad q>1$

where, here and below, $\Phi(\cdot)$ represents the cdf of the standard normal distribution. The proof of this results is based on the following identity,

$$\int_{-\infty}^{\infty} \phi(z)\Phi(a+bz)dz = \Phi\left(\frac{a}{\sqrt{1+b^2}}\right),$$

where $a, b \in \mathbb{R}$ and $\Phi(z)$ represents the pdf of the standard normal distribution.

10.2.2 *Other Inequality Measures*

In this section we obtain expressions for some additional inequality measures for members of the McDonald family. First of all, the inequality measures expressed in terms of the raw moments can be obtained using the formulas contained in Table 10.1. We continue with the Theil indices $T_0(X)$ and $T_1(X)$ defined in Chap. 5 which are given by

$$T_0(X) = -E\left(\log\frac{X}{\mu}\right),$$

and

$$T_1(X) = E\left(\frac{X}{\mu}\log\frac{X}{\mu}\right),$$

respectively, where $X \in \mathcal{L}$ and $E(X) = \mu$. The corresponding formulas for $T_0(X)$ and $T_1(X)$ for the three subclasses of the McDonald family of distributions

Table 10.5 Theil index $T_0(X)$ for the members of the McDonald family

Distribution	$T_0(X)$
GG	$-\dfrac{\psi(p)}{a}+\log\dfrac{\Gamma(p+\frac{1}{a})}{\Gamma(p)}$
GB1	$-\dfrac{\psi(p)-\psi(p+q)}{a}+\log\dfrac{\Gamma(p+q)\Gamma(p+\frac{1}{a})}{\Gamma(p)\Gamma(p+q+\frac{1}{a})}$
GB2	$-\dfrac{\psi(p)-\psi(q)}{a}+\log\dfrac{\Gamma(p+\frac{1}{a})\Gamma(q-\frac{1}{a})}{\Gamma(p)\Gamma(q)},\quad q>\dfrac{1}{a}$

Note: We denote by $\psi(z)=\frac{d\log\Gamma(z)}{dz}=\frac{\Gamma'(z)}{\Gamma(z)}$ the digamma function

Table 10.6 Theil index $T_1(X)$ for the members of the McDonald family

Distribution	$T_1(X)$
GG	$\dfrac{\psi(p+\frac{1}{a})}{a}-\log\dfrac{\Gamma(p+\frac{1}{a})}{\Gamma(p)}$
GB1	$\dfrac{\psi(p+\frac{1}{a})-\psi(p+q+\frac{1}{a})}{a}-\log\dfrac{\Gamma(p+q)\Gamma(p+\frac{1}{a})}{\Gamma(p)\Gamma(p+q+\frac{1}{a})}$
GB2	$\dfrac{\psi(p+\frac{1}{a})-\psi(q-\frac{1}{a})}{a}-\log\dfrac{\Gamma(p+\frac{1}{a})\Gamma(q-\frac{1}{a})}{\Gamma(p)\Gamma(q)},\quad q>\dfrac{1}{a}$

Note: We denote by $\psi(z)=\frac{d\log\Gamma(z)}{dz}=\frac{\Gamma'(z)}{\Gamma(z)}$ the digamma function

are included in Tables 10.5 and 10.6, respectively. For the lognormal distribution $X\sim LN(\mu,\sigma^2)$ we have

$$T_0(X)=T_1(X)=\frac{\sigma^2}{2}.$$

The Theil indices for the GB2 distribution were obtained by Jenkins (2009). The T_1 index for the classical gamma distribution was obtained by McDonald and Ransom (2008) and the formula for T_1 for the classical beta distribution was obtained by McDonald (1981) and Pham-Gia and Turkkan (1992). The rest of the formulas were obtained by Sarabia et al. (2017a).

We continue with the Pietra index. We recall that if X is a non-negative random variable in $\mathcal{L}$, the Pietra index is defined as the maximal vertical deviation between the Lorenz curve and the egalitarian line, that is,

$$P(X)=\max_{0\le p\le 1}\{p-L_X(p)\}.$$

If $X\in\mathcal{L}$, an alternative simple expression is available for the Pietra index in terms of the cdf of X and the first moment distribution $X_{(1)}$, i.e.,

$$P(X)=F_X(\mu_X)-F_X^{(1)}(\mu_X). \tag{10.22}$$

Table 10.7 Special cases of the Pietra index for the GG family

Distribution	Pietra Index
Half normal GG(2, 1/2)	$\mathrm{erfc}\left(\frac{1}{\sqrt{\pi}}\right) - 1 + \exp\left(-\frac{1}{\pi}\right)$
Classical gamma GG(1, p)	$\frac{p^p e^{-p}}{\Gamma(p)}$
Exponential GG(1, 1)	$\frac{1}{e} = 0,367879$
Weibull GG(a, 1)	$1 - \exp(-\mu_W^a) - G\left(\mu_W^a; 1 + \frac{1}{a}\right)$

Note: Notation, $\mathrm{erfc}(x)$ is the complementary error function defined by $\mathrm{erfc}(x) = \frac{2}{\sqrt{\pi}} \int_x^\infty e^{-t^2} dt$ and $\mu_W = \Gamma\left(1 + \frac{1}{a}\right)$ the mean of the Weibull distribution

The corresponding formulas have been obtained in a unified way by Sarabia and Jordá (2014b) . First, if $X \sim \mathrm{GG}(a, p)$, i.e., has a generalized gamma distribution, where $a, p > 0$, then its Pietra index is given by

$$P_{\mathrm{GG}}(a, p) = G\left(\mu_{\mathrm{GG}}^a; p\right) - G\left(\mu_{\mathrm{GG}}^a; p + \frac{1}{a}\right), \tag{10.23}$$

where

$$\mu_{\mathrm{GG}} = \frac{\Gamma(p + \frac{1}{a})}{\Gamma(p)},$$

and $G(x; p)$ is defined in (10.8). The proof is direct using Eq. (10.22) and the elements contained in Table 10.1. Some special cases of the Pietra index for the GG distribution are given in Table 10.7.

If $X \sim \mathrm{GB1}(a, p, q)$ has a generalized beta distribution of the first kind, where $a, p, q > 0$, then its Pietra index is given by

$$P_{\mathrm{GB1}}(a, p, q) = B\left(\mu_{\mathrm{GB1}}^a; p, q\right) - B\left(\mu_{\mathrm{GB1}}^a; p + \frac{1}{a}, q\right), \tag{10.24}$$

where

$$\mu_{\mathrm{GB1}} = \frac{B(p + q, \frac{1}{a})}{B(p, \frac{1}{a})},$$

and $B(x; p, q)$ is defined in (10.10). Some special cases of the Pietra index for the $GB1$ distribution are given in Table 10.8.

Table 10.8 Special cases of the Pietra index for the GB1 family

Distribution	Pietra index
Power GB1$(a, p, 1)$	$\frac{(ap)^{ap}}{(1+ap)^{1+ap}}$
Classical beta GB1$(1, p, q)$	$B\left(\frac{p}{p+q}; p, q\right) - B\left(\frac{p}{p+q}; p+1, q\right)$
Kumaraswamy GB1$(a, 1, q)$	$1 - (1-\mu_K^a)^q - B\left(\mu_K^a; 1+\frac{1}{a}, q\right)$

Note: Here $\mu_K = \frac{B\left(1+q, \frac{1}{a}\right)}{B\left(1, \frac{1}{a}\right)}$ is the mean of the Kumaraswamy distribution

Table 10.9 Special cases of the Pietra index for the GB2 family

Distribution	Pietra Index
Second kind beta GB2$(1, p, q)$	$B\left(\frac{p}{p+q-1}; p, q\right) - B\left(\frac{p}{p+q-1}; p+1, q-1\right)$
Singh–Maddala GB2$(a, 1, q)$	$1 - \frac{1}{(1+\mu_{\text{SM}}^a)^q} - B\left(\frac{\mu_{\text{SM}}^a}{1+\mu_{\text{SM}}^a}; 1+\frac{1}{a}, q-\frac{1}{a}\right)$
Dagum GB2$(a, p, 1)$	$\frac{1}{(1+\mu_D^{-a})^p} - B\left(\frac{\mu_D^a}{1+\mu_D^a}; p+\frac{1}{a}, 1-\frac{1}{a}\right)$

Note: We denote as $\mu_{\text{SM}} = \frac{\Gamma(1+\frac{1}{a})\Gamma(q-\frac{1}{a})}{\Gamma(q)}$ the mean of the Singh–Maddala distribution and $\mu_D = \frac{\Gamma(p+\frac{1}{a})\Gamma(1-\frac{1}{a})}{\Gamma(p)\Gamma(q)}$ the mean of the Dagum distribution

Finally, if $X \sim \text{GB2}(a, p, q)$, i.e., has a generalized beta distribution of the second kind, where $a, p, q > 0$, then the Pietra index is given by

$$P_{\text{GB2}}(a, p, q) = B\left(\frac{\mu_{\text{GB2}}^a}{1+\mu_{\text{GB2}}^a}; p, q\right) - B\left(\frac{\mu_{\text{GB2}}^a}{1+\mu_{\text{GB2}}^a}; p+\frac{1}{a}, q-\frac{1}{a}\right), \tag{10.25}$$

where $q > 1/a$ and

$$\mu_{\text{GB2}} = \frac{\Gamma(p+\frac{1}{a})\Gamma(q-\frac{1}{a})}{\Gamma(p)\Gamma(q)},$$

and $B(x; p, q)$ is defined in (10.10). Some relevant special cases of the Pietra index for the GB2 distribution are included in Table 10.9.

For the lognormal distribution, since its first moment distribution is again lognormal, using Eq. (10.22) we obtain

$$P_{LN} = F_X(\mu) - F_{X^{(1)}}(\mu) = 2\Phi\left(\frac{\sigma}{2}\right) - 1. \tag{10.26}$$

10.3 The Generalized Pareto Distributions

In this section we will study the hierarchy of Pareto distributions proposed by Arnold (2015b).

The different members of this family are all heavy-tailed distributions. In addition, they are easy to compute because they are defined in terms of simple survival functions. Many of the main properties of this family including: distributional properties, order statistics, record values, residual life, asymptotic, characterizations, related distributions, measures of inequality, multivariate versions, and statistical inference have been studied in detail by Arnold (1983, 2015b).

The starting point of the hierarchy is the classical Pareto distribution (which we have used in several parts of the book), called the Pareto (I) distribution. Its survival function is of the form

$$\bar{F}(x) = \left(\frac{x}{\sigma}\right)^{-\alpha}, \quad x > \sigma, \tag{10.27}$$

where $\bar{F} = 1 - F$, $\sigma > 0$ is the scale parameter and $\alpha > 0$ is the Pareto index of inequality. In practice, we assume that $\alpha > 1$, so that distribution has a finite mean. A random variable X with survival function (10.27) will be denoted by $X \sim P(I)(\sigma, \alpha)$.

The second member of the hierarchy is the Pareto(II) distribution. This distributions is also defined in terms of the survival function by

$$\bar{F}(x) = \left[1 + \left(\frac{x - \mu}{\sigma}\right)\right]^{-\alpha}, \quad x > \mu, \tag{10.28}$$

where $\mu \in \mathbb{R}$ is the location parameter, $\sigma > 0$ is positive, and $\alpha > 0$ is the shape parameter. In many applications μ is positive. We will denote this distribution by $X \sim P(II)(\mu, \sigma, \alpha)$.

The next member of the hierarchy, which provides tail behavior similar to that corresponding to the survival function (10.28) is the Pareto(III) family with survival function,

$$\bar{F}(x) = \left[1 + \left(\frac{x - \mu}{\sigma}\right)^{1/\gamma}\right]^{-1}, \quad x > \mu, \tag{10.29}$$

where $\mu \in \mathbb{R}$ and $\sigma, \gamma > 0$. We will call γ the inequality parameter. If $\mu = 0$ and $\gamma \leq 1$, the shape parameter γ turns out to be the Gini index of inequality. If X has (10.29) as its survival function, we will denote this by $X \sim P(III)(\mu, \sigma, \gamma)$.

The last member of the hierarchy is the Pareto(IV) distribution defined by

$$\bar{F}(x) = \left[1 + \left(\frac{x - \mu}{\sigma}\right)^{1/\gamma}\right]^{-\alpha}, \quad x > \mu, \tag{10.30}$$

Table 10.10 Quantile functions for the generalized Pareto distributions

Pareto Distribution	Quantile function
Pareto (I)(σ, α)	$F_I^{-1}(u) = \sigma(1-u)^{-1/\alpha}$
Pareto (II)(μ, σ, α)	$F_{II}^{-1}(u) = \mu + \sigma\left[(1-u)^{-1/\alpha} - 1\right]$
Pareto (III)(μ, σ, γ)	$F_{III}^{-1}(u) = \mu + \sigma\left[(1-u)^{-1} - 1\right]^{\gamma}$
Pareto (IV)$(\mu, \sigma, \gamma, \alpha)$	$F_{IV}^{-1}(u) = \mu + \sigma\left[(1-u)^{-1/\alpha} - 1\right]^{\gamma}$

Table 10.11 Moments for the generalized Pareto distributions

Pareto Distribution	Moments
Pareto (I)(σ, α)	$E(X^k) = \dfrac{\alpha\sigma^k}{\alpha - k}, \quad \alpha > k$
Pareto (II)$(0, \sigma, \alpha)$	$E(X^k) = \dfrac{\sigma^k\Gamma(\alpha-k)\Gamma(1+k)}{\Gamma(\alpha)}, \quad \alpha > k$
Pareto (III)$(0, \sigma, \gamma)$	$E(X^k) = \sigma^k\Gamma(1-k\gamma)\Gamma(1+k\gamma), \quad -\dfrac{1}{\gamma} < k < \dfrac{1}{\gamma}$
Pareto (IV)$(0, \sigma, \gamma, \alpha)$	$E(X^k) = \dfrac{\sigma^k\Gamma(\alpha-k\gamma)\Gamma(1+k\gamma)}{\Gamma(\alpha)}, \quad -\dfrac{1}{\gamma} < k < \dfrac{\alpha}{k}$

with $\mu \in \mathbb{R}$ and σ, γ, and α positive. If a random variable has the survival function (10.30), we will denote $X \sim P(IV)(\mu, \sigma, \gamma, \alpha)$.

The Pareto (IV) family provides a convenient vehicle for computing distributional results for the three more specialized families. All three of the previous distributions may be identified as special cases of the Pareto (IV) family as follows:

$$\begin{aligned} P(I)(\sigma, \alpha) &= P(IV)(\sigma, \sigma, 1, \alpha), \\ P(II)(\mu, \sigma, \alpha) &= P(IV)(\mu, \sigma, 1, \alpha), \\ P(III)(\mu, \sigma, \gamma) &= P(IV)(\mu, \sigma, \gamma, 1). \end{aligned}$$

The quantile function and the raw moments of the generalized Pareto distributions appear in Tables 10.10 and 10.11, respectively. Samples of the different families of the generalized Pareto distributions can be obtained easily by simulation using formulas in Table 10.10.

Some of the distributions in this hierarchical family of Pareto models can be identified with other known families of distributions. For example, the Pareto III distribution corresponds to the Fisk distribution (Fisk 1961a,b), also called the log-logistic distribution. If fact, if X has a logistic distribution with cdf,

$$F(x) = \left[1 + e^{-(x-\mu)/\sigma}\right]^{-1}, \quad x \in \mathbb{R}$$

we have

$$\exp(X) \sim P(III)(0, e^{\mu}, \sigma).$$

On the other hand, if $X \sim P(II)(\mu, \sigma, \alpha)$, i.e., is a Pareto (II) variable, and if $Y_1 \sim \Gamma(1, 1)$ is an exponential distribution and $Y_\alpha \sim \Gamma(\alpha, 1)$ a gamma distribution with shape parameter α and independent of Y_1, we have

$$X = \mu + \sigma \frac{Y_1}{Y_\alpha}. \tag{10.31}$$

The stochastic representation (10.31) has been used by Sarabia et al. (2016) to generate multivariate Pareto distribution and then to study some aggregated distributions under dependence. See also Chap. 6 of Arnold (2015b) where several multivariate Pareto models are defined in terms of independent gamma distributed components. One of these is mentioned in Sect. 12.9 below. The representation (10.31) can also be written in terms of mixture distributions.

The Pareto IV distribution was proposed by Arnold and Laguna (1976); Ord (1975) and Cronin (1977, 1979). It is related to the Feller–Pareto (Feller 1971) distribution. If $B \sim B(\gamma_1, \gamma_2)$, i.e., it has a beta distribution of the first kind, then a Feller–Pareto variable is defined by

$$X = \mu + \sigma \left(\frac{1}{B} - 1 \right)^{\gamma}, \tag{10.32}$$

where $\mu \in \mathbb{R}$ and σ, γ, γ_1 and γ_2 are positive. This distribution will be denoted as $X \sim FP(\mu, \sigma, \gamma, \gamma_1, \gamma_2)$. The probability density function corresponds to (10.32) is

$$f(x) = \frac{1}{\gamma \sigma B(\gamma_1, \gamma_2)} \frac{[(x-\mu)/\sigma]^{\gamma_2/\gamma_1 - 1}}{\left\{1 + [(x-\mu)/\sigma]^{1/\gamma}\right\}^{\gamma_1+\gamma_2}}, \quad x > \mu.$$

In the case $\mu = 0$, Kalbfleisch and Prentice (1980) called this density the generalized F density. The Pareto (IV) distributions can be identified with the Feller–Pareto distributions as

$$P(IV)(\mu, \sigma, \gamma, \alpha) = FP(\mu, \sigma, \gamma, \alpha, 1) \tag{10.33}$$

and if $\mu = 0$, the GB2 and Feller–Pareto distributions are related by

$$FP(0, \sigma, \gamma, \gamma_1, \gamma_2) = GB2(\gamma^{-1}, \gamma_1, \gamma_2, \sigma). \tag{10.34}$$

The Pareto (IV) distribution has a representation as a mixture of Weibull distributions. If

$$P(X > x | Y = y) = e^{-y[(x-\mu)/\sigma]^{1/\gamma}}, \quad x > \mu,$$

that is, a translated Weibull distribution and if $Y \sim \Gamma(\alpha, 1)$, we have $X \sim P(IV)(\mu, \sigma, \gamma, \alpha)$.

10.3.1 Lorenz Curves and Gini Indices

In this section we include the Lorenz curves and the Gini indices for the Pareto (I)–Pareto (IV) distributions. The expressions for the Lorenz curves can be found in Arnold (2015b) and are the following.

- Lorenz curve for the Pareto (I)(σ, α) distribution:

$$L(u) = 1 - (1-u)^{(\alpha-1)/\alpha}, \quad 0 < u < 1,$$

  with $\alpha > 1$.
- Lorenz curve for the Pareto (II)$(0, \sigma, \alpha)$ distribution:

$$L(u) = 1 - B((1-u)^{1/\alpha}; \alpha - 1, 2), \quad 0 < u < 1,$$

  with $\alpha > 1$.
- Lorenz curve for the Pareto (III)$(0, \sigma, \gamma)$ distribution:

$$L(u) = 1 - B(1-u; 1-\gamma, \gamma+1), \quad 0 < u < 1,$$

  with $\gamma < 1$.
- Lorenz curve for the Pareto (IV)$(\mu, \sigma, \gamma, \alpha)$ distribution:

$$L(u) = \frac{\mu u + \sigma\alpha B(\alpha-\gamma, \gamma+1)[1 - B((1-u)^{1/\alpha}; \alpha-\gamma, \gamma+1)]}{\mu + \sigma\alpha B(\alpha-\gamma, \gamma+1)},$$

  with $0 < u < 1$ and $\alpha > \gamma$

10.3.2 Inequality Measures

In this section we obtain several inequality measures for the Pareto (I)–Pareto (IV) distributions. Obviously, many inequality measures expressed in terms of the raw moments and quantiles can be obtained using the moment formulas contained in Tables 10.10 and 10.11. Table 10.12 contains the Gini indices for these distributions.

Table 10.12 Gini indices for the generalized Pareto distributions

Pareto Distribution	Gini index
Pareto (I)(σ, α)	$\frac{1}{2\alpha-1}, \quad \alpha > 1$
Pareto (II)$(0, \sigma, \alpha)$	$\frac{\alpha}{2\alpha-1}, \quad \alpha > 1$
Pareto (III)$(0, \sigma, \gamma)$	$\gamma, \quad \gamma < 1$
Pareto (IV)$(\mu, \sigma, \gamma, \alpha)$	$1 - \frac{\mu + 2\sigma\alpha B(2\alpha-\gamma, \gamma+1)}{\mu + \sigma\alpha B(\alpha-\gamma, \gamma+1)}, \quad \alpha > \gamma$

Table 10.13 Theil index $T_0(X)$ for the generalized Pareto distributions

Distribution	$T_0(X)$
Pareto (I)(σ, α)	$-\frac{1}{\alpha} + \log \frac{\alpha}{\alpha-1}, \quad \alpha > 1$
Pareto (II)$(0, \sigma, \alpha)$	$\gamma + \psi(\alpha) - \log(\alpha - 1), \quad \alpha > 1$
Pareto (III)$(0, \sigma, a)$	$\log[\Gamma(1-a)\Gamma(1+a)], \quad a < 1$
Pareto (IV)$(0, \sigma, a, \alpha)$	$a[\gamma + \psi(\alpha)] + \log\left[\frac{\Gamma(1+a)\Gamma(\alpha-a)}{\Gamma(\alpha)}\right], \quad \alpha > a$

Note: We denote the Euler's constant as $\gamma = 0.5772$ and $\psi(z) = \frac{d \log \Gamma(z)}{dz} = \frac{\Gamma'(z)}{\Gamma(z)}$ the digamma function

Table 10.14 Theil index $T_1(X)$ for the generalized Pareto distributions

Distribution	$T_1(X)$
Pareto (I)(σ, α)	$\frac{1}{\alpha-1} - \log \frac{\alpha}{\alpha-1}, \quad \alpha > 1$
Pareto (II)$(0, \sigma, \alpha)$	$1 - \gamma - \psi(\alpha - 1) + \log(\alpha - 1), \quad \alpha > 1$
Pareto (III)$(0, \sigma, a)$	$a[\psi(1+a) - \psi(1-a)] - \log[\Gamma(1-a)\Gamma(1+a)], \quad a < 1$
Pareto (IV)$(0, \sigma, a, \alpha)$	$a[\psi(1+a) - \psi(\alpha-a)] - \log\left[\frac{\Gamma(1+a)\Gamma(\alpha-a)}{\Gamma(\alpha)}\right], \quad \alpha > a$

Note: We denote the Euler's constant as $\gamma = 0.5772$ and $\psi(z) = \frac{d \log \Gamma(z)}{dz} = \frac{\Gamma'(z)}{\Gamma(z)}$ the digamma function

Tables 10.13 and 10.14 contain the Theil indices of the generalized Pareto distributions. These expressions appear in Sarabia et al. (2017a,b) .

The Pietra indices for the Pareto(I)(σ, α) and the Pareto(II)$(0, \sigma, \alpha)$ distributions are (see Arnold 2015b)

$$P_I = \frac{(\alpha-1)^{\alpha-1}}{\alpha^\alpha},$$

$$P_{II} = \left(1 - \frac{1}{\alpha}\right)^{\alpha-1},$$

respectively. The Pietra indices for the other two members can also be obtained using the formulas in Chap. 5.

10.4 Stochastic Orderings Within the McDonald Family

10.4.1 Introduction and the Orderings to be Used

In this section we study the comparison of income distributions in terms of inequality and relative deprivation for distributions in the McDonald family. We will consider the following three stochastic orderings: the Lorenz ordering, the star-shaped ordering, and the order based on the expected proportional shortfall (EPS) function. The first two orderings were introduced in Chaps. 3 and 8. The order based on the expected proportional shortfall function is an intermediate ordering between the Lorenz and the star-shaped orderings, with an interesting economic interpretation.

The different orderings for the Pareto(IV) and the Feller–Pareto distributions can be obtained directly using its relation with the GB2 distribution given by the identities (10.33) and (10.34).

First, we introduce the EPS function, proposed by Belzunce et al. (2012).

Definition 10.4.1 Let $X \in \mathcal{L}$. The expected proportional shortfall function of X is defined as

$$\mathrm{EPS}_X(u) = E\left[\left(\frac{X - F_X^{-1}(u)}{F_X^{-1}(u)}\right)_+\right],$$

for all $u \in (0, 1)$ with $F_X^{-1}(u) > 0$, where $(x)_+ = x$ if $x \geq 0$ and $(x)_+ = 0$ if $x < 0$.

This measure can be interpreted as a measure of the relative deprivation felt by an individual with rank u in the distribution F_X. Since the quantile function F_u^{-1} of $\left(\frac{X-F_X^{-1}(u)}{F_X^{-1}(u)}\right)_+$ is,

$$F_u^{-1}(t) = \frac{F_X^{-1}(t) - F_X^{-1}(u)}{F_X^{-1}(u)}, \quad t \geq u,$$

and 0 if $t < u$, the EPS function can be computed as

$$\mathrm{EPS}_X(u) = \int_0^1 F_u^{-1}(t)dt = \frac{\int_u^1 (F_X^{-1}(t) - F_X^{-1}(u))dt}{F_X^{-1}(u)}. \tag{10.35}$$

We consider a simple example.

Example 10.4.1 In the case in which X has a standard exponential distribution with cdf $F_X(x) = 1 - e^{-x}$, if $x \geq 0$ and quantile function $F_X^{-1}(u) = -\log(1-u)$, if $u \in (0, 1)$, using Formula (10.35), the EPS function of X is

$$\mathrm{EPS}_X(u) = \frac{\int_u^1 (-\log(1-t) + \log(1-u))dt}{-\log(1-u)} = \frac{1-u}{-\log(1-u)}.$$

Using the EPS function, we introduce the expected proportional shortfall ordering.

Definition 10.4.2 For $X_1, X_2 \in \mathcal{L}$, we say that X_1 exhibits less relative deprivation than X_2 in the expected proportional shortfall sense, to be denoted by $X_1 \leq_{ps} X_2$ if,

$$\mathrm{EPS}_{X_1}(u) \leq \mathrm{EPS}_{X_2}(u)$$

for all $u \in (0, 1)$.

Recently, Belzunce et al. (2012) have shown that the expected proportional shortfall ordering is intermediate between the Lorenz and the star-shaped orderings, that is,

$$X_1 \leq_* X_2 \Rightarrow X_1 \leq_{ps} X_2 \Rightarrow X_1 \leq_L X_2.$$

10.4.2 *Comparisons for Two Distributions in the Same Subfamily of McDonald Distributions*

In this section we will study comparison between two distributions inside the families GG, GB1, and GB2.

We begin with the GG family. The Lorenz ordering in the GG family was initially studied by Taillie (1981). In a rigorous contribution, Wilfling (1996b) provided the necessary and sufficient conditions for Lorenz ordering in the GG family. Subsequently, Belzunce et al. (2013) proved that these conditions were also necessary and sufficient for the star-shaped and the expected proportional shortfall orderings. Thus we have:

Theorem 10.4.1 *Let $X_i \sim \mathrm{GG}(a_i, p_i)$, $i = 1, 2$ be two random variables with GG distributions. Then,*

$$X_1 \leq_* X_2 \Longleftrightarrow X_1 \leq_{ps} X_2 \Longleftrightarrow X_1 \leq_L X_2 \Longleftrightarrow a_1 \geq a_2 \text{ and } a_1 p_2 \geq a_2 p_2. \tag{10.36}$$

In the following corollary we consider the necessary and sufficient conditions for the three orderings in two subfamilies (Gamma and Weibull distributions) of the GG distribution.

Corollary 10.4.1 *Assume that one of the following conditions holds:*

1. $X_i \sim \Gamma(p_i, 1)$, $i = 1, 2$ *with* $p_1 \geq p_2$,
2. $X_i \sim W(a_i)$, $i = 1, 2$ *with* $a_1 \geq a_2$.

Then,

$$X_1 \leq_* X_2,\ X_1 \leq_{ps} X_2 \text{ and } X_1 \leq_L X_2.$$

Now, we consider the GB1 family. The following theorem provides sufficient conditions for the Lorenz ordering of two GB1 distributions with different parameters (Wilfling 1996b).

Theorem 10.4.2 *Let* $X \sim GB1(a_1, p_1, q_1)$ *and* $Y \sim GB1(a_2, p_2, q_2)$ *with* $a_1, a_2, p_1, p_2, q_2 > 0$ *and* $q_1 > 1$. *We have* $X \leq_L Y$ *if:*

(a) $a_1 = a_2 =: a \geq 1$ *and* $p_1 = q_1$ *and* $p_2 = q_2 = p_1 - 1$.
(b) $a_1 = a_2 =: a$ *and* $p_1 > p_2$ *and* $q_1 = q_2 =: q \geq 1$.
(c) $a_1 = a_2 =: a$ *and* $p_1 > p_2$ *and* $q_2 > \xi > 0$, *with* ξ *satisfying* $E(X) = E(Z)$ *for* $Z_1 \sim GB1(a, p_2, \xi)$.
(d) $a_1 \geq a_2$ *and* $p_1 = p_2$ *and* $q_1 = q_2$.

Remark The constant ξ in part (c) of the previous theorem has to be determined numerically. For fixed a and $p = p_2$, it results from varying q until the corresponding expectation coincides with the mean of a GB1(a, p_1, q_1) distribution (see Wilfling 1996b).

Next, we consider distributions in the GB2 family. For this family, Wilfling (1996a) has proved the following result for Lorenz ordering.

Theorem 10.4.3 *Let* $X_i \sim GB2(a_i, p_i, q_i)$, $i = 1, 2$ *be two random variables with GB2 distributions. Then,*

(a) If $X_1 \leq_L X_2$, *then* $a_1 p_1 \geq a_2 p_2$ *and* $a_1 q_1 \geq a_2 q_2$.
(b) If $a_1 \geq a_2$, $p_1 > p_2$ *and* $q_1 > q_2$, *then* $X_1 \leq_L X_2$.

Later, Kleiber (1999) proved that the Lorenz order holds assuming weaker conditions on the parameters. Subsequently, Belzunce et al. (2013) showed that the sufficient conditions considered by Kleiber (1999) are sufficient conditions for the starshaped order and, as a consequence, for the expected proportional shortfall order also. Thus

Theorem 10.4.4 *Let* $X_i \sim GB2(a_i, p_i, q_i)$, $i = 1, 2$ *be two random variables with GB2 distributions. If* $a_1 \geq a_2$, $a_1 p_1 \geq a_2 p_2$, *and* $a_1 q_1 \geq a_2 q_2$, *then* $X_1 \leq_* X_2$.

In the following corollary we consider the sufficient conditions for the three orderings in five relevant subfamilies of the GB2 class of distributions.

Corollary 10.4.2 *Assume that one of the following conditions holds:*

1. $X_i \sim B2(p_i, q_i)$ *with* $q_i > 1$, $i = 1, 2$ *and* $p_1 \geq p_2$, $q_1 \geq q_2$,
2. $X_i \sim \mathrm{SM}(a_i, q_i)$ *with* $a_i q_i > 1$, $i = 1, 2$ *and* $a_1 \geq a_2$, $a_1 q_1 \geq a_2 q_2$,
3. $X_i \sim D(a_i, p_i)$ *with* $a_i > 1$, $i = 1, 2$ *and* $a_1 \geq a_2$, $a_1 p_1 \geq a_2 p_2$,
4. $X_i \sim L(q_i)$ *with* $q_i > 1$, $i = 1, 2$ *and* $q_1 \geq q_2$,
5. $X_i \sim F(a_i)$ *with* $a_i > 1$, $i = 1, 2$ *and* $a_1 \geq a_2$.

It follows that $X_1 \leq_* X_2$ *and so also that* $X_1 \leq_{ps} X_2$ *and* $X_1 \leq_L X_2$..

10.4.3 Comparisons for Two Distributions in Different McDonald Subfamilies

In this section we provide some results on comparisons between two distribution from different McDonald subfamilies. We will begin by considering the Lorenz order and then deal with comparison in terms of the star-shaped order and as a consequence, in terms of the expected proportional shortfall order.

Theorem 10.4.5 *Let* $X_1 \sim GB1(a, p, q)$, $X_2 \sim GG(a, p)$ *and* $X_3 \sim GB2(a, p, q)$ *with* $aq > 1$. *Then,*

$$X_1 \leq_L X_2 \leq_L X_3. \tag{10.37}$$

Proof The proof of these results can be found in Theorems 4, 5 and 6 in Sarabia et al. (2002). Here, we only include the proof for $X_2 \leq_L X_3$. Let $X_2 \sim \text{GG}(a, p)$ and $\tilde{X}_2 \sim \text{GG}(a, q)$ be independent GG random variables. We have

$$X_3 = \frac{X_2}{\tilde{X}_2} \sim \text{GB2}(a, p, q),$$

and then

$$E(X_3|X_2) = E\left(\frac{X_2}{\tilde{X}_2}|X_2\right) = kX_2,$$

where $k = E(\tilde{X}_2^{-1})$, which exists if $aq > 1$. Consequently, $X_2 = E(k^{-1}X_3|X_2)$ and then

$$X_2 \leq_L k^{-1}X_3 \Rightarrow X_2 \leq_L X_3,$$

since the Lorenz ordering is invariant with respect to changes of scale. ■

The following theorem is a stronger version of the previous theorem, this time for the star-shaped order.

Theorem 10.4.6 *Let* $X_1 \sim GB1(a, p, q)$, $X_2 \sim GG(a, p)$ *and* $X_3 \sim GB2(a, p, q)$. *Then,*

$$X_1 \leq_* X_2 \leq_* X_3.$$

Proof For proofs of these results, see Theorems 4.1 and 4.3 in Belzunce et al. (2013). ■

The following corollary identifies some pairs of income distributions that are star-shaped ordered as consequences of the above theorem.

Corollary 10.4.3 *Assume that one of the following conditions holds:*

1. $X_1 \sim GG(a_1, p_1)$ *and* $X_2 \sim GG(a_2, p_2, q_2)$ *with* $a_1 \geq a_2$ *and* $a_1 p_1 \geq a_2 p_2$,
2. $X_1 \sim GG(a_1, p_1)$ *and* $X_2 \sim D(a_2, p_2)$ *with* $a_2 > 1$ *and* $a_1 \geq a_2$ *and* $a_1 p_1 \geq a_2 p_2$,
3. $X_1 \sim G(p_1)$ *and* $X_2 \sim B2(p_2, q_2)$ *with* $q_2 > 1$ *and* $p_1 \geq p_2$,
4. $X_1 \sim W(a_1)$ *and* $X_2 \sim SM(a_2, q_2)$ *with* $a_2 q_2 > 1$ *and* $a_1 \geq a_2$,
5. $X_1 \sim W(a_1)$ *and* $X_2 \sim F(a_2)$ *with* $a_2 > 1$ *and* $a_1 \geq a_2$,
6. $X_1 \sim E(1)$, $X_2 \sim L(q)$ *with* $q > 1$
7. $X_1 \sim GB1(a_1, p_1, q_1)$ *and* $X_2 \sim GB2(a_2, p_2, q_2)$ *with* $a_1 \geq a_2$ *and* $a_1 p_1 \geq a_2 p_2$,

then,

$$X_1 \leq_* X_2.$$

10.5 Exercises

1. For the McDonald family, prove the following results:

   (a) Let $X \sim \text{GB2}(a, p, q)$, then,

   $$Y = \left(\frac{X^a}{1 + X^a}\right)^{1/a} \sim \text{GB1}(a, p, q).$$

   (b) Let $X_1 \sim \text{GG}(a, p)$ and $X_2 \sim \text{GG}(a, q)$ be independent random variables, then,

   $$Y = \frac{X_1}{X_2} \sim \text{GB2}(a, p, q).$$

2. Let $X \sim \text{GB2}(a, p, q, b)$ be a random variable with a GB2 distribution. Prove that:

   (a) The Singh–Maddala distribution $\text{SM}(a, q, b) = \text{GB2}(a, 1, q, b)$ has the simple cdf,

   $$F(x; a, q, b) = 1 - \left[1 + \left(\frac{x}{b}\right)^a\right]^{-q}, \quad x \geq 0.$$

   (b) The Dagum distribution $D(a, p, b) = \text{GB2}(a, p, 1, b)$ has the following simple cdf,

   $$F(x; a, p, b) = \left[1 + \left(\frac{x}{b}\right)^{-a}\right]^{-p}, \quad x \geq 0.$$

(c) If $X \sim \text{SM}(a, q, b)$, then $\frac{1}{X} \sim D(a, q, \frac{1}{b})$.

3. Use R software, to write code for computing the Lorenz curves for the GG, GB1, and GB2 distributions.

4. A random variable X is said to have an extended half normal distribution if its pdf is given by

$$f(x; a) = \frac{\exp(-x^a/2)}{2^{1/a}\Gamma(1 + \frac{1}{a})}, \quad x \geq 0,$$

and $f(x; a) = 0$ if $x < 0$, where $a > 0$.

(a) Check that X has a distribution that belongs to the GG family, identifying the corresponding parameters.
(b) Prove that the Pietra index of $X$is given by

$$G\left(\mu^a; \frac{1}{a}\right) - G\left(\mu^a; \frac{2}{a}\right),$$

where $\mu = \frac{\Gamma(2/a)}{\Gamma(1/a)}$.

5. The Lamé class of Lorenz curves (Sarabia et al. 2017b) includes Lorenz curves of the following two types,

$$L_1(u) = [1 - (1 - u)^a]^{1/a}, \quad 0 \leq u \leq 1,$$

where $0 < a \leq 1$ and

$$L_2(u) = 1 - (1 - u^a)^{1/a}, \quad 0 \leq u \leq 1,$$

where $a \geq 1$.

(a) Identify the cdf's corresponding to the Lorenz curves L_1 and L_2. Verify that both cdfs are special cases of the GB2 family, identifying the corresponding parameters.
(b) Prove that the Gini indices are

$$G_1(a) = 1 - \frac{\Gamma(1/a)^2}{a\Gamma(2/a)}, \quad 0 < a \leq 1,$$

and

$$G_2(a) = \frac{\Gamma(1/a)^2}{a\Gamma(2/a)} - 1, \quad a \geq 1,$$

respectively for L_1 and L_2.

(c) Prove that the Pietra indices for L_1 and L_2 are given by

$$P_1(a) = 1 - \frac{1}{2^{1/a-1}}, \quad 0 < a \leq 1,$$

and

$$P_2(a) = \frac{1}{2^{1/a-1}} - 1, \quad a \geq 1,$$

respectively.

6. Let $X_1 \sim \text{GB1}(a, p, q)$ and $X_2 \sim \text{GB2}(a, p, q)$ be GB1 and GB2 random variables with the same parameters. Use Theorem 4.1.1 to prove that $X_1 \leq_L X_2$.

7. Consider a Weibull distribution with cdf $F(x) = 1 - e^{-x^n}$ if $x \geq 0$. Verify that the corresponding expected proportional shortfall function is given by

$$-(1-u) - \frac{\Gamma\left(1 + \frac{1}{n}, -\log(1-u)\right)}{\log(1-u)},$$

with $u \in (0, 1)$, where $\Gamma(x; \nu)$ denotes the incomplete gamma function.

8. Consider a Truncated Pareto distribution with survival function,

$$\bar{F}(x) = \frac{(x/\sigma)^{-\alpha} - b^{-\alpha}}{a^{-\alpha} - b^{-\alpha}}, \quad a\sigma < x < b\sigma,$$

where $\sigma, \alpha > 0$ and $1 < a < b$. Prove that the Lorenz curve is given by (Arnold 2015b),

$$L(u) = \frac{1 - [1 - (1 - (a/b)^{\alpha})u]^{1-1/\alpha}}{1 - (a/b)^{\alpha-1}}, \quad 0 < u < 1,$$

when $\alpha \neq 1$. If $\alpha = 1/2$, prove that the previous curve can be written as

$$L(u) = \frac{u}{\sqrt{b/a} + (\sqrt{b/a} - 1)u}, \quad 0 < u < 1.$$

Chapter 11
Some Applications

A glance at the table of contents of Marshall, Olkin, and Arnold's (2011) book will indicate that approximately 40% of the book (i.e., 327 pages) is devoted to applications. Clearly, we will make no attempt to duplicate such an exhaustive list. We will content ourselves in this and the following chapter with a selection of examples which, it is hoped, will hint at the breadth of the areas of possible application. For more, of course, the reader will need to consult Marshall et al. (2011, Chapters 7–10 and 12–13).

11.1 A Geometric Inequality of Cesaro

Let ℓ_1, ℓ_2, and ℓ_3 denote the lengths of the three sides of a triangle. Cesaro is credited with the observation that, for any triangle, its side lengths satisfy

$$\ell_1\ell_2\ell_3 \leq \frac{1}{8}(\ell_1+\ell_2)(\ell_2+\ell_3)(\ell_3+\ell_1). \tag{11.1}$$

To prove this, we merely demonstrate that the left-hand side and right-hand side are just a particular Schur convex function evaluated at two points in $\mathbb{R}^3$ which are related by majorization. To this end, consider

$$\underline{\tilde{\ell}} = \begin{pmatrix} 1/2 & 1/2 & 0 \\ 0 & 1/2 & 1/2 \\ 1/2 & 0 & 1/2 \end{pmatrix} \underline{\ell}. \tag{11.2}$$

Since the matrix in (11.2) is doubly stochastic, we have $\underline{\tilde{\ell}} \leq_M \underline{\ell}$. Now consider the function $g(\underline{x}) = -\prod_{i=1}^{3} x_i$. It is easy to verify that g is Schur convex (apply

B. C. Arnold, J. M. Sarabia, *Majorization and the Lorenz Order with Applications in Applied Mathematics and Economics*, Statistics for Social and Behavioral Sciences,
https://doi.org/10.1007/978-3-319-93773-1_11

Schur's condition, Theorem 2.2.3). Inequality (11.1) then follows readily. But, in retrospect, it had nothing to do with the fact that ℓ_1, ℓ_2, ℓ_3 were lengths of sides of a triangle! If indeed the ℓ_i's were lengths of sides of a triangle, we would know that the largest of the three would have to be no larger than the sum of the other two, i.e.,

$$\ell_{3:3} \leq \ell_{1:3} + \ell_{2:3}, \tag{11.3}$$

and, of course, $\ell_{1:3} \geq 0$. It follows that if the ℓ_i's are sides of a triangle, we have

$$(\ell_1, \ell_2, \ell_3) \leq_M \left(0, \frac{p}{2}, \frac{p}{2}\right) \tag{11.4}$$

where $p = \ell_1+\ell_2+\ell_3$ is the perimeter of the triangle. Equation (11.4) does not hold for any vector (ℓ_1, ℓ_2, ℓ_3), only for vectors $\underline{\ell}$ which satisfy (11.3). If we evaluate the Schur convex function

$$h(\underline{x}) = -(x_1 + x_2)(x_2 + x_3)(x_3 + x_1)$$

at each of the vectors in (11.4), we conclude that

$$(\ell_1 + \ell_2)(\ell_2 + \ell_3)(\ell_3 + \ell_1) \geq p^3/4 \tag{11.5}$$

for any triangle (actually the inequality is strict if we consider only non-degenerate triangles). In addition, for any vector (ℓ_1, ℓ_2, ℓ_3) whose coordinates are lengths of the sides of a triangle we have

$$\left(\frac{p}{3}, \frac{p}{3}, \frac{p}{3}\right) \leq_M (\ell_1, \ell_2, \ell_3) \tag{11.6}$$

where again $p = \ell_1 + \ell_2 + \ell_3$. Evaluating $h(\underline{x})$ at each of the points in (11.6) yields

$$(\ell_1 + \ell_2)(\ell_2 + \ell_3)(\ell_3 + \ell_1) \leq 8p^3/27 \tag{11.7}$$

with equality in the case of an equilateral triangle.

11.2 Matrices with Prescribed Characteristic Roots

The relationship between the diagonal elements of an $n \times n$ matrix and the vector of its n eigenvalues was early discovered to involve majorization. Schur showed that if A is an $n \times n$ Hermitian matrix, then its diagonal vector is necessarily majorized by the vector of its eigenvalues. A more recent theorem discovered independently by Horn (1954) and Mirsky (1958) is the following.

Theorem 11.2.1 *Let $\underline{a}$ and $\underline{w}$ be two vectors in $\mathbb{R}^n$. If $\underline{a} \leq_M \underline{w}$, then there exists a real symmetric $n \times n$ matrix A with $\underline{a}$ as its diagonal vector and $\underline{w}$ as its vector of eigenvalues.*

To capture the basic idea of the proof without too much algebraic complexity, we will consider the two preliminary lemmas only in the case $n = 3$. First we have

Lemma 11.2.1 *If* $\underline{a}, \underline{b} \in \mathbb{R}^3$ *are such that* $\underline{a} \leq_M \underline{b}$, *then there exist numbers* c_2 *and* c_3 *such that* $b_{1:3} \leq c_2 \leq b_{2:3} \leq c_3 \leq b_{3:3}$ *and* $(a_{2:3}, a_{3:3}) \leq_M (c_2, c_3)$.

Proof Without loss of generality $a_1+a_2+a_3 = b_1+b_2+b_3 = 1$ and $a_1 \leq a_2 \leq a_3$, $b_1 \leq b_2 \leq b_3$. Since $\underline{a} \leq_M \underline{b}$ we have $a_1 \geq b_1$, $a_1 + a_2 \geq b_1 + b_2$ and $b_3 \geq a_3$. Obviously $b_1 \leq a_2$ and $a_3 \leq b_3$. We consider three cases.

Case (i). If $a_2 \leq b_2 \leq a_3$, we may take $c_2 = a_2$ and $c_3 = a_3$ and we are done.

Case (ii). Suppose $a_2 \leq a_3 < b_2$. We seek c_2, c_3 such that $c_2 + c_3 = a_2 + a_3$, $a_2 \geq c_2$ and $b_1 \leq c_2 \leq b_2 \leq c_3 \leq b_3$. Thus we wish to choose $c_2 \in (b_1, a_2)$. An acceptable choice is any c_2 in the interval $(b_1, a_1 + a_2 - b_3)$ (and $c_3 = a_2 + a_3 - c_2$).

Case (iii). Suppose $b_2 < a_2 \leq a_3$. In this case an acceptable choice is $c_2 = b_2$ and $c_3 = a_2 + a_3 - b_2$. ■

Lemma 11.2.2 *If the real numbers* (w_1, w_2, w_3) *and* (α_2, α_3) *satisfy*

$$w_1 \leq \alpha_2 \leq w_2 \leq \alpha_3 \leq w_3, \tag{11.8}$$

then there exists a real symmetric matrix of the form,

$$A = \begin{pmatrix} \alpha_3 & 0 & p_3 \\ 0 & \alpha_2 & p_2 \\ p_3 & p_2 & p_1 \end{pmatrix} \tag{11.9}$$

whose eigenvalues are w_1, w_2, *and* w_3.

Proof If we successively substitute w_1, w_2 and w_3 for λ in the determinantal equality $|A - \lambda I| = 0$, we obtain three linear equations in three unknowns p_1, p_2^2 and p_3^2. Subject to the constraint (11.8), it is not difficult to verify that a solution of the form $p_1 \in \mathbb{R}$, $p_2^2 \geq 0$, $p_3^2 \geq 0$ does exist. Hence, we can find p_1, p_2, and p_3 with the desired properties. ■

n-dimensional versions of Lemmas 11.2.1 and 11.2.2 were obtained by Mirsky (1958). Using them, an inductive proof of Theorem 11.2.1 is readily obtainable.

11.3 Variability of Sample Medians and Means

Let $X_1, X_2, \ldots, X_n$ be i.i.d. non-negative random variables with corresponding order statistics $X_{1:n}, X_{2:n}, \ldots, X_{n:n}$. If $n = 2m + 1$ is odd, then the sample median is $X_{m+1:2m+1}$. As m increases, it is plausible that the sample median will tend to concentrate more and more closely near to the population median, i.e., $F_X^{-1}(\frac{1}{2})$. For convenience we will assume that the common distribution of the X_i's is absolutely continuous with corresponding probability density function $f_X(x)$. Yang (1982) showed that $\text{var}(X_{m+1:2m+1})$ is less than or equal to $\text{var}(X)$. In the case of a symmetric density, i.e., f_X of the form

$$f_X(x) = \begin{cases} f_X(c-x), & 0 < x < c, \\ 0, & x \notin (0,c), \end{cases} \tag{11.10}$$

we can make an even stronger statement. In fact, we know that if the common density of the X_i's is of the form (11.10), we have

$$X_{m+2:2m+3} \leq_L X_{m+1:2m+1} \tag{11.11}$$

for every $m = 0, 1, 2, \ldots$ (this was Example 9.1.2). Since $E(X_{m+1:2m+1}) = c, \forall m$, we can conclude that

$$\text{var}(X_{m+1:2m+1}) \downarrow \text{ as } m \uparrow \tag{11.12}$$

(since $u(x) = x^2$ is convex).

If we turn to consider sample means rather than medians, then, assuming second moments exist, but not assuming symmetry, we have for each n,

$$\text{var}(\bar{X}_{n+1}) \leq \text{var}(\bar{X}_n) \tag{11.13}$$

(where $\bar{X}_n = \frac{1}{n}\sum_{i=1}^n X_i$). In fact $\bar{X}_n$ and $\bar{X}_{n+1}$ are Lorenz ordered.

To see this, we argue as follows. By exchangeability of the X_i's we have

$$E(X_i|X_1 + \cdots + X_{n+1}) = \frac{X_1 + \cdots + X_{n+1}}{n+1},$$

for $i = 1, 2, \ldots, n+1$. Thus,

$$\begin{aligned} E\left(\bar{X}_n|\bar{X}_{n+1}\right) &= \frac{1}{n}E\left(\sum_{i=1}^n X_i|\bar{X}_{n+1}\right) \\ &= \frac{1}{n}\sum_{i=1}^n E\left(X_i|X_1 + \cdots + X_{n+1}\right) \\ &= \bar{X}_{n+1} \end{aligned} \tag{11.14}$$

Since $E(\bar{X}_{n+1}) = E(\bar{X}_n)$, we can apply Theorem 3.2.3 and conclude from (11.14) that $\bar{X}_{n+1} \leq_L \bar{X}_n$.

Thus, if we were considering use of $\bar{X}_n$ or $\bar{X}_{n+1}$ as estimates of $E(X_i) = \mu$ and had a loss function which was a convex function of the error of estimation, i.e., of the form $g(T - \mu)$ where T is the estimate and g is convex, then no matter what choice of convex g is deemed appropriate we would prefer $\bar{X}_{n+1}$ to $\bar{X}_n$.

Under what circumstances is the sample median less unequal in the Lorenz order sense than the sample mean? An answer to this question would be of interest in that it would identify the forms of the common distribution of the X_i's for which the median would be generally preferred to the mean for estimation of the center of the distribution. For example, in the case in which the common distribution of the X_i's is uniform $(0, \sigma)$, one may verify by a density crossing argument (Theorem 9.1.4) that

$$\bar{X}_3 \leq_L X_{2:3} \tag{11.15}$$

Equation (11.15) does not always hold. For example, if the common distribution of the X_i's is discrete with

$$P(X=0) = P(X=2) = \frac{1}{7}, \quad P(X=1) = \frac{5}{7}, \tag{11.16}$$

then $\bar{X}_3$ and $X_{2:3}$ are not Lorenz comparable. Note that (11.16) defines a symmetric distribution, so symmetry alone is not enough to guarantee (11.15).

Intuitively, (11.15) should hold for light tailed distributions.

See Arnold and Villaseñor (1986) for further discussion.

11.4 Reliability

Consider a complex system involving n independent components. Each component either functions or does not. Denote by p_i the reliability of component i, i.e., the probability that the i'th component functions, $i = 1, 2, \ldots, n$. The probability that the system functions will be some function of $\underline{p}$ say $h(\underline{p})$, called the reliability of the system. For example, a series system which functions only if all components function will have as its reliability function $h(\underline{p}) = \prod_{i=1}^{n} p_i$. A k out of n system functions if any k of its n components are functioning. For such a system we have

$$h(p) = P\left(\max_{\pi} \sum_{j=1}^{k} X_{\pi(j)} = k\right) \tag{11.17}$$

where the summation is over all permutations of length k of the integers $1, 2, \ldots, n$ and X_i is an indicator random variable of the event that component i is functioning $(i = 1, 2, \ldots, n)$. The expression on the right of (11.17) clearly is a function of $\underline{p}$ only, albeit a complicated function.

Suppose that we have limited resources and must use them to construct components. For a given system, should we try to construct all components to have equal reliabilities, or are there key components on which we should concentrate our resources? If $h(\underline{p})$ is a symmetric function of $\underline{p}$ (as in the series, parallel and k out of n systems), intuition suggests that equal values of the p_i's would be reasonable. A majorization result may be sought. Roughly, we might expect that increased inequality among the component reliabilities, the p_i's, will decrease overall system reliability. Or, maybe, it is the other way round; it depends on your intuition. Can we prove such a result? It turns out that a majorization result can be proved. But the appropriate parameterization does not involve p_i's, rather we should let

$$\lambda_i = -\log p_i, \quad i = 1, 2, \ldots, n. \tag{11.18}$$

Pledger and Proschan (1971) show that, indeed, in any k out of n system ($k = 1, 2, \ldots, n$), the system reliability is a Schur convex function of $\underline{\lambda}$.

To see that this is true, we argue as follows. Consider two vectors $(\lambda_1, \lambda_2, \ldots, \lambda_n)$ and $(\lambda_1', \lambda_2', \ldots, \lambda_n')$ which differ only in their first two coordinates, and without loss of generality assume $\lambda_1 < \lambda_2$, $\lambda_1' = \lambda_1 + \epsilon$, $\lambda_2' = \lambda_2 - \epsilon$ where ϵ is small. We must show $h(\underline{\lambda}) \geq h(\underline{\lambda}')$. Let δ_k denote the probability that at least k of components 3 through n are functioning, and let $\delta_{k-1}, \delta_{k-2}$ denote the probability that exactly $k-1$ (respectively $k-2$) of components 3 though n are functioning. Conditioning on the number of components functioning among components 3 through n, we find

$$\begin{aligned} h(\underline{\lambda}) - h(\underline{\lambda}') = \delta_k(1-1) & \\ &+\delta_{k-1}[(p_1 + p_2 - p_1 p_2) - (p_1' + p_2' - p_1' p_2')] \\ &+\delta_{k-2}(p_1 p_2 - p_1' p_2') \end{aligned}$$

where $\lambda_i = -\log p_i$ and $\lambda_i' = -\log p_i'$. Since $\lambda_1 + \lambda_2 = \lambda_1' + \lambda_2'$, we have $p_1 p_2 = p_1' p_2'$. So we can write

$$\begin{aligned} h(\underline{\lambda}) - h(\underline{\lambda}') &= \delta_{k-1}[p_1 + p_2 - p_1' - p_2'] \\ &= \delta_{k-1}[e^{-\lambda_1} + e^{-\lambda_2} - e^{-(\lambda_1+\epsilon)} - e^{-(\lambda_2-\epsilon)}] \end{aligned}$$

which is positive, since e^{-x} is a decreasing convex function.

Note that in order to compare λ and λ' in the sense of majorization, we must have $\sum_{i=1}^n \lambda_i = \sum_{i=1}^n \lambda_i'$ or equivalently $\prod_{i=1}^n p_i = \prod_{i=1}^n p_i'$ (where $p_i' = e^{-\lambda_i'}$). Thus, the Pledger–Proschan result tells us about orderings of system reliabilities of k out of n systems as functions of individual reliabilities subject to the constraint that the reliability of a series system constructed with the components is some fixed quantity.

11.5 Genetic Selection

Suppose that we are in the cattle breeding business, and we wish to develop a strain of cattle with a high beef yield. A reasonable approach to this problem involves culling in each generation and saving for further breeding purposes only the most meaty cattle. It is frequently deemed appropriate to model such a situation with a random effects linear model. Thus, in a particular generation we will have, say, m families of cattle of sizes $k_1, k_2, \ldots, k_m$. The meat yield of the i'th member of the j'th family is represented by

$$X_{ij} = \mu + A_i + E_{ij}, \quad i = 1, 2, \ldots, m; \quad j = 1, 2, \ldots, k_i \tag{11.19}$$

where the A_i's are assumed to be i.i.d. normal $(0, \sigma_A^2)$ random variables, and the E_{ij} are i.i.d. normal $(0, \sigma^2)$ random variables. This is a classical intraclass correlation model. The culling scheme involves retaining the m' animals corresponding to the m' largest observed values of the X_{ij}'s. See Rawlings (1976), Hill (1976, 1977), and Tong (1982) for a more precise and detailed description of the phenomena in question. Rawlings and Hill assumed $k_i = k, i = 1, 2, \ldots, m$. We are here focussing on Tong's results which are concerned with the effect of heterogeneity of the k_i's. Tong restricts attention to the largest X_{ij} which we here denote by Z. He first observes that for every z, the quantity $P(Z \leq z)$ is a monotone increasing function of σ_A^2/σ^2. This is intuitively plausible. If we think of the extreme case when σ_A^2 is very large, then we are really only effectively dealing with the maximum of m rather than $\sum_{i=1}^m k_i = n$ variables, and Z will tend to be smaller.

The other result obtained by Tong is that increasing the "variability" of the vector of family sizes also tends to make Z stochastically smaller. In this setting we measure "variability" in the sense of majorization. Specifically, consider two data sets of the form (11.19) denoted by $\{X_{ij}\}$ and $\{X'_{ij}\}$. Let the corresponding vectors of family sizes be $\underline{k} = (k_1, \ldots, k_m)$ and $\underline{k}' = (k'_1, \ldots, k'_m)$, respectively. Assume that the total number of observations is the same in both data sets, i.e., $\sum_{i=1}^m k_i = \sum_{i=1}^m k'_i = n$ say. Denote the corresponding maxima by Z and Z'. Tong's result may then be stated in the form:

Lemma 11.5.1 *If $\underline{k} \leq_M \underline{k}'$, then for every z, $P(Z \leq z) \leq P(Z' \leq z)$.*

Proof Referring back to the model (11.19), we may compute for any z,

$$P(X_{ij} \leq z | A_i = a) = \phi_Z(a) \quad \text{say}$$

Also note that, given $A_1, A_2, \ldots, A_m$, the X_{ij}'s are conditionally independent. It follows that

$$\begin{aligned} P(Z \leq z) &= P(X_{ij} \leq z \;\forall\, i, j) \\ &= E(P(X_{ij} \leq z \;\forall\, i, j | A_1, A_2, \ldots, A_m)) \\ &= E\left\{\prod_{i=1}^m [\phi_Z(A_i)]^{k_i}\right\} \\ &= E\left\{\sum_\pi \prod_{i=1}^m [\phi_Z(A_{\pi(i)})]^{k_i}\right\} / m! \end{aligned} \tag{11.20}$$

where the summation is over all permutations of the integers $1, 2, \ldots, m$. Analogously,

$$P(Z' \le z) = E\left\{\sum_{\pi}\prod_{i=1}^{m}[\phi_Z(A_{\pi(i)})]^{k'_i}\right\}/m! \tag{11.21}$$

If $\underline{k} \le_M \underline{k}'$, then the expressions inside the expectations in (11.20) and (11.21) are ordered by Muirhead's theorem (Theorem 2.2.7). The result then follows. ■

In the above Lemma it is evidently sufficient to have the joint distribution of $(A_1, A_2, \ldots, A_m)$ be exchangeable, the A_i's do not have to be independent. It is also evident that normality plays no role in the result.

An interesting open question is whether this majorization result can be extended to cover the case of the largest k of the X_{ij}'s rather than just the largest one. Other possible extensions are mentioned by Tong.

11.6 Large Interactions

Bechhofer et al. (1977) encountered a majorization relationship in the study of $2 \times c$ two-way analysis of variance. They use the following constrained maximization result provided by Kemperman (1973).

Lemma 11.6.1 *Let $\underline{x} \in \mathbb{R}^n$ satisfy $\ell \le x_i \le u$, $i = 1, 2, \ldots, n$. Then*

$$\underline{x} \le_M (\ell, \ell, \ell, \ldots, \ell, \tau, u, u, \ldots, u) = \underline{y}$$

where there are k ℓ's and $n-k-1$ u's and k and $\tau \in [\ell, u]$ are uniquely determined by the requirement that $\sum_{i=1}^n x_i = \sum_{i=1}^n y_i$.

Proof Since $\ell \le x_i \le u$, it follows that, for $j \le k$, $\sum_{i=1}^j x_{i:n} \ge \sum_{i=1}^j y_{i:n}$ and for $j > k$, that $\sum_{i=j+1}^n x_{i:n} \le \sum_{i=j+1}^n y_{i:n}$. Thus, for all $j \le n$, $\sum_{i=1}^j x_{i:n} \ge \sum_{i=1}^j y_{i:n}$, i.e. $\underline{x} \le_M \underline{y}$. ■

Now consider a two-way fixed effects analysis of variance situation with 2 rows and J columns with K observations per cell. Thus, our data is of the form

$$X_{ijk} = \mu + \alpha_i + \beta_j + \gamma_{ij} + E_{ij},\ i = 1, 2;\ j = 1, 2, \ldots, J;\ k = 1, 2, \ldots, K$$

where the E_{ijk}'s are i.i.d. $N(0, \sigma^2)$ and σ^2 is assumed known. As usual, we assume our parameters satisfy certain constraints:

$$0 = \sum_{i=1}^{2}\alpha_i = \sum_{j=1}^{J}\beta_j = \sum_{i=1}^{2}\gamma_{ij} = \sum_{j=1}^{J}\gamma_{ij}. \tag{11.22}$$

Suppose that we are interested in identifying the cell whose interaction parameter γ_{ij} is largest. We wish to derive a selection procedure which will identify the largest interaction cell with high probability when there is some discernible interaction and when one cell interaction is discernibly larger than the others. Specifically, we want a rule that will select the cell corresponding to the largest interaction parameter which without loss of generality will be the $(1, 1)$ cell with probability at least P^* when $\gamma_{11} \geq \Delta^*$ and $\gamma_{ij} < \Delta^* - \delta^*$, $(i, j) \neq (1, 1)$, where P^*, Δ^* and δ^* are specified in advance by the experimenter. This is a typical formulation of a ranking and selection problem (see, e.g., Bechhofer et al. 1968).

A plausible decision rule is one which selects as the cell with largest interaction that cell whose observed interaction, namely,

$$Z_{ij} = X_{ij.} - X_{i..} - X_{.j.} + X_{...} \tag{11.23}$$

is largest. As usual, in (11.23) the dots indicate averaging over missing subscripts. For a configuration of interaction parameters $\underline{\gamma}$ satisfying

$$\gamma_{11} \geq \Delta^*$$

and

$$\gamma_{ij} \leq \Delta^* - \delta^*, \quad (i, j) \neq (1, 1), \tag{11.24}$$

the probability of correct selection using the above described rule is simply

$$P_{\underline{\gamma}}(Z_{1,1} > Z_{i,j} \ \forall\ (i, j) \neq (1, 1)). \tag{11.25}$$

We want to determine the sample size necessary to assure us that the probability of correct selection given by (11.25) is at least equal to P^*. It is evident that we can easily achieve this goal if K is enormous. What is the smallest value of K which will suffice? Let Γ denote the set of all interaction arrays $\underline{\gamma}$ which satisfy (11.24). Is there a least favorable $\underline{\gamma}$ in Γ (i.e., one for which the probability of correct selection (11.25) is smallest)? There is, and it can be identified by a majorization argument. Observe that an array will belong to Γ provided that

$$\gamma_{11} \geq \Delta^*$$

and

$$|\gamma_{ij}| \leq \gamma_{11} - \delta^*, \quad j = 2, \ldots, J. \tag{11.26}$$

Also note that $\sum_{j=2}^{J} \gamma_{1j} = -\gamma_{11}$. Expression (11.26) describes a coordinatewise bounded collection of vectors of dimension $(J - 1)$ to which Kemperman's result (Lemma 11.6.1) can be applied. The vector $\tilde{\gamma} = (\gamma_{12}, \gamma_{13}, \ldots, \gamma_{1J})$ is majorized by the vector

$$\underline{\tilde{\gamma}}^* = (\ell, \ell, \ldots, \ell, \tau, u, u, \ldots, u)$$

described in Lemma 11.6.1, with

$$\ell = -(\gamma_{11} - \delta^*),$$
$$u = (\gamma_{11} - \delta^*)$$

and k and τ determined such that $\sum_{j=2}^{J} \gamma_{1j} = k\ell + \tau + (n-k-1)u$. The vector $\underline{\tilde{\gamma}}^*$ can be used in conjunction with the constraints (11.22) to determine an array $\underline{\tilde{\gamma}}^*$ in Γ (with $\gamma_{11}^* = \gamma_{11}$). A non-trivial application of Theorem 2.2.4 yields the result that the probability of correct selection for a fixed value of γ_{11} is a Schur concave function of $\underline{\tilde{\gamma}} = (\gamma_{12}, \ldots, \gamma_{1J})$. It follows then that

$$P_{\underline{\gamma}}(Z_{11} > Z_{ij} \;\; \forall\, (i,j) \neq (1,1))$$
$$\geq P_{\underline{\tilde{\gamma}}^*}(Z_{11} > Z_{ij} \;\; \forall\, (i,j) \neq (1,1)).$$

It remains only to verify that the probability of correct selection is a monotone increasing function of γ_{11}, and then we conclude that the least favorable configuration is of the form $\underline{\tilde{\gamma}}^*$ with $\gamma_{11}^* = \Delta^*$. In principle, we can then determine by numerical integration the minimal value of K necessary to achieve the desired level P^* for the probability of correct selection for any $\underline{\gamma}$ in Γ.

11.7 Unbiased Tests

Suppose that we have a data set $\underline{X}$ whose distribution depends on a k-dimensional parameter $(\theta_1, \theta_2, \ldots, \theta_k)$. Not infrequently we are interested in testing the hypothesis of homogeneity, i.e., $H : \theta_1 = \theta_2 = \cdots = \theta_k$. The test will often be of the form: reject H if $\phi(\underline{X}) \geq c$. Consequently, the power function will be

$$\beta(\underline{\theta}) = P_{\underline{\theta}}(\phi(\underline{X}) \geq c).$$

It is often the case that ϕ is a Schur convex function of $\underline{x}$, and consequently, that $\beta(\underline{\theta})$ is a Schur convex function of $\underline{\theta}$. This information may permit us to conclude that our test is unbiased. A specific case in which this program works perfectly is that in which $\underline{X}$ has a multinomial $(N, \underline{\theta})$ distribution and the hypothesis to be tested is $H : \theta_1 = \theta_2 = \cdots = \theta_k = 1/k$. In this setting the likelihood ratio test is of the form: reject H if $\sum_{i=1}^{k} X_i^2 \geq c$. This is clearly a Schur convex function of $\underline{X}$ and the test is verified to be unbiased (cf. Perlman and Rinott 1977). The program is also feasible to verify unbiasedness of the likelihood ratio test of sphericity of a multivariate normal population and certain invariant tests of equality of mean vectors in multivariate analysis of variance.

11.8 Summation Modulo *m*

Suppose X and Y are independent random variables each with possible values $0, 1, 2, \ldots, m-1$. Define $Z = X \oplus Y$ where the symbol $\oplus$ denotes addition modulo m. Define vectors $\underline{p}$, $\underline{q}$ and $\underline{r}$ by

$$\begin{aligned} p_i &= P(X = i), \\ q_i &= P(Y = i), \\ r_i &= P(Z = i), \end{aligned}$$

where $i = 0, 1, 2, \ldots, m - 1$.

Lemma 11.8.1 *$\underline{r} \leq_M \underline{p}$ and $\underline{r} \leq_M \underline{q}$.*

Proof By conditioning we can verify that

$$\underline{r} = P\underline{q} \tag{11.27}$$

where P is the circulant matrix

$$\begin{pmatrix} p_0 & p_{m-1} & p_{m-2} & \cdots & p_2 & p_1 \\ p_1 & p_0 & p_{m-1} & \cdots & p_3 & p_2 \\ \vdots & & & & & \\ p_{m-1} & p_{m-2} & \cdots & \cdots & p_1 & p_0 \end{pmatrix}$$

Evidently, P is doubly stochastic and, thus, by the HLP Theorem 2.1.1, we conclude $\underline{r} \leq_M \underline{q}$. Analogously, $\underline{r} \leq_M \underline{p}$. ■

Lemma 11.8.1 can be generalized to cover random variables assuming values in an arbitrary finite group (rather than just the group of non-negative integers under summation modulo m). Marshall et al. (2011, pp. 509–510) give the general expression. Brown and Solomon (1976) focused on the case in which the group consisted of vectors of non-negative integers under coordinate wise summation modulo m. The problem is of interest in the context of pseudo-random number generators. If our pseudo random number generator generates integers $0, 1, 2, \ldots, 9$ with non-uniform probabilities, our Lemma tells us we may improve things (i.e., obtain more uniformly distributed random digits) by summing successively generated digits modulo 10. Or we might combine outputs from distinct random number generators. If this is done by addition modulo 10, our Lemma guarantees that the output will be at least as uniformly distributed and, most likely, more uniformly distributed than was the output of either individual generator.

The improvement can be striking. For example, if

$$\underline{p} = (0.07,\ 0.19,\ 0.02,\ 0.11,\ 0.06,\ 0.13,\ 0.01,\ 0.18,\ 0.09,\ 0.14),$$

and

$$\underline{q} = (0.13,\ 0.07,\ 0.02,\ 0.11,\ 0.17,\ 0.13,\ 0.07,\ 0.02,\ 0.11,\ 0.17),$$

then from (11.27) we find

$$\underline{r} = (0.1,\ 0.1,\ 0.1,\ 0.1,\ 0.1,\ 0.1,\ 0.1,\ 0.1,\ 0.1,\ 0.1).$$

Examples such as this are discussed in Arnold (1979) in a related context.

11.9 Forecasting

Every night on TV Channel 5, Bert announces the probability of rain for the next day. Every night on Channel 11, Wilbur gives out his probability of rain. Is Bert a better forecaster than Wilbur? How should we appropriately determine if one is the better forecaster? De Groot and Fienberg have studied this phenomenon in a series of papers. A representative reference addressed to the general reader is De Groot and Fienberg (1983). The concept of majorization plays a central role. In the present section the scenario will only be sketched along the lines suggested by De Groot and Fienberg. A particular forecaster is asked each day to give us his subjective probability say x that there will be a measurable amount of precipitation at a certain location. At the end of each day we can observe whether or not it did rain. Denote by X the random prediction of our forecaster (most easily visualized in a long run frequency sense). Denote by Y an indicator random variable. Y assumes the value 1 if it rains, 0 otherwise. The performance of our forecaster is summarized by the density function $\nu(x)$ of the random variable X and for each possible value x of X, the conditional probability of rain given that the forecaster's prediction is x, denoted by $\rho(x)$.

Thus

$$P(X \in A) = \int_A \nu(x)\,dx \tag{11.28}$$

and

$$P(Y = 1|X = x) = \rho(x). \tag{11.29}$$

The forecaster is said to be well-calibrated if $\rho(x) \equiv x$ (such a forecaster is sometimes said to be perfectly reliable). If he says there is a 50% chance of rain, then indeed there is a 50% chance of rain. Any coherent forecaster who updates his subjective probability based on experience using Bayes theorem will be asymptotically well calibrated (see, e.g., Dawid 1982). It is usually deemed appropriate to restrict attention to well-calibrated forecasters. Clearly, however, there are differences between well-calibrated forecasters. If in a given town rain occurs on about 15% of the days during the year, then a forecaster who every day announces $x = 0.15$ will be well calibrated but somewhat boring and of questionable utility. At the other extreme, a forecaster who always announces either $x = 0$ or $x = 1$ and who is always right is also well-calibrated (and displays amazing skill).

Comparisons between well-calibrated forecasters are made in terms of a partial ordering known as refinement. Consider two well-calibrated forecasters A and B whose forecasts X_A and X_B have corresponding density functions ν_A and ν_B (as in (11.28)). Since both are well-calibrated we have (cf. Eq. (11.28))

$$\rho_A(x) = \rho_B(x) = x, \quad \forall x \in [0, 1]$$

A stochastic transformation is a function $h(y|x)$ defined on $[0, 1] \times [0, 1]$ satisfying

$$\int_0^1 h(y|x)\, dy = 1, \quad \forall x \in [0, 1]. \tag{11.30}$$

Forecaster A is said to be at least as refined as forecaster B if there exists a stochastic transformation $h(y|x)$ for which

$$\nu_B(y) = \int_0^1 h(y|x)\nu_A(x)\, dx, \quad \forall y \in [0, 1] \tag{11.31}$$

and

$$y\nu_B(y) = \int_0^1 h(y|x)x\nu_A(x)\, dx, \quad \forall y \in [0, 1]. \tag{11.32}$$

We denote by μ, the long run frequency of rainy days. For any well-calibrated forecaster we must have

$$\int_0^1 x\nu(x)\, dx = \mu.$$

Any density ν with mean μ can be identified with a well-calibrated forecaster. Thus our refinement partial order may be thought of as being defined on the class of all random variables with range (0,1) and mean μ. In this context we can allow the prediction random variables X to be discrete with obvious modifications in our earlier discussion which assumed absolutely continuous prediction variables (just replace integral signs by summation signs).

Inspection of conditions (11.31) and (11.32), in either the discrete or absolutely continuous case, yields the interpretation that forecaster A (with prediction variable X_A) is at least as refined as B (with prediction variable X_B) if and only if there exist random variables X'_A and X'_B defined on some convenient probability space with

$$X'_A \overset{d}{=} X_A,$$
$$X'_B \overset{d}{=} X_B \tag{11.33}$$

and

$$E(X'_A|X'_B) = X'_B. \tag{11.34}$$

Thus the refinement ordering, on random variables with support [0, 1] and mean μ, is exactly identifiable with Lorenz ordering on that restricted class of random variables (recall Theorem 3.2.3). Since our random variables have identical means it can also be identified, if desired, with the second order stochastic dominance partial order (as defined in Chap. 9).

If (11.33) and (11.34) hold, then we know that

$$E(u(X_A)) \geq E(u(X_B)) \tag{11.35}$$

for any convex function u. In the forecasting setting this has an interpretation in terms of what are known as strictly proper scoring rules.

A scoring rule is a device for comparing forecasters (even it they are not well-calibrated). Suppose that the forecaster's prediction is x. If rain does occur, he receives a score of $g_1(x)$, if it does not rain he receives a score of $g_2(x)$. The expected score is, using (11.28) and (11.29), thus

$$S = E[\rho(X)g_1(X) + [1 - \rho(X)]g_2(X)] \tag{11.36}$$

It is reasonable to assume that g_1 is an increasing function and g_2 is a decreasing function. A scoring rule is said to be strictly proper if for each value of y in the unit interval the function

$$yg_1(x) + (1 - y)g_2(x) \tag{11.37}$$

is maximized only when $x = y$. Such a rule encourages the predictor to announce his true subjective probability of rain. It is not difficult to verify that if g_1 is increasing, g_2 is decreasing and the scoring rule is strictly proper then the function

$$xg_1(x) + (1 - x)g_2(x) \tag{11.38}$$

is strictly convex. Referring to (11.36), if we consider two well-calibrated forecasters A and B (for whom $\rho_A(x) = \rho_B(x) = x$) where A is at least as refined as B, then using any proper scoring rule forecaster A will receive a higher score than forecaster B (unless $X_A \overset{d}{=} X_B$ in which case they clearly receive identical scores). Subsequently De Groot and Fienberg (1986) extended these concepts to allow for comparison of "multivariate" forecasters. Thus instead of predicting (rain)-(no-rain) the forecaster might predict a finite set of temperature ranges. The prediction random variable X becomes s dimensional (assuming values in the simplex $x_i \geq 0$, $\sum_{i=1}^{s} x_i = 1$). The concepts of refinement, well calibration, and scoring rules continue to be meaningful in this more complex setting.

11.10 Ecological Diversity

Consider a closed ecological community (an island in the Pacific, for example). Following an exhaustive study, representatives of s different species of insects are found on the island. For $i = 1, 2, \ldots, s$ let n_i denote the abundance of species i, the number of insects of that species on the island. For $i = 1, 2, \ldots, s$ define the relative abundance π_i by

$$\pi_i = n_i / \sum_{j=1}^{s} n_j. \tag{11.39}$$

The community can and will be identified with its relative abundance profile $(\pi_1, \ldots, \pi_s)$. Evidently some communities are more diverse than others. What is an acceptable ordering or partial ordering of relative abundance profiles which will capture the spirit of the concept of diversity? Several diversity measures have been proposed in the literature. Patil and Taillie (1982), taking on the ecological role played in the economics context by Dalton, sought basic principles of diversity which might lead to a widely accepted diversity order. They were led in this fashion to what they call the intrinsic diversity ordering. Suppose that $\underline{\pi}^{(1)}$ and $\underline{\pi}^{(2)}$ are species abundance profiles of two communities involving (without loss of generality) the same s species in both communities. They propose that it is appropriate to determine whether $\underline{\pi}^{(1)}$ is more diverse than $\underline{\pi}^{(2)}$ by considering the corresponding intrinsic diversity profiles.

The diversity of an abundance profile $\underline{\pi}$ is deemed to be a function of the ordered components of $\underline{\pi}$, viz. $\pi_{1:s}, \pi_{2:s}, \ldots, \pi_{s:s}$. Thus a population with 25% butterflies, 60% ants, and 15% beetles is deemed to be equally diverse as one containing 60% butterflies, 15% ants, and 25% beetles. The intrinsic diversity profile (IDP) is actually defined in terms of the decreasing order statistics of $\underline{\pi}$, i.e. $\pi_{(1:s)} \geq \pi_{(2:s)} \geq \cdots \geq \pi_{(s:s)}$. It is a plot of the points $\{(\sum_{j=1}^{i} \pi_{(j:s)}, i/s)\}$. The curve is completed by linear interpolation. If all the species are equally represented the IDP will be a 45° line. This is taken to be the case of maximum diversity. If one species is extremely numerous [i.e., $\pi_{(1:s)} \approx 1$], the curve will be very high and diversity is judged to be small.

Community $\underline{\pi}^{(1)}$ will be judged to be intrinsically more diverse than community $\underline{\pi}^{(2)}$ if the IDP of $\underline{\pi}^{(1)}$ is uniformly below the IDP of $\underline{\pi}^{(2)}$ and we may write

$$\underline{\pi}^{(1)} \geq_{IMD} \underline{\pi}^{(2)}$$

It is interesting to view how this intrinsic diversity ordering ranks uniform distributions over different numbers of species. Suppose $\underline{\pi}_1$ has K_1 non-zero components all equal to $1/K_1$ while $\underline{\pi}_2$ has K_2 non-zero components, all equal to $1/K_2$. If $K_1 > K_2$, then $\underline{\pi}_1$ surely exhibits more diversity than $\underline{\pi}_2$. Inspection of the corresponding IDP's confirms that in this case $\underline{\pi_1} \geq_{IMD} \underline{\pi_2}$.

Referring to the definition of an intrinsic diversity profile and the definition of majorization (in $\mathbb{R}^s$) we see that they are intimately related. In fact

$$\underline{\pi}^{(1)} \geq_{IMD} \underline{\pi}^{(2)} \Leftrightarrow \underline{\pi}^{(1)} \leq_M \underline{\pi}^{(2)}. \tag{11.40}$$

In the light of (11.40), summary measures of diversity should be Schur concave functions of $\underline{\pi}$ in order to respect the intrinsic diversity ordering. Typical examples of such measures are:

- Shannon's measure

$$D_1(\underline{\pi}) = -\sum_{i=1}^{s} \pi_i \log \pi_i, \tag{11.41}$$

and

- Simpson's measure

$$D_2(\underline{\pi}) = 1 - \sum_{j=1}^{s} \pi_i^2 \tag{11.42}$$

Many diversity measures are interpretable as coefficients of expected rarity. Imagine that each species has a measure of rarity attached to it. The simplest case occurs when there exists a rarity function $R(\pi)$ defined on [0, 1] such that the rarity of an individual belonging to a species with relative abundance π in the community is equal to $R(\pi)$. Imagine we randomly select an individual from the community. The expected rarity of the individual so chosen is then given by

$$(\text{ER})(\underline{\pi}) = \sum_{i=1}^{s} \pi_i R(\pi_i) \tag{11.43}$$

Such a measure will be Schur concave, and hence will respect the intrinsic diversity order, if R is concave. Both the Shannon (11.41) and the Simpson (11.42) indices are of the form (11.43) for suitable concave functions $R(\pi)$.

Patil and Taillie (1982) suggest several alternative diversity profiles. They also present interpretations of many diversity indices in terms of random encounters (inter-species and intraspecies).

11.11 Covering a Circle

Suppose that n arcs of lengths $\ell_1, \ell_2, \ldots, \ell_n$ are placed independently and uniformly on the unit circle (a circle with unit circumference). With $\underline{\ell} = (\ell_1, \ldots, \ell_n)$, let $P(\underline{\ell})$ denote the probability that the unit circle is completely covered by these arcs. To make the problem interesting assume that the total length of the arcs, $L = \sum_{i=1}^{n} \ell_i$, exceeds 1, otherwise coverage would be impossible, and that none of the arcs has length greater than or equal to 1, otherwise coverage would be certain. Stevens (1939) gave the following explicit expression for this coverage probability when all arcs are of equal length, i.e., when each arc is of length $\overline{\ell} = L/n$:

$$P(\overline{\ell}, \ldots, \overline{\ell}) = \sum_{k=0}^{n} (-1)^k \tbinom{n}{k} [(1 - k\overline{\ell})^+]^{n-1}. \tag{11.44}$$

If one arc is of length L assumed to be greater than 1, and all other arcs are of length 0, then coverage would be certain. This suggests that increasing the variability among the ℓ_i's, subject to the requirement that the sum of the lengths is L, might be expected to increase the coverage probability. Thinking along these lines, Proschan conjectured that $P(\underline{\ell})$ is a Schur convex function of $\underline{\ell}$. Huffer and Shepp (1987) show that (11.44) represents an extremal case by verifying the conjecture that $P(\ell)$ is a Schur-convex function. Thus, for a given total sum of arc lengths, increased inequality among the lengths of the arcs yields a greater coverage probability. It suffices to consider the effect on $P(\underline{\ell})$ of making a small change in two unequal ℓ_i's (to make them more alike), holding the other ℓ_i's fixed. The result turns out however to be more troublesome to verify than might have been hoped. See Huffer and Shepp (1987) for details.

11.12 Waiting for a Pattern

Imagine that a monkey seated in front of a keyboard types a sequence of letters, spaces, and punctuation marks at random. We come into the room and are amazed to see that he has just typed "To be or not to be, that is the question." Will he do it again, and if so, how long will we have to wait to see the second performance. A long time, that's for sure. We can formulate this problem mathematically as follows.

Suppose that $X_1, X_2, \ldots$ is a sequence of independent, identically distributed random variables with possible values $1, 2, \ldots, k$ and associated positive probabilities $p_1, p_2, \ldots, p_k$. Let N denote the waiting time until a particular string of consecutive values is observed, or until one of a collection of particular strings is observed. If we are waiting for the string $t_1, t_2, \ldots, t_\ell$ where each t_i is number chosen from the set $1, 2, \ldots, k,$, variability can affect the distribution of N in several ways.

The distribution of the random variable N will be affected by variability among the p_i's, the probabilities of the possible values of the X_i's. For example, an extreme case in which $p_1 \approx 1$ will result in long waits for strings $\underline{t}$ which include entries which are not all equal to 1. The waiting time also, in general, will be affected by variability among the t_j's appearing in the string for whose appearance we are waiting. For example, we might expect to wait longer for the string 1, 1, 1, 1, 1 than for the string 2, 2, 1, 1, 1. There are several possible aspects of this problem that might be amenable to an analysis involving majorization.

In particular, Ross (1999) considers the waiting time until a string of k outcomes occurs that includes all of the k possible outcomes (in any order), i.e.,

$$N = \min\{n \geq k : X_{n-k+1}, X_{n-k+1}, \ldots, X_n \quad \text{are all distinct}\}.$$

As a function of $\underline{p}$, he shows that, for every n, $P_{\underline{p}}(N > n)$ is a Schur convex function of $\underline{p}$. From this, it follows that $E_{\underline{p}}(N)$ is also a Schur convex function of $\underline{p}$, with the shortest waiting time corresponding to the case in which the p_j's are all equal to $1/k$.

The reader might wish to consider whether the waiting time to observe a string of k outcomes $t_1, t_2, \ldots, t_k$ (in no particular order) is affected by the intrinsic diversity profile of the string (as defined in Sect. 11.10).

11.13 Paired Comparisons

Paired comparison experiments involving food preferences have a long history. Blind tests of the preference for, say, Coca Cola when compared with other soft drinks are familiar to all, as are recent tests comparing turkey hamburgers to traditional beef hamburgers. Paired comparisons also occur in sporting events and it is on this arena that we will focus attention, while recognizing that the analysis will also be meaningful in taste testing and other product comparison settings. In our discussion, we will rely heavily on Joe (1988) for both content and notation.

In a league consisting of k teams, during a season each team will typically play each other team a total of m times. To simplify the presentation, we ignore home field advantage and we assume that ties do not occur (typically the league will have specific rules for breaking any ties that do occur).

To model the outcome of a typical season, consider a $k \times k$ matrix $P = (p_{ij})$ in which, for $i \neq j$, p_{ij} denotes the probability that team i beats team j in a particular game. Since we assume that ties do not occur, we have $p_{ij} + p_{ji} = 1$. The diagonal elements of P are left empty, so P has $k(k-1)$ non-negative (usually positive) elements. For each i, define $p_i = \sum_{j \neq i} p_{ij}$. This row total can be viewed as providing a measure of the strength of team i. For a given vector of team strengths $\underline{p} = (p_1, \ldots, p_k)$ we define $\mathscr{P}(\underline{p})$ to be the class of all probability matrices P(with only off-diagonal elements defined) and with row totals given by $\underline{p}$.

Joe (1988) defines a variability ordering on the members of the class $\mathscr{P}(\underline{p})$ based on majorization. For $P, Q \in \mathscr{P}(\underline{p})$, the matrix P is majorized by the matrix Q ($P \leq_M Q$) if and only if $P^* \leq_M Q^*$ in the usual sense of majorization, where P^* (respectively Q^*) is the $k(k-1)$-dimensional vector whose entries are all the defined elements of P (respectively Q).

Joe (1988) describes the following application of this variability ordering on $\mathscr{P}(\underline{p})$. We will say that team i is better than team j if $p_{ij} > 0.5$. It is reasonable to assume that if team i is better than team j and if team j is better than team ℓ, then it should be the case that team i is better than team ℓ.

P is said to be weakly transitive if $p_{ij} \geq 0.5$ and $p_{j\ell} \geq 0.5$ imply $p_{i\ell} \geq 0.5$ and is said to be strongly transitive if $p_{ij} \geq 0.5$ and $p_{j\ell} \geq 0.5$ imply $p_{i\ell} \geq \max\{p_{ij}, p_{j\ell}\}$. The link with the variability ordering on $\mathscr{P}(\underline{p})$ is as follows. A matrix $P \in \mathscr{P}(\underline{p})$ is said to be minimal if $Q \leq_L P$ implies $Q^* = P^*$ up to rearrangement. Joe (1988) shows that any strongly transitive P is minimal.

See Joe (1988) for discussion in which ties are permitted and home field advantage is considered.

11.14 Phase Type Distributions

Consider a continuous-time Markov chain with $n+1$ states, in which states $1, 2, \ldots, n$ are transient and state $n+1$ is absorbing. The time, T, until absorption in state $(n+1)$ is said to have a *phase-type distribution* (Neuts 1975). This distribution is determined by an initial distribution over the transient states denoted by $\underline{\alpha} = (\alpha_1, \alpha_2, \ldots, \alpha_n)$ (assume that the chain has probability 0 of being initially in the absorbing state). The intensity matrix Q for transitions among the transient states has elements satisfying $q_{ii} < 0$ and $q_{ij} \geq 0$ for $j \neq i$. In this setting, the time, T, to absorption in state $(n+1)$ is said to have a phase-type distribution with parameters $\underline{\alpha}$ and Q. A particularly simple case is one in which $\underline{\alpha} = \underline{\alpha}^* = (1, 0, \ldots, 0)$ and $Q = Q^*$, where $q^*_{ii} = -\delta$ for each i and $q^*_{ij} = \delta$ for $j = i+1$, $q^*_{ij} = 0$ otherwise. In this case, the Markov chain begins in state 1 with probability 1, and spends an exponential time with mean $1/\delta$ in each state before moving to the next state. The corresponding time to absorption, T^*, is thus a sum of n independent and identically distributed random variables having an exponential distribution with intensity parameter δ; thus, T^* has a gamma distribution with scale parameter $1/\delta$ and shape (convolution) parameter n.

There are multiple possible representations of phase-type distributions. The same distribution of time to absorption can be associated with more than one choice of n, α, and Q.

A phase-type distribution is said to be of order n if n is the smallest integer such that the distribution can be identified as an absorption time for a chain with n transient states and one absorbing state. In some sense the variable T^* clearly exhibits the least variability among phase-type distributions of order n. Aldous and Shepp (1987) show that the T^* having a gamma distribution with parameters $1/\delta$

and n has the smallest coefficient of variation among phase-type distributions of order n. In fact, O'Cinneide (1991) shows that T^* exhibits the least variability in a more fundamental sense. For any phase-type variable T of order n, $T^* \leq_L T$.

11.15 Gaussian Correlation

Suppose that the random vector $\underline{Z}$ has a normal distribution with mean zero and identity covariance matrix, and that A and B are symmetric convex sets in $\mathscr{R}^n$. It was conjectured that in such a case,

$$P\left(\underline{Z} \in A \cap B\right) \geq P\left(\underline{Z} \in A\right) P\left(\underline{Z} \in B\right).$$

The conjecture was verified in the case $n = 2$ by Pitt (1977). Vitale (1999) verified that, in n dimensions, the conjecture is true when A and B are what are called *Schur cylinders*. A set C in $\mathscr{R}^n$ is a Schur cylinder if $\underline{x} \in C$ implies $\underline{x} + k\underline{1} \in C$ for every $k \in \mathscr{R}$ and the indicator function of C is Schur-concave.

Chapter 12
More Applications

Here we provide brief coverage of a further selection of examples in which majorization and/or the Lorenz order makes an appearance.

12.1 Catchability

Assume that an island community contains an unknown number ν of butterfly species. Butterflies are trapped sequentially until n individuals have been captured. Let r denote the number of distinct species represented among the n butterflies that have been trapped. On the basis of r and n, we wish to estimate ν. A popular model for this scenario involves the assumption that butterflies from species $j, \quad j = 1, 2, \ldots, \nu$, enter the trap according to a Poisson (λ_j) process and that these Poisson processes are independent. If $p_j = \lambda_j / \sum_{i=1}^{\nu} \lambda_i$, then p_j denotes the probability that a particular trapped butterfly is of species j. The p_j's reflect the relative catchabilities of the various species. Under the assumption of equal catchability ($p_j = 1/\nu, \quad j = 1, 2, \ldots, \nu$), and under the (somewhat restrictive) assumption that $\nu \leq n$, there exists a minimum variance unbiased estimate $\widetilde{\nu}$ of ν based on r, n, namely,

$$\widetilde{\nu} = S(n + 1, r)/S(n, r), \tag{12.1}$$

where $S(n, x)$ denotes a Stirling number of the second kind [see, e.g., Abramowitz and Stegun (1972, p. 835)]. How will this estimate fare in the presence of unequal catchability? It can fare badly. For example, if one species is "trap-happy," i.e., easy to trap, while all the other species are very difficult to trap, then we will often observe $r = 1$ and we will usually seriously underestimate ν. Nayak and Christman (1992) show that this phenomenon is widespread. They investigate the effect of unequal catchability on the performance of the estimate (12.1). They observe that

B. C. Arnold, J. M. Sarabia, *Majorization and the Lorenz Order with Applications in Applied Mathematics and Economics*, Statistics for Social and Behavioral Sciences,
https://doi.org/10.1007/978-3-319-93773-1_12

the random number, R, of species captured has a distribution function that is a Schur-convex function of $\underline{p}$ and conclude that the estimate (12.1) is negatively biased in the presence of any unequal catchability.

12.2 Server Assignment Policies in Queueing Networks

Consider a queueing setup with M stations. Assume that all stations are single-server stations with service rate μ_i for station i, $i = 1, 2, \ldots, M$. Let λ_i denote the total arrival rate from external and internal sources to station i. Let $\rho_i = \lambda_i/\mu_i$ and assume that $\rho_i < 1$ for every i. ρ_i is referred to as the loading of station i. Yao (1987) discusses optimal assignment of servers to stations in this context. Using majorization and arrangement orderings, he shows that a better loading policy is one which distributes the total work more uniformly to the stations, i.e., one that makes the ρ_i's more uniform, and that better server assignment policies are those which, not surprisingly, assign faster servers to busier stations.

12.3 Disease Transmission

Eisenberg (1991) introduced a simple model of disease transmission that was further investigated by Lefévre (1994). In a later contribution, Tong (1997) identified an aspect of the model involving majorization. The Eisenberg model may be described as follows. Consider a closed population of $n + 1$ individuals. One individual (individual number $n + 1$) is susceptible to the disease but as yet is uninfected. The other n individuals, numbered $1, 2, \ldots, n$, are carriers of the disease. Let p_i denote the probability of avoiding infection after a single contact with individual i, $i = 1, 2, \ldots, n$. It is assumed that individual $n + 1$ makes a total of J contacts with the other individuals in the population governed by a preference vector $\underline{\alpha} = (\alpha_1, \alpha_2, \ldots, \alpha_n)$, where $\alpha_i > 0$ and $\sum_{i=1}^{n} \alpha_i = 1$. Individual $n + 1$ also has an associated lifestyle vector $\underline{k} = (k_1, k_2, \ldots, k_J)$ where the k_i's are non-negative integers summing to J. For given vectors $\underline{\alpha}$ and $\underline{k}$, individual $n+1$ selects a "partner" among the n carriers according to the preference distribution $\underline{\alpha}$. Thus individual 1 is selected with probability α_1, individual 2 is selected with probability α_2, etc. Individual $n + 1$ then makes k_1 contacts with this partner. He or she then selects a second partner (which could be the same one as previously selected) independently again according to the preference distribution $\underline{\alpha}$ and has k_2 contacts with this partner. This continues until $J = \sum_{\ell=1}^{J} k_\ell$ contacts have been made.

The probability of escaping infection under this model is denoted by $H(\underline{k}, \underline{\alpha}, \underline{p})$, to indicate its dependence on the preference vector $\underline{\alpha} = (\alpha_1, \ldots, \alpha_n)$, on the nontransmission probabilities $\underline{p} = (p_1, \ldots, p_n)$, and on the lifestyle vector $\underline{k} = (k_1, \ldots, k_j)$. Majorization could be of interest here in more than one way.

Variability among the coordinates of any or all of $\underline{p}$, $\underline{k}$, and $\underline{\alpha}$ can be expected to influence $H(\underline{k}, \underline{\alpha}, \underline{p})$. Tong, however, only considers the effect of variability among the k_i's for fixed choices of $\underline{p}$ and $\underline{\alpha}$.

Two extreme lifestyles are readily identified, they are associated with the vectors, $(J, 0, \ldots, 0)$ and $(1, 1, \ldots, 1)$. In the first case, which could be called the monogamous lifestyle, the individual $n + 1$ chooses one partner according to the preference vector $\underline{\alpha}$, and stays with that partner for all J contacts. In the second case, which we call the random lifestyle, each contact is made with an individual chosen at random independently according to $\underline{\alpha}$. It is not difficult to verify that the probability of escaping infection in these two cases is given by $\sum_{i=1}^{n} \alpha_i p_i^J$ and $(\sum_{i=1}^{n} \alpha_i p_i)^J$, respectively. It follows from Jensen's inequality that the probability of escaping infection is larger with the monogamous lifestyle $(J, 0, \ldots, 0)$ than it is with random lifestyle $(1, 1, \ldots, 1)$. This result holds for every $\underline{\alpha}$ and every $\underline{p}$. It will be observed that these two particular lifestyle vectors are extreme cases with respect to majorization. It is consequently quite plausible that the probability of escaping infection, $H(\underline{k}, \underline{\alpha}, \underline{p})$, is a Schur-convex function of the lifestyle vector $\underline{k}$, for each $\underline{\alpha}$ and each $\underline{p}$. Tong (1997) confirms this conjecture. Discussion is also provided to cover some cases in which a random number of contacts are made. Note that the value assumed by $H(\underline{k}, \underline{\alpha}, \underline{p})$ when $\underline{\alpha}$ takes an extreme value, either $(1, 0, 0, \ldots, 0)$ or $(1/n, 1/n, \ldots, 1/n)$, may suggest other possible majorization results related to this model.

12.4 Apportionment in Proportional Representation

Proportional representation seeks to assign to each political party a proportion of seats that closely reflects the proportion of votes obtained by that party, in order to approximate the ideal of one man-one vote. Thus if there are N seats available and if a political party received $100q\%$ of the votes, then ideally that party should be assigned Nq seats. But fractional seats are not assigned (though they could be) and some "rounding" rule must be used to arrive at integer valued assignments of seats to the various parties in a manner that will closely approximate proportional representation. Since individual seats in a legislative body are potentially highly influential in subsequent decision making there has historically been considerable discussion of alternative proposals regarding which method of rounding should be used. Several well-known American politicians have proposed rounding methods for use in this situation. For more a detailed general discussions of voting issues, see Saari (1995). Alternatively, Balinski and Young (2001) provide a survey of rounding methods that are used and/or proposed. There are five apportionment schemes that have received considerable attention in the United States, which are named after their well-known proposers: John Quincy Adams, James Dean (no, not the actor), Josef A. Hill, Daniel Webster, and Thomas Jefferson. In the order given, they move from a method (Adams) kinder to small parties to the method (Jefferson) which most favors large parties.

We can describe these apportionment methods in terms of a sequence of sign-posts that govern rounding decisions. The sign-posts $s(k)$ are numbers in the interval $[k, k+1]$ such that $s(k)$ is a strictly increasing function of k, and the associated rounding rule is that a number in the interval $[k, k+1]$ is rounded down if the number is less than $s(k)$ and rounded up if greater than $s(k)$. If the number exactly equals $s(k)$, we are allowed to round up or down. Power-mean sign-post sequences have received much attention. They are of the form

$$s_p(k) = \left(\frac{k^p + (k+1)^p}{2} \right)^{1/p}, \quad -\infty \leq p \leq \infty. \tag{12.2}$$

The five popular apportionment methods named above can all be interpreted as being based on a power-mean sign-post sequence: (Adams) $p = -\infty$, which involves rounding up; (Dean) $p = -1$; (Hill) $p = 0$; (Webster) $p = 1$; (Jefferson) $p = \infty$, which involves rounding down.

Marshall et al. (2002) show that for two sign-post sequences $s(k)$ and $s'(k)$, a sufficient condition to ensure that the seating vector produced by the method using s is always majorized by the seating vector produced by the method using s' is that the sequence of sign-post ratios $s(k)/s'(k)$ is strictly increasing in k. It follows that the result of a power-mean rounding of order p is always majorized by the corresponding power-mean rounding of order p' if and only if $p \leq p'$. Consequently, among the five popular apportionment procedures, a move from Adams toward Jefferson is a move toward favoring large parties in the sense of majorization. In a particular case, the move from an Adams apportionment to a Jefferson apportionment can be achieved by a series of single seat reassignments from a poorer party (with fewer votes) to a relatively richer party (with more votes), i.e., a sequence of "reverse" Robin Hood operations.

12.5 Connected Components in a Random Graph

Ross (1981) considers a random graph with n nodes numbered $1, 2, \ldots, n$. Let $X_1, X_2, \ldots, X_n$ be independent, identically distributed random variables with distributions determined by the probability vector $\underline{p} = (p_1, p_2, \ldots, p_n)$, where

$$P(X_i = j) = p_j, \quad j = 1, 2, \ldots, n. \tag{12.3}$$

A random graph is constructed by drawing n random arcs that connect i to X_i, $i = 1, 2, \ldots, n$, so that one arc emanates from each node. But, of course, several arcs can share the same termination node. Let M denote the number of connected components of this random graph. A connected component is a set of nodes such that any pair of nodes in the graph are connected by an arc, and there are no arcs joining a node in the set to a node outside the set. The distribution of M will be affected by the probability distribution vector $\underline{p}$ appearing in (12.3). For example, if $\underline{p} = (1, 0, \ldots, 0)$, then all arcs will terminate at node 1 and there will be a single connected subset of nodes, so that $M = 1$ with probability 1. In this type of random graph, the expected number of connected components is equal to the expected number of cycles (i.e., closed paths without repetitions) in the graph. Consequently,

$$E(M) = \sum_{S} (|S| - 1)! \prod_{j \in S} p_j, \tag{12.4}$$

where the summation extends over all nonempty subsets of $\{1, 2, \ldots, n\}$ and $|S|$ denotes the cardinality of S. Ross proves that $E(M)$ is a Schur-concave function of $\underline{p}$. Consequently, the expected number of connected components in the random graph is maximized when $p_j = 1/n, \quad j = 1, 2, \ldots, n$.

12.6 A Stochastic Relation Between the Sum and the Maximum of Two Random Variables

Dalal and Fortini (1982), in the context of constructing confidence intervals for a difference between the means of two normal distributions with unknown unequal variances (the so-called Behrens–Fisher setting), derived the following inequality relating the distribution of the sum of two nonnegative random variables to the distribution of the maximum of the two variables. Schur convexity plays a key role in the result.

Theorem 12.6.1 *If X_1 and X_2 are nonnegative random variables with a symmetric joint density $f(x_1, x_2)$, such that $f(\sqrt{x_1}, \sqrt{x_2})$ is a Schur-convex function of $\underline{x}$, then*

$$P(X_1 + X_2 \leq c) \geq P\left(\sqrt{2} \max(X_1, X_2) \leq c\right)$$

for every $c > 0$.

The theorem is proved by conditioning on $X_1^2 + X_2^2$, and using the fact that, on any circle, the density $f(\underline{x})$ increases as $\underline{x}$ moves away from the line $x_1 = x_2$.

An important example of a nonnegative random vector with a joint density f such that $f(\sqrt{x_1}, \sqrt{x_2})$ is Schur-convex is one of the form $(X_1, X_2) = (|Y_1|, |Y_2|)$, where Y_1 and Y_2 are normally distributed with zero means, common variance σ^2, and correlation ρ.

Dalal and Fortini (1982) also derived a related n-dimensional result, as follows.

Theorem 12.6.2 *If $X_1, X_2, \ldots, X_n$ are independent identically distributed positive random variables with a common density function f that satisfies:*

(1) $\log f(\sqrt{x})$ *is concave*
and
(2) $f(x)/x$ *is nonincreasing,*

then

$$\sum_{i=1}^{n} X_i \leq_{\text{st}} \sqrt{n} \max \{X_1, X_2, \ldots, X_n\}.$$

Finally, we mention a result of Dalal and Fortini (1982) that specifically involves majorization.

Theorem 12.6.3 *If $X_1, X_2, \ldots, X_n$ are independent identically distributed non-negative random variables with a common density function f with the property that $\log f(\sqrt{x})$ is concave, and if $\underline{a} \leq_M \underline{b}$, then*

$$\sum_{i=1}^{n} \sqrt{a_i} X_i \leq_{\text{st}} \sum_{i=1}^{n} \sqrt{b_i} X_i.$$

12.7 Segregation

The key reference for this section is Arnold and Gokhale (2014). Segregation is an issue of social and political concern. Efforts to reduce segregation in educational settings continue to be of interest. How should one measure the degree of segregation in a given system at a given time?

We will focus on the degree of ethnic segregation in a given urban school district. The data can be arrayed in a contingency table in which rows correspond to individual schools in the district and columns correspond to ethnic groups. Segregation is judged to be in evidence if certain schools contain a disproportionately low or high number of students from particular ethnic groups. Lack of segregation would be associated with the situation in which the ethnic mix is roughly the same in all schools in the district. The concept of segregation is seen to be one which has to do with the degree of dependence between the rows and columns of a contingency table. There exist a variety of dependency orderings and dependency measures for contingency tables. However, one dependency partial ordering in particular (the Scarsini order) appears to be tailor made for use in segregation studies. Almost all of the segregation measures that have been proposed in the literature respect the Scarsini order. It is intimately related to what we call progressive exchanges, wherein one student of an over-represented group at a particular school is exchanged for a student at a second school who belongs to an ethnic group that is overrepresented at the second school. Such exchanges are quite generally viewed as ones which reduce segregation. Such exchanges leave the row and column totals in the table unchanged. Consequently we initially consider dependence ordering within tables with given marginals. Subsequently attention will be given to problem of comparing dependence (i.e., degree of segregation) in tables with different marginal totals and indeed with possibly different numbers of rows and columns. This will be essential for allowing us to compare, for example, segregation in a Los Angeles school district with segregation in a New Orleans district. The two districts will have different numbers of schools (so the number of rows will be different) and different numbers of ethnic groups (so the number of columns will be different).

To begin with, assume that the number of schools and the capacity of the schools is fixed and the number of students in each of the ethnic groups is fixed. What can change is the way in which the students are distributed among the schools. Segregation is evidenced by different ethnic distributions in different schools. Lack of segregation corresponds to (roughly) the same ethnic distribution in all schools. We are then dealing with an $I \times J$ contingency table $N = (n_{ij})_{i=1,j=1}^{I,J}$ where the rows correspond to the I schools in the district and the columns correspond to the J ethnic groups present in the student population. With this notation, n_{ij} denotes the number of students of the j-th ethnic group who are attending the i-th school. Marginal totals, which are assumed to be fixed, are denoted by

$$n_{i+} = \sum_{j=1}^{J} n_{ij}, \quad i = 1, 2, \ldots, I,$$

and

$$n_{+j} = \sum_{i=1}^{I} n_{ij}, \quad j = 1, 2, \ldots, J.$$

Thus, n_{2+}, the sum of the entries in row 2 denotes the number of students in school number 2, the capacity of that school. In parallel fashion n_{+4}, the total of column 4 is the number of students in the district who are members of the 4th ethnic group. The district capacity will be denoted by n_{++}, defined by

$$n_{++} = \sum_{i=1}^{I} n_{i+} = \sum_{j=1}^{J} n_{+j}.$$

We introduce four other $I \times J$ matrices that are determined by the elements of N and which will be useful in our efforts to quantify and modify segregation in the table.

Define $I \times J$ matrices $P = P(N)$, $Q = Q(N)$, $R = R(N)$, and $\widetilde{N}$ to have corresponding elements as follows:

$$p_{i,j} = \frac{n_{ij}}{n_{++}}, \qquad i = 1, 2, \ldots, I, \quad j = 1, 2, \ldots, J, \tag{12.5}$$

$$q_{i,j} = \frac{n_{i+}n_{+j}}{n_{++}n_{++}}, \qquad i = 1, 2, \ldots, I, \quad j = 1, 2, \ldots, J, \tag{12.6}$$

$$r_{i,j} = \frac{p_{ij}}{q_{ij}} = \frac{n_{ij}n_{++}}{n_{i+}n_{+j}}, \quad i = 1, 2, \ldots, I, \quad j = 1, 2, \ldots, J, \tag{12.7}$$

$$\widetilde{n}_{ij} = n_{++}q_{ij}, \qquad i = 1, 2, \ldots, I, \quad j = 1, 2, \ldots, J. \tag{12.8}$$

The matrix P has an (i, j)-element which represents the proportion of the total number of students in the district that are found in cell (i, j). The matrix Q has an (i, j)-element which represents the proportion of the total number of students in the district that would be found in cell (i, j) if the rows and columns were independent (i.e., no segregation). The matrix R is a local dependence matrix. If an element $r_{ij} > 1$, then ethnic group j is over-represented in school i. If $r_{ij} < 1$, then there is under-representation. Segregation will be in evidence if there are many elements of R with values that are noticeably different from 1. Successful measures to mitigate segregation should result in an R matrix whose elements are closer to 1 in value. Finally the matrix $\widetilde{N}$ has elements such that $\widetilde{n}_{ij}$corresponds to the number of students in school i of ethnic group j when there is no segregation. The elements of $\widetilde{N}$ will often not be integers, indicating that, without assigning fractions of students to certain schools, complete independence in the table will not be achievable for the given row and column totals. The optimal assignment in such cases will be that array of integers with the correct marginal totals which is closest to $\widetilde{N}$ using some measure of distance between matrices (perhaps $d(A, B) = \max\{|a_{ij} - b_{ij}| : i = 1, 2, \ldots, I, \ j = 1, 2, \ldots, J\}$). However, in our discussion of segregation and segregation orders it is convenient to sometimes consider moving fractional students from one school to another. In this way, complete independence will always be attainable.

An exchange consists of moving one student(or fraction of a student) from school i_1 to school i_2 and in return moving a different student (or fraction thereof) from school i_2 to school i_1, thus preserving marginal totals. Not all such exchanges will reduce segregation. Many will make it worse. Those that reduce segregation are what we call progressive exchanges and are characterized as follows:

Definition 12.7.1 (Progressive Exchange) An exchange which moves an individual (or fraction of an individual) of ethnic group j_1 from school i_1 to school i_2 and in return moves an individual(or fractional individual) of ethnic group j_2 from school i_2 to school i_1 will be called a progressive exchange if $r_{i_1j_1} > r_{i_2j_1}$ and $r_{i_2j_2} > r_{i_1j_2}$.

A progressive exchange will, in general, be most effective in reducing segregation if

$$r_{i_1j_1} > 1 > r_{i_2j_1} \text{ and } r_{i_2j_2} > 1 > r_{i_1j_2} \tag{12.9}$$

Definition 12.7.2 (Strongly Progressive Fractional Exchange) An exchange which moves a fraction δ of an individual of ethnic group j_1 from school i_1 to school i_2 and in return moves the same fraction δ of an individual of ethnic group j_2 from school i_2 to school i_1 will be called a strongly progressive fractional exchange if before the exchange $r_{i_1j_1} > 1 > r_{i_2j_1}$ and $r_{i_2j_2} > 1 > r_{i_1j_2}$ and after the exchange $r_{i_1j_1} \geq 1 \geq r_{i_2j_1}$ and $r_{i_2j_2} \geq 1 \geq r_{i_1j_2}$.

The reason for introducing fractional exchanges is that, when the table is close to having independent rows and columns but with $\widetilde{N}$ having some entries which are not non-negative integers, a progressive exchange involving movement of a whole student may fail to reduce segregation because it will result in "overshooting" independence.

In some tables which exhibit some degree of segregation it may not be possible to implement a strongly progressive fractional exchange. For example, if R is as follows:

$$R = \begin{pmatrix} + & - & 1 \\ 1 & + & - \\ - & 1 & + \end{pmatrix},$$

where a $+$-sign indicates an element greater than 1 and a $-$-sign indicates an element less than 1. However in this case a "chained" series of strongly progressive fractional exchange is possible to clearly reduce segregation. This chained exchange will move a fraction δ of a student in ethnic group 1 from school 1 to school 3, then move a fraction δ of a student of ethnic group 3 from school 3 to school 2 and finally move a fraction δ of a student of ethnic group 2 from school 2 to school 1. Since, after this exchange, all the entries in R have been moved closer to 1 and the marginals have been preserved, this has reduced segregation. Note that, one can choose δ to be just large enough to ensure that the number of 1's in R is increased by at least one after the chained exchange.

In any matrix R that does not contain all entries equal to 1, it is possible to identify a chained series of strongly progressive fractional exchange of order k for some $k \leq min\{I, J\}$. Such an exchange will be associated with an upper left hand corner submatrix of a possibly row and/or column rearrangement of R of dimension $k \times k$ with diagonal elements all greater than 1, with $r_{i,i+1} < 1$, $i = 1, 2, \ldots, k-1$, and with $r_{k,1} < 1$. Thus any R which exhibits segregation can have its segregation reduced by some possibly chained strongly progressive fractional exchange. Note that a chained exchange of order k involves individuals from k different ethnic groups and in k different schools. However, it can be verified that a chained series of strongly progressive fractional exchange can be instead implemented by a finite series of fractional progressive changes each of which involves only two schools.

An attractive possible partial ordering on the class of segregation matrices N with fixed marginals may then be defined as follows.

Definition 12.7.3 (Progressive Exchange Ordering) Let $\mathcal{N}$ denote the class of all $I \times J$ matrices with non-negative real elements with fixed marginal totals n_{i+}, $i = 1, 2, \ldots, I$ and n_{+j}, $j = 1, 2, \ldots, J$. For $N_1, N_2 \in \mathcal{N}$, N_1 is said to exhibit no more segregation in the progressive exchange ordering sense than N_2, written $N_1 \leq_{PE} N_2$ if N_1 can be obtained from N_2 by a finite string of possibly fractional progressive exchanges.

In order to extend this segregation order to deal with matrices of different dimensions and with different marginals, another interpretation is desirable.

Scarsini (1990) introduced a partial ordering on $\mathcal{N}$ to reflect dependence in the tables. But, in the segregation setting, dependence and segregation are synonymous. We can thus adopt the Scarsini partial order as a suitable segregation ordering. But to argue that it is the "right" partial order requires us to relate it to the Progressive Exchange ordering that was defined above. The Scarsini order, which we will call the Scarsini Segregation order is defined as follows.

Definition 12.7.4 (Scarsini Segregation Ordering) For any $N \in \mathcal{N}$ define a random variable X_N by

$$P(X_N = r_{ij}) = q_{ij}, \quad i = 1, 2, \ldots, I, \quad j = 1, 2, \ldots, J, \tag{12.10}$$

and denote the Lorenz curve of this random variable by $S_N(u) = L_{X_N}(u)$. For $N_1, N_2 \in \mathcal{N}$, N_1 is said to exhibit no more segregation in the Scarsini Segregation sense than N_2, written $N_1 \leq_S N_2$ if $S_{N_1}(u) \geq S_{N_2}(u), \;\; 0 \leq u \leq 1$.

The curve $S_N(u)$ will be called the segregation curve of the matrix N. It can be verified that if N' is obtained from N by a small fractional progressive exchange, then $X_{N'} \leq_S X_N$.

On the basis of this result, it can be argued that the segregation order $\leq_S$ is the appropriate order to be used in comparing tables with regard to the segregation that they exhibit.

Just as in the Wealth and Income literature where it is often felt to be desirable to have a single number index of inequality, in Segregation discussions the use of a single number index of segregation is often deemed to be desirable. Based on the present discussion it appears to be appropriate to only use indices of segregation that are monotone with respect to the segregation order $\leq_S$. Thus an index $S(N)$, say, will be acceptable provided that whenever $N_1 \leq_S N_2$ it is the case that $S(N_1) \leq S(N_2)$. The majority of the segregation indices that have been proposed in the literature do satisfy this requirement. We will review several of these next.

Two general classes of segregation measures can be usefully identified: (1) Those which are of the form $S_h(N) = E(h(X_N))$ for some particular convex function h, and (2) Those which are based on geometric characteristics of the segregation curve $S_N(u)$.

For a particular continuous convex function h, the corresponding segregation measure is defined by

$$S_h(N) = E(h(X_N)) = \sum_{i=1}^{I}\sum_{j=1}^{J} h(r_{ij}) q_{ij} = \sum_{i=1}^{I}\sum_{j=1}^{J} h\left(\frac{n_{ij}n_{++}}{n_{i+}n_{+j}}\right)\frac{n_{i+}}{n_{++}}\frac{n_{+j}}{n_{++}}. \tag{12.11}$$

The following simple continuous convex choices for h can be considered.

$$h_1(y) = y^2, \qquad h_2(y) = (y-1)^2,$$

$$h_3(y) = 1/(1+y), \qquad h_4(y) = y \ln y,$$

with corresponding segregation indices S_1, S_2, S_3, and S_4.

Using (12.11) one finds for h_1,

$$S_1(N) = \sum_{i=1}^{I} \sum_{j=1}^{J} \frac{n_{ij}^2}{n_{i+}n_{+j}}, \tag{12.12}$$

which is an attractively simple expression for measuring segregation. The index corresponding to h_2 is clearly closely related. It can be simplified to the form

$$S_2(N) = \frac{1}{n_{++}} \sum_{i=1}^{I} \sum_{j=1}^{J} \frac{(n_{ij} - \widetilde{n}_{ij})^2}{\widetilde{n}_{ij}}, \tag{12.13}$$

which is a constant multiple of the classical Pearson chi-squared statistic for testing for independence in the table. Note that the relation between S_1 and S_2 is

$$S_1(N) = S_2(N) + 1, \tag{12.14}$$

since $E(X_N) = 1$. Consequently the index S_1 is bounded below by 1 since S_2 is clearly non-negative.

The index corresponding to h_3 simplifies to the form

$$S_3(N) = \sum_{i=1}^{I} \sum_{j=1}^{J} \frac{\widetilde{n}_{ij}^2}{(\widetilde{n}_{ij} + n_{ij})n_{++}}. \tag{12.15}$$

The index S_3 is bounded below by 1/2 since, using Jensen's inequality, $E(1/(1 + X_N)) \geq 1/E(1 + X_N) = 1/2$.

Using h_4 we obtain

$$\begin{aligned} S_4(N) &= \sum_{i=1}^{I} \sum_{j=1}^{J} \frac{n_{ij}}{n_{++}} [\ln n_{ij} - \ln \widetilde{n}_{ij}] \\ &= \sum_{i=1}^{I} \sum_{j=1}^{J} p_{ij} [\ln p_{ij} - \ln q_{ij}]. \end{aligned} \tag{12.16}$$

Note that, in (12.16), any summand of the form $n_{ij}[\ln n_{ij} - \ln \widetilde{n}_{ij}]/n_{++}$ with $n_{ij} = 0$ is to be interpreted as zero. This measure is recognizable as the Kullback–Leibler divergence from P to Q, or distance between P and Q.

The segregation curve $S_N(u)$ is a Lorenz curve and, as such, is a convex curve joining the points $(0, 0)$ and $(1, 1)$ in the unit square lying below the egalitarian curve which is the straight line joining $(0, 0)$ to $(1, 1)$. The egalitarian curve corresponds to a matrix N with independent rows and columns, i.e., with $r_{ij} = 1$ for every (i, j). Undoubtedly the most commonly used measure of inequality for Lorenz curves is the Gini index which corresponds to twice the area between the

Lorenz curve and the egalitarian line. A convenient formula for the Gini index of the segregation curve $S_N(u)$, which we will call the Gini segregation index, is the following.

$$S_G(N) = \frac{E(|X_N^{(1)} - X_N^{(2)}|)}{2E(X_N)}, \tag{12.17}$$

where $X_N^{(1)}$ and $X_N^{(2)}$ are independent copies of X_N. Since, for any N, $E(X_N) = 1$, the Gini segregation index can be expressed in the form

$$S_G(N) = \frac{1}{2n_{++}^2} \sum_{i=1}^{I} \sum_{j=1}^{J} \sum_{k=1}^{I} \sum_{\ell=1}^{J} |n_{ij}\widetilde{n}_{k\ell} - n_{k\ell}\widetilde{n}_{ij}|. \tag{12.18}$$

An alternative geometric inequality measure is the Pietra index. A convenient formula for the Pietra index of the segregation curve, which we will call the Pietra segregation index is

$$S_P(N) = \frac{E(|X_N - E(X_N)|)}{2E(X_N)}. \tag{12.19}$$

Since, for any N, $E(X_N) = 1$ the Pietra segregation index can be expressed in the form

$$S_P(N) = \frac{1}{2n_{++}} \sum_{i=1}^{I} \sum_{j=1}^{J} |n_{ij} - \widetilde{n}_{ij}|. \tag{12.20}$$

Thus the Pietra index is a scalar multiple of the $\mathcal{L}_1$-distance between N and N^*. It is perhaps the simplest of all the indices that have been described and, on that basis, might be recommended for general use. Note that the Pietra index also admits a representation in the form

$$S_P(N) = E(h_P(X_N)) \tag{12.21}$$

where $h_P(y) = |y - 1|$, a continuous convex function.

A third geometric inequality measure, suggested by Amato (1968), is the length of the Lorenz curve. For our segregation curves, which are piecewise linear functions, it is not difficult to obtain a formula for the length of the Segregation curve, to be called the Amato segregation index and denoted by $S_A(N)$.

$$\begin{aligned} S_A(N) &= \sum_{i=1}^{I} \sum_{j=1}^{J} \sqrt{q_{ij}^2 + p_{ij}^2} \\ &= \frac{1}{n_{++}} \sum_{i=1}^{I} \sum_{j=1}^{J} \sqrt{\widetilde{n}_{ij}^2 + n_{ij}^2} \end{aligned} \tag{12.22}$$

Note that the possible values for the Amato index range between $\sqrt{2}$, in the independent case, and 2 the maximal length of Lorenz curve in the unit square. If a segregation index ranging from 0 to 1 is desired, one simply needs to subtract $\sqrt{2}$ from the Amato index and divide by $2 - \sqrt{2}$. We may rewrite $S_A(N)$in the following form

$$S_A(N) = \sum_{i=1}^{I} \sum_{j=1}^{J} \left[\sqrt{1 + \left(\frac{p_{ij}}{q_{ij}} \right)^2} \right] q_{ij}$$
$$= E(\sqrt{1 + X_N^2}) = E(h_A(X_N)), \tag{12.23}$$

where $h_A(y) = \sqrt{1 + y^2}$, a continuous convex function. Consequently the Amato index has a double interpretation. It is the length of the Lorenz curve, but also it is a segregation measure based on a particular choice of the continuous convex function h_A.

On the basis of mathematical simplicity, the Pietra index is appealing. The Amato index is an intriguing alternative. Familiarity with the classical Pearson measure of divergence between N and N^* and the Kullback–Leibler divergence measure might be used to argue in favor of using S_2 or S_4.

Quite obviously it will be appropriate to compare different school districts with regard to segregation and not just restrict attention to the fixed marginal case, which effectively only applies to movement of students within a district. Thus we might wish to compare two tables $N^{(1)}$ which is of dimension $I_1 \times J_1$, and $N^{(2)}$ which is of dimension $I_2 \times J_2$. For $i = 1, 2$, table $N^{(i)}$ has an associated matrix $R^{(i)}$ and an associated random variable $X_{N^{(i)}}$. It is then quite natural to extend the segregation order to allow comparison of matrices of arbitrary dimensions by defining $N^{(1)} \leq_S N^{(2)}$ if $S_{N^{(1)}}(u) \geq S_{N^{(2)}}(u)$ for every $u \in [0, 1]$. We no longer have an available interpretation in terms of progressive exchanges, but the comparison of tables of differing dimensions on the basis of summary measures of segregation has a long pedigree and, absent counterexamples in which the extended segregation order is demonstrably misleading, there seems to be no strong argument against the extension. The available segregation indices will be used as summary measures for comparing tables of different dimensions, just as they were for comparing tables with the same dimensions and the same marginals.

Of course, the definitions of ethnic groups can vary from school district to school district, and even from time to time within a school district. Thus for example, the Asian category might or might not be broken down into sub-categories labeled Chinese, Korean, etc. The opposite of this disaggregation would be aggregation or combining previously distinguished categories. How will such combining of categories affect the segregation curve and segregation measures ? Without loss of generality we can consider a case in which columns 1 and 2 are combined. The dimension of the table is then reduced from $I \times J$ to $I \times (J - 1)$ but most of the marginal totals are unchanged. To compare the segregation in the original table

N with the segregation in the new table N_C (with columns 1 and 2 combined) we need to compare $E(g(X_N))$ with $E(g(X_{N_C}))$ for an arbitrary convex function g. It is a straightforward algebraic exercise to verify that the convexity of g implies that $E(g(X_{N_C})) \leq E(g(X_N))$. Consequently, $N_C \leq_S N$, i.e., combining categories reduces (or, better, does not increase) segregation. Conversely, disaggregation will generally increase segregation. For example, if categories are defined in terms of ethnic group and gender, there will be more segregation than if only ethnic groups were used to define categories.

As a consequence of this aggregation effect, it is quite possible that district 1 may have less segregation than district 2 before an aggregation of categories, but that, after aggregation, district 2 has less segregation. Of course, disaggregation can also change the ordering between districts.

Note that the segregation order is only a partial order. When segregation curves cross, it clearly becomes difficult to make decisions regarding which array is "more desirable." This complication causes lively discussion involving additional criteria to be invoked to permit preferential judgments. Nevertheless, a central role for the Lorenz curve based segregation order appears to be defensible.

12.8 Lorenz Order with Common Finite Support

In this section we focus on the problem of characterizing the Lorenz order restricted to a class of random variables with common finite support. In addition, we identify operations that reduce inequality in this class. These are called Robin Hood exchanges and they play a parallel role in the finite support setting to that played by Robin Hood (or progressive) transfers in the general Lorenz ordering case.

For a fixed positive integer n and a fixed set of n distinct numbers $0 < x_1 < x_2 < \cdots < x_{n-1} < x_n$, consider the class $\mathscr{L}_{\underline{x}}^{(n)}$ of all random variables with support $\{x_1, x_2, \ldots, x_n\}$. A random variable $X(\underline{p})$ in this class can be associated with a probability vector $\underline{p} = (p_1, p_2, \ldots, p_n)$ where $p_i = P(X(\underline{p}) = x_i)$, $i = 1, 2, \ldots, n$. We address the question of identifying how two probability vectors, $\underline{p}$ and $\underline{q}$, must be related in order that $X(\underline{p}) \leq_L X(\underline{q})$. The first somewhat surprising result in this direction is the following.

Theorem 12.8.1 *For $X(\underline{p}), X(\underline{q}) \in \mathscr{L}_{\underline{x}}^{(n)}$, if $X(\underline{p}) \leq_L X(\underline{q})$ then $E(X(\underline{p})) = E(X(\underline{q}))$.*

See Arnold and Gokhale (2017) for a proof of this assertion.

Now for two random variables with equal means, such as these, a convenient characterization of the Lorenz order is available involving "angle functions."

We have: $X(\underline{p}) \leq_L X(\underline{q})$ (with equal means) iff

$$E[(X(\underline{p}) - c)^+] \leq E[(X(\underline{q}) - c)^+] \text{ for every } c \in (0, \infty).$$

Because $X(\underline{p})$ and $X(\underline{q})$ only take on the values $x_1, x_2, \ldots, x_n$ we have

$$X(\underline{p}) \leq_L X(\underline{q}) \text{ iff } E[(X(\underline{p}) - x_i)^+] \leq E[(X(\underline{q}) - x_i)^+] \text{ for } i = 1, 2, \ldots, n-1,$$

i.e., if $\sum_{j=i+1}^{n}(x_j - x_i)p_j \leq \sum_{j=i+1}^{n}(x_j - x_i)q_j$ for $i = 1, 2, \ldots, n-1$, or equivalently if $\sum_{j=i+1}^{n}(x_j - x_i)(p_j - q_j) \leq 0$, for $i = 1, 2, \ldots, n-1$. This can be written in the form $A(\underline{x})(\underline{p} - \underline{q}) \leq 0$ for a suitable matrix $A(\underline{x})$.

So our final conclusion is

$$X(\underline{p}) \leq_L X(\underline{q}) \text{ iff } E(X(\underline{p})) = \sum_{j=1}^{n} x_i p_i = \sum_{j=1}^{n} x_i q_i = E(X(\underline{q}))$$

and

$$A(\underline{x})(\underline{p} - \underline{q}) \leq 0.$$

Of course, the case $n = 2$ is not very interesting because in that case if $E(X(\underline{p})) = E(X(\underline{q}))$ then necessarily $X(\underline{p}) \equiv X(\underline{q})$.

In a standard Lorenz scenario the activities of Robin Hood play a key role. These are simple inequality attenuating operations that effectively define the Lorenz order. In the common finite support setting, we need to identify analogs of these Robin Hood operations.

Thus we seek to identify simple changes in the coordinates of a probability vector $\underline{p}$ that will yield a new probability vector, say $\underline{p^*}$ such that $X(\underline{p^*}) \leq_L X(\underline{p})$. Note that the new vector $\underline{p^*}$ must be such that $E(X(\underline{p^*})) = E(X(\underline{p}))$, i.e., the inequality reducing operation must preserve the mean.

We introduce the concept of an exchange to be applied to a probability vector $\underline{p}$. A vector $\underline{\delta}$ will be called an exchange if it satisfies $\sum_{i=1}^{n} \delta_i = 0$ and $\underline{p} + \underline{\delta} \geq \underline{0}$. The result of the application of an exchange $\underline{\delta}$ to probability vector $\underline{p}$ is a new probability vector $\underline{p^*} = \underline{p} + \underline{\delta}$. An exchange $\underline{\delta} \neq \underline{0}$ is inequality attenuating if $X(\underline{p^*}) \leq_L X(\underline{p})$.

In order to be a mean-preserving exchange it is necessary that $\underline{\delta}$ has at least 3 non-zero coordinates. Exchanges of the simplest form, those with exactly 3 non-zero coordinates, will be called Robin Hood Exchanges, provided that they are mean-preserving and inequality attenuating. A Robin Hood exchange will have, for some indices $j < k < \ell$, $\delta_k = \psi > 0$, $\delta_j = -(1-\alpha)\psi$ and $\delta_\ell = -\alpha\psi$ where α is selected to preserve the mean and ψ is not too large. Thus we must have $p_j \geq (1-\alpha)\psi$ and $p_\ell \geq \alpha\psi$ so that the post-exchange vector is a probability vector.

Suppose that $\underline{p} \neq \underline{q}$ and $X(\underline{p}) \leq_L X(\underline{q})$. Our goal is to identify a finite sequence of Robin Hood exchanges applied successively to $\underline{q}$ that will yield the vector $\underline{p}$. First note that if $X(\underline{p}) \leq_L X(\underline{q})$ with $\underline{p} \neq \underline{q}$, then it must be the case that there exist integers r and m such that (i) $p_i = q_i$, $i = 1, 2, \ldots, r-1$ and $p_r < q_r$, (ii) $p_j = q_j$, $j = m+1, m+2, \ldots, n$, and $p_m < q_m$, and (iii) there exists k such that $r < k < m$ and $p_k > q_k$. Define $N(\underline{q}, \underline{p}) = \sum_{i=1}^{n} I(q_i \neq p_i)$, i.e., the number of indices i where q_i and p_i differ. Note that $N(\underline{q}, \underline{p})$ might be zero, but it cannot be 1 nor can it be equal to 2 since $E(X(\underline{q})) = E(X(\underline{p}))$. To prove our claim, it will suffice to show that for any pair $\underline{q}, \underline{p}$ with $\underline{p} \neq \underline{q}$ and $X(\underline{p}) \leq_L X(\underline{q})$, there exists a Robin Hood exchange which will reduce $N(\underline{q}, \underline{p})$.

However this can be seen to be the case. Because we have assumed that $\underline{p} \neq \underline{q}$ and $X(\underline{p}) \leq_L X(\underline{q})$, there may exist several indices t for which $p_t = q_t$. Note that if the first few indices ℓ are such that $p_\ell = q_\ell$, then the first index, say ℓ^*, for which equality does not occur must be such that $p_{\ell^*} < q_{\ell^*}$. Moreover, if the last few indices are such that the corresponding coordinates of p and q are equal then the last index, say ℓ^{**}, for which equality does not occur must be such that $p_{\ell^{**}} > q_{\ell^{**}}$. It is then possible to choose indices $r < k < m$ such that $p_r < q_r$, $p_m < q_m$, and $p_k > q_k$, and with no indices t with $r < t < m$ and $p_t = q_t$. Consider then, a Robin Hood exchange defined by

$$\alpha = \frac{p_k x_k - p_r x_r}{p_m x_m - p_r x_r}$$

to preserve the mean, and then choose $\psi = min\{(p_k - q_k), (1 - \alpha)^{-1}(q_r - p_r), \alpha^{-1}(q_m - p_m)\}$. Application of this Robin Hood exchange to $\underline{q}$ will yield a new probability vector $\underline{q*}$ with

$$X(\underline{p}) \leq_L X(\underline{q*}) \leq_L X(\underline{q}),$$

and $N(\underline{q*}, \underline{p}) < N(\underline{q}, \underline{p})$. A finite number of such exchanges will bring us to $\underline{p}$.

For further details and examples, see Arnold and Gokhale (2017).

12.9 The Scarsini Dependence Order

The partial order introduced by Scarsini (1990) for discrete bivariate distributions, continues to make sense in the absolutely continuous case. Thus, for any bivariate random variable (X, Y) with density, say $f_{12}(x, y)$ with marginals $f_1(x)$ and $f_2(y)$, we may define a dependence variable Z by

$$Z = \frac{f_1(X) f_2(Y)}{f_{12}(X, Y)}$$

with corresponding Lorenz curve $L_Z(u)$, which will be called the dependence curve corresponding to the density f_{12}, or equivalently, to the random variable (X, Y). Note that it is always the case that $E(Z) = 1$. We will say that a joint density $f_{12}^{(1)}$ exhibits at least as much dependence as the density $f_{12}^{(2)}$, with corresponding dependence variables $Z^{(1)}$ and $Z^{(2)}$, if $L_{Z^{(1)}}(u) \leq L_{Z^{(2)}}(u)\ \forall u \in [0; 1]$.

Example 12.9.1 (Basic FGM Distributions) The joint densities in this case will be of the form

$$f_{12}^{(i)}(x, y) = 1 + \theta^{(i)}(1 - 2x)(1 - 2y), \quad 0 < x, y < 1,$$

with uniform marginals (here $\theta^{(i)} \in [-1, 1]$, $i = 1, 2$). Thus

$$f_1^{(i)}(x) = I(0 < x < 1), \text{ and } f_2^{(i)}(y) = I(0 < y < 1), \quad i = 1, 2.$$

The corresponding dependence variables are

$$Z^{(i)} = [1 + \theta^{(i)}(1 - 2X)(1 - 2Y)]^{-1}.$$

It is natural to consider more general FGM distributions, obtained via marginal transformations. However, the following theorem shows that there is no need to consider them. In fact, the theorem shows that the dependence ordering is in fact a copula ordering.

Theorem 12.9.1 *Suppose that the random variable (X, Y) has dependence variable $Z_{X,Y}$ and that the random variable (U, V) is obtained from (X, Y) by invertible marginal transformations, i.e., $U = h_1(X)$ and $V = h_2(Y)$, then the two corresponding dependence variables $Z_{X,Y}$ and $Z_{U,V}$ are identically distributed.*

Proof For simplicity assume that the marginal transformations have differentiable inverses. It is then possible to represent the joint and marginal densities of (U, V) as

$$f_{U,V}(u, v) = \left|\frac{d}{du}h_1^{-1}(u)\right| \left|\frac{d}{dv}h_2^{-1}(v)\right| f_{X,Y}(h_1^{-1}(u), h_2^{-1}(v))$$

and

$$f_U(u) = \left|\frac{d}{du}h_1^{-1}(u)\right| f_X(h_1^{-1}(u)) \text{ and } f_V(v) = \left|\frac{d}{dv}h_2^{-1}(v)\right| f_Y(h_2^{-1}(v)).$$

Consequently, after simplification, we have

$$\begin{aligned} Z_{U,V} &= \frac{f_X(h_1^{-1}(U) f_Y(h_2^{-1}(V)}{f_{X,Y}(h_1^{-1}(U), h_2^{-1}(V)} \\ &= \frac{f_X(X) f_Y(Y)}{f_{X,Y}(X, Y)} = Z_{X,Y}. \end{aligned}$$

In particular, one can select $h_1(x) = F_X(x)$ and $h_2(y) = F_Y(y)$, to get

$$Z_{X,Y} = Z_{F_X(X), F_Y(Y)}$$

which is the dependence variable of the copula of (X, Y). ■

However, notwithstanding this last observation, in order to identify properties of the distribution of the dependence variable of (X, Y), it may be best to consider it as a function of (X, Y) rather than as a function of the copula.

Note that, as a consequence of the theorem, the dependence variable is not changed by marginal scale changes. Location changes will also not affect it.

Example 12.9.2 (The Bivariate Normal Distributions) From the above discussion, we may, without loss of generality, assume zero means and unit variances. The bivariate normal density is then of the following form, where $-1 < \rho < 1$:

$$f_{X,Y}(x, y) = (2\pi)^{-1}(1-\rho^2)^{-1/2} \exp\left[-\frac{1}{2(1-\rho^2)}(x^2 - 2\rho xy + y^2)\right].$$

Since the marginal densities are standard normal, the corresponding dependence variable $Z_{X,Y}$ is of the form:

$$Z_{X,Y} = (1-\rho^2)^{1/2} \exp\left\{[2(1-\rho^2)]^{-1}[\rho^2 X^2 + \rho^2 Y^2 - 2\rho XY]\right\}.$$

Identification of the distribution of this random variable appears to be difficult. However it is not difficult to simulate. If we wish to evaluate the corresponding Gini index it will be perhaps simplest to make use of the following representation of the Gini index.

$$G_{Z_{X,Y}} = \frac{E(\max[Z^{(1)}_{X,Y}, Z^{(2)}_{X,Y}])}{E(Z_{X,Y})} - 1,$$

which involves i.i.d. copies of $Z_{X,Y}$ and can be approximated easily via simulation.

Example 12.9.3 (The Mardia Bivariate Pareto Distribution) In this model X and Y are defined as functions of independent gamma random variables, thus:

$$(X, Y) = \left(\frac{U}{W}, \frac{V}{W}\right),$$

where U, V, and W are independent random variables with

$$U \sim \Gamma(1, 1), \quad V \sim \Gamma(1, 1) \text{ and } W \sim \Gamma(\gamma, 1).$$

The joint survival function for this distribution is of the form:

$$\overline{F}_{X,Y}(x, y) = (1 + x + y)^{-\gamma}, \quad x, y > 0,$$

where $\gamma > 0$. The corresponding dependence variable is

$$Z_{X,Y} = \frac{\gamma^2(1 + X + Y)^{\gamma+2}}{\gamma(\gamma+1)(1+X)^{\gamma+1}(1+Y)^{\gamma+1}}.$$

Alternatively this can be written in terms of the basic gamma variables, U, V, and W (which may be preferable for simulation). Thus

$$Z_{X,Y} = \frac{\gamma^2 W^\gamma (U + V + W)^{\gamma+2}}{\gamma(\gamma+1)(U+W)^{\gamma+1}(V+W)^{\gamma+1}}.$$

A more general model with second kind beta marginals might be considered, of the form:

$$(X, Y) = \left(\frac{U}{W}, \frac{V}{W}\right),$$

where U, V, and W are independent random variables with

$$U \sim \Gamma(\alpha, 1), \quad V \sim \Gamma(\beta, 1) \text{ and } W \sim \Gamma(\gamma, 1).$$

However, this can be recognized as a simple marginal transformation of the Olkin–Liu bivariate beta model, to be discussed next, and will have the same dependence variable.

Example 12.9.4 (The Olkin–Liu Bivariate Beta Distribution) In this model X and Y are also defined as functions of independent gamma random variables. Thus

$$(X, Y) = \left(\frac{U}{U+W}, \frac{V}{V+W}\right),$$

where U, V, and W are independent random variables with

$$U \sim \Gamma(\alpha, 1), \quad V \sim \Gamma(\beta, 1) \text{ and } W \sim \Gamma(\gamma, 1).$$

The corresponding joint density (Olkin and Liu 2003) is of the form

$$f_{X,Y}(x, y) = \frac{\Gamma(\alpha+\beta+\gamma)}{\Gamma(\alpha)\Gamma(\beta)\Gamma(\gamma)} \frac{x^{\alpha-1}y^{\beta-1}(1-x)^{\beta+\gamma-1}(1-y)^{\alpha+\gamma-1}}{(1-xy)^{\alpha+\beta+\gamma}}.$$

Taking into account the fact that X and Y have beta densities, the dependence variable for this bivariate distribution is of the form:

$$Z_{X,Y} = \frac{\Gamma(\alpha+\gamma)\Gamma(\beta+\gamma)}{\Gamma(\alpha+\beta+\gamma)\Gamma(\gamma)} \frac{(1-XY)^{\alpha+\beta+\gamma}}{(1-X)^{\beta}(1-Y)^{\alpha}},$$

or in terms of the gamma distributed components,

$$Z_{X,Y} = \frac{\Gamma(\alpha+\gamma)\Gamma(\beta+\gamma)}{\Gamma(\alpha+\beta+\gamma)\Gamma(\gamma)} \frac{\left(1 - \frac{U}{U+W}\frac{V}{V+W}\right)^{\alpha+\beta+\gamma}}{\left(1-\frac{U}{U+W}\right)^{\beta}\left(1-\frac{V}{V+W}\right)^{\alpha}}.$$

12.9.1 Extension to k Dimensions

Instead of a two-dimensional random variable (X, Y) we may consider a k-dimensional random variable $(X_1, X_2, \ldots, X_k)$ with corresponding joint and marginal densities:

$$f_{X_1,X_2,\ldots,X_k}(x_1, x_2, \ldots, x_k)$$

and

$$f_{X_i}(x_i), \quad i = 1, 2, \ldots, k.$$

We then can define the corresponding dependence variable as

$$Z_{X_1,X_2,\ldots,X_k} = \frac{\prod_{i=1}^{k} f_{X_i}(X_i)}{f_{X_1,X_2,\ldots,X_k}(X_1, X_2, \ldots, X_k)}.$$

As in two dimensions, we use the Lorenz order to compare dependence variables.

How should we interpret comparisons between variables of differing dimensions?

For example, how are $Z_{X_1,X_2,\ldots,X_k}$ and $Z_{X_1,X_2,\ldots,X_{k-1}}$ related? Note that we can simplify the ratio of these two variables as follows.

$$\frac{Z_{X_1,X_2,\ldots,X_k}}{Z_{X_1,X_2,\ldots,X_{k-1}}} = \frac{f_{X_k}(X_k)}{f_{X_k|X_1,X_2,\ldots,X_{k-1}}(X_k|X_1, X_2, \ldots, X_{k-1})}$$

In fact, we can argue as follows that these two variables, $Z_{X_1,X_2,\ldots,X_k}$ and $Z_{X_1,X_2,\ldots,X_{k-1}}$, are Lorenz ordered.

Consider

$$\begin{aligned}
&E(Z_{X_1,X_2,\ldots,X_k}|(X_1, X_2, \ldots, X_{k-1}) = (x_1, x_2, \ldots, x_{k-1})) \\
&= \int_{-\infty}^{\infty} \frac{\prod_{i=1}^{k} f_{X_i}(x_i)}{f_{X_1,X_2,\ldots,X_k}(x_1, x_2, \ldots, x_k)} \frac{f_{X_1,X_2,\ldots,X_k}(x_1, x_2, \ldots, x_k)}{f_{X_1,X_2,\ldots,X_{k-1}}(x_1, x_2, \ldots, x_{k-1})} dx_k \\
&= \frac{\prod_{i=1}^{k-1} f_{X_i}(x_i)}{f_{X_1,X_2,\ldots,X_{k-1}}(x_1, x_2, \ldots, x_{k-1})} \int_{-\infty}^{\infty} f_{X_k}(x_k) dx_k \\
&= \frac{\prod_{i=1}^{k-1} f_{X_i}(x_i)}{f_{X_1,X_2,\ldots,X_{k-1}}(x_1, x_2, \ldots, x_{k-1})} \times 1.
\end{aligned}$$

Thus

$$E(Z_{X_1,X_2,\ldots,X_k}|X_1, X_2, \ldots, X_{k-1})$$
$$= \frac{\prod_{i=1}^{k-1} f_{X_i}(X_i)}{f_{X_1,X_2,\ldots,X_{k-1}}(X_1, X_2, \ldots, X_{k-1})} = Z_{X_1,X_2,\ldots,X_{k-1}}.$$

It follows that $Z_{X_1,X_2,\ldots,X_k}$ is a balayage of $Z_{X_1,X_2,\ldots,X_{k-1}}$ and, consequently,

$$Z_{X_1,X_2,\ldots,X_{k-1}} \leq_L Z_{X_1,X_2,\ldots,X_k}.$$

By induction, we may conclude that if $\underline{X} = (\underline{X}^{(1)}, \underline{X}^{(2)})$ then

$$Z_{\underline{X}^{(1)}} \leq_L Z_{\underline{X}}.$$

Example 12.9.5 (The Normal Case Revisited, This Time in k Dimensions) Consider a k-dimensional normal random variable $(X_1, X_2, \ldots, X_k)$ with, without loss of generality, standard normal marginals, i.e., $\underline{X} \sim N^{(k)}(\underline{0}, R)$ where R is a correlation matrix. The corresponding dependence variable is then:

$$Z_{\underline{X}} = \frac{(2\pi)^{-k/2} \exp[-(1/2)\underline{X}'\underline{X}]}{(2\pi)^{-k/2}|R|^{-1/2} \exp[-(1/2)\underline{X}'R^{-1}\underline{X}]}$$
$$= |R|^{1/2} \exp[-(1/2)\underline{X}'(I - R^{-1})\underline{X}].$$

It is then possible to identify the moment generating function of $\log[Z_{\underline{X}}]$ and, from it, determine the distribution of $Z_{\underline{X}}$.

Bibliography

Aaberge, R. (2000). Characterization of Lorenz curves and income distributions. *Social Choice and Welfare, 17*, 639–653.

Aaberge, R. (2001). Axiomatic characterization of the Gini coefficient and Lorenz curve orderings. *Journal of Economic Theory, 101*, 115–132. Erratum: 140, e356.

Abramowitz, M., & Stegun, I. A. (1972). *Handbook of mathematical functions*. New York: Dover Publications.

Aggarwal, V. (1984). On optimal aggregation of income distribution data. *Sankhya B, 46*, 343–335.

Aggarwal, V., & Singh, R. (1984). On optimum stratification with proportional allocation for a class of Pareto distributions. *Communications in Statistics - Theory and Methods, 13*, 3107–3116.

Aldous, D., & Shepp, L. (1987). The least variable phase type distribution is Erlang. *Communications in Statistics. Stochastic Models, 3*, 467–473.

Alker, H. R. (1965). *Mathematics and politics*. New York: Macmillan.

Alker, H. R., & Russet, B. M. (1966). Indices for comparing inequality. In R. L. Merritt & S. Rokkan (Eds.), *Comparing nations: The use of quantitative data in cross-national research* (pp. 349–372). New Haven, CT: Yale University Press.

Amato, V. (1968). *Metodologia Statistica Strutturale* (Vol. 1). Bari: Cacucci.

Arnold, B. C. (1979). Non-uniform decompositions of uniform random variables under summation modulo *m*. *Bollettino Unione Matematica Italiana, 5*(16-A), 100–102.

Arnold, B. C. (1983). *Pareto distributions*. Fairland, MD: International Cooperative Publishing House.

Arnold, B. C. (1986a). *Inequality Attenuating and Inequality Preserving Weightings*. Technical Report 146, University of California, Riverside.

Arnold, B. C. (1986b). A class of hyperbolic Lorenz curves. *Sankhya B, 48*, 427–436.

Arnold, B. C. (1987). *Majorization and the Lorenz order: A brief introduction. Lecture notes in statistics* (Vol. 43). Berlin: Springer.

Arnold, B. C. (2012). On the Amato inequality index. *Statistics and Probability Letters, 82*, 1504–1506.

Arnold, B. C. (2015a). On Zenga and Bonferroni curves. *Metron, 73*, 25–30.

Arnold, B. C. (2015b). *Pareto distributions* (2nd ed.). Boca Raton, FL: CRC Press, Taylor & Francis Group.

Arnold, B. C., Balakrishnan, N., & Nagaraja, H. N. (1992). *A first course in order statistics*. New York: Wiley.

B. C. Arnold, J. M. Sarabia, *Majorization and the Lorenz Order with Applications in Applied Mathematics and Economics*, Statistics for Social and Behavioral Sciences,
https://doi.org/10.1007/978-3-319-93773-1

Arnold, B. C., & Gokhale, D. V. (2014). On segregation: Ordering and measuring. *Sankhya B, 76*, 141–166.

Arnold, B. C., & Gokhale, D. V. (2017). Lorenz order with common finite support. *Metron, 75*, 215–226.

Arnold, B. C., & Laguna, L. (1976). A stochastic mechanism leading to asymptotically Paretian distributions. In *Business and Economic Statistics Section, Proceedings of the American Statistical Association* (pp. 208–210).

Arnold, B. C., & Laguna, L. (1977). *On generalized Pareto distributions with applications to income data. International studies in economics, monograph* (Vol. 10). Department of Economics. Ames, Iowa: Iowa State University.

Arnold, B. C., Robertson, C. A., Brockett, P. L., & Shu, B.-Y. (1987). Generating ordered families of Lorenz curves by strongly unimodal distributions. *Journal of Business and Economic Statistics, 5*, 305–308.

Arnold, B. C., & Sarabia, J. M. (2017). *Explicit expressions for the Amato inequality index* (Unpublished Manuscript)

Arnold, B. C., & Sarabia, J. M. (2018). *Analytic expressions for multivariate Lorenz surfaces* (submitted)

Arnold, B. C., & Villaseñor, J. A. (1984). *Some Examples of Fitted General Quadratic Lorenz Curves*. Technical Report 130, University of California, Riverside.

Arnold, B. C., & Villaseñor, J. A. (1985). *Inequality Preserving and Inequality Attenuating Transformations*. Technical Report, Colegio de Postgraduados, Chapingo, Mexico.

Arnold, B. C., & Villaseñor, J. A. (1986). Lorenz ordering of means and medians. *Statistics and Probability Letters, 4*, 47–49.

Arnold, B. C., & Villaseñor, J. A. (1998). Lorenz ordering of order statistics and record values. In N. Balakrishnan & C. R. Rao (Eds.), *Order statistics: Theory and methods. Handbook of statistics* (Vol. 16, pp. 75–87). Amsterdam: North-Holland.

Arnold, B. C., & Villaseñor, J. A. (2015). Bias of the sample Lorenz curve. In *Ordered data analysis, modeling and health research methods* (Springer proceedings in mathematics and statistics, pp. 3–15). 149, Cham: Springer.

Atkinson, A. B. (1970). On the measurement of inequality. *Journal of Economic Theory, 2*, 244–263.

Bairamov, I., & Kotz, S. (2003). On a new family of positive quadrant dependent bivariate distributions. *International Journal of Mathematics, 3*, 1247–1254.

Balinski, M. L., & Young, H. P. (2001). *Fair representation - Meeting the ideal of one man one vote* (2nd ed.). Washington, DC: Brookings Institute Press.

Basmann, R. L., Hayes, K. L., Slottje, D. J., & Johnson, J. D. (1990). A general functional form for approximating the Lorenz curve. *Journal of Econometrics, 43*, 77–90.

Beach, C. M., & Davidson, R. (1983). Distribution-free statistical inference with Lorenz curves and income shares. *The Review of Economic Studies, 50*, 723–735.

Bechhofer, R. E., Kiefer, J., & Sobel, M. (1968). *Sequential identification and ranking procedures*. Chicago, IL: University of Chicago Press.

Bechhofer, R. E., Santer, T. J., & Turnbull, B. W. (1977). Selecting the largest interaction in two-factor experiment. In S. S. Gupta & D. S. Moore (Eds.), *Statistical decision theory and related topics II* (pp. 1–18). New York: Academic.

Belzunce, F., Pinar, J. F., Ruiz, J. M., & Sordo, M. A. (2012). Comparisons of risks based on the expected proportional shortfall. *Insurance: Mathematics and Economics, 51*, 292–302.

Belzunce, F., Pinar, J. F., Ruiz, J. M., & Sordo, M. A. (2013). Comparison of concentration for several families of income distributions. *Statistics and Probability Letters, 83*, 1036–1045.

Birkhoff, G. (1946). Tres observaciones sobre el algebra lineal. Univ. Nac. Tucuman Rev. Ser. A, S, (pp. 147–151).

Bonferroni, C. E. (1930). *Elementi di Statistica Generale*. Firenze: Libreria Seber.

Bourguignon, F. (1979). Decomposable income inequality measures. *Econometrica, 47*, 901–920.

Brown, M., & Solomon, H., (1976). *On a Method of Combining Pseudorandom Number Generators*. Technical Report No. 233. Department of Statistics, Stanford University, Stanford, CA.

Butler, R. J., & McDonald, J. B. (1987). Interdistributional income inequality. *Journal of Business and Economic Statistics, 5*, 13–18.

Butler, R. J., & McDonald, J. B. (1989). Using incomplete moments to measure inequality. *Journal of Econometrics, 42*, 109–119.

Chandra, M., & Singpurwalla, N. D. (1978). *The Gini Index, the Lorenz Curve and the Total Time on Test Transforms*. Technical Report T-368, George Washington University, School of Engineering and Applied Science, Washington, DC.

Chandra, M., & Singpurwalla, N. D. (1981). Relationships between some notions which are common to reliability theory and economics. *Mathematics of Operations Research, 6*, 113–121.

Chotikapanich, D. (1993). A Comparison of alternative functional forms for the Lorenz curve. *Economic Letters, 41*, 129–138.

Chotikapanich, D., Griffiths, W. E., & Rao, D. S. P. (2007). Estimating and combining national income distributions using limited data. *Journal of Business and Economic Statistics, 25*, 97–109.

Chotikapanich, D., Valenzuela, R., & Rao, D. S. P. (1997). Global and regional inequality in the distribution of income: Estimation with limited and incomplete data. *Empirical Economics, 22*, 533–546.

Cobham, A., & Sumner, A. (2014). Is inequality all about the tails? The Palma measure of income inequality. *Significance, 11*, 10–13.

Colombi, R. (1990). A new model of income distribution: The Pareto-lognormal distribution. In C. Dagum & M. Zenga (Eds.), *Studies in contemporary economics* (pp. 18–32). Heidelberg: Springer.

Cowell, F. A. (1980). On the structure of additive inequality measures. *Review of Economic Studies, 47*, 521–531.

Cowell, F. A. (2011). *Measuring inequality* (3rd ed.). LSE perspectives in economic analysis. Oxford: Oxford University Press.

Cowell, F. A., & Kuga, K. (1981). Additivity and the entropy concept: An axiomatic approach to inequality measurement. *Journal of Economic Theory, 25*, 131–143.

Cronin, D. C. (1977). *Theory and applications of the log-logistic distribution*. Unpublished M.A. thesis. Polytechnic of Central London.

Cronin, D. C. (1979). A function for size distribution of incomes: A further comment. *Econometrica, 47*, 773–774.

Dagum, C. (1977). A new model of personal income distribution: Specification and estimation. *Economie Appliqueée, 30*, 413–437.

Dalal, S. R., & Fortini, P. (1982). An inequality comparing sums and maxima with application to Behrens-Fisher type problem. *Annals of Statistics, 10*, 297–301.

Dalton, H. (1920).The measurement of the inequality of incomes. *Economic Journal, 30*, 348–361.

Darbellay, G. A., & Vajda, I. (2000). Entropy expressions for multivariate continuous distributions. *IEEE Transactions on Information Theory, 46*, 709–712.

Darbellay, G. A., & Vajda, I. (2000). Entropy expressions for multivariate continuous distributions. *IEEE Transactions on Information Theory, 46*, 709–712.

David, H. A. (1968). Gini's mean difference rediscovered. *Biometrika, 55*, 573–574.

Davies, J. B., Green, D. A., & Paarsch, H. J. (1998). Economic statistics and social welfare comparisons. A review. In *Handbook of applied economic statistics* (pp. 1–38). New York: Marcel Dekker.

Davies, J. B., & Hoy, M. (1994). The normative significance of using third-degree stochastic dominance in comparing income distributions. *Journal of Economic Theory, 64*, 520–530.

Dawid, A. P. (1982). The well-calibrated Bayesian. *Journal of American Statistical Association, 77*, 605–610.

De Groot, M. H., & Fienberg, S. E. (1983). The comparison and evaluation of forecasters. *The Statistician, 32*, 12–22.

De Groot, M. H., & Fienberg, S. E. (1986). Comparing probability forecasters: Basic binary concepts and multivariate extensions. In P. K. Goel & A. Zellner (Eds.), *Bayesian inference and decision techniques* (pp. 247–264). Essays in honor of B. de Finetti. Amsterdam: North Holland.

Dempster, A. P., & Kleyle, R. M. (1968). Distributions determined by cutting a simplex with hyperplanes. *Annals of Mathematical Statistics, 39*, 1473–1478.

Donaldson, D., & Weymark, J. A. (1980). A single parameter generalization of the Gini indices of inequality. *Journal of Economic Theory, 22*, 67–86.

Dorfman, R. (1979). A formula for the Gini coefficient. *Review of Economics and Statistics, 61*, 146–149.

Ebrahimi, N., Maasoumi, E., & Soofi, E. S. (1999). Ordering univariate distributions by entropy and variance. *Journal of Econometrics, 90*, 317–336.

Eisenberg, B. (1991). The effect of variable infectivity on the risk of HIV infection. *Statistics in Medicine, 10*, 131–139.

Elteto, O., & Frigyes, E. (1968). New income inequality measures as efficient tools for causal analysis and planning. *Econometrica, 36*, 383–396.

Elton, J., & Hill, T. P. (1992). Fusions of a probability distribution. *Annals of Probability, 20*, 421–454.

Emlen, J. M. (1973). *Ecology: An evolutionary approach*. Reading, MA: Addison-Wesley.

Farahat, H. K., & Mirsky, L. (1960). Permutation endomorphisms and refinement of a theorem of Birkhoff. *Proceedings of the Cambridge Philological Society, 56*, 322–328.

Feller, W. (1971). *An introduction to probability theory and its applications* (2nd ed., Vol. 2). New York: Wiley.

Fellman, J. (1976). The effect of transformations on Lorenz curves. *Econometrica, 44*, 823–824.

Fishburn, P. C. (1976). Continua of stochastic dominance relations for bounded probability distributions. *Journal of Mathematical Economics, 3*, 295–311.

Fishburn, P. C. (1980). Continua of stochastic dominance relations for unbounded probability distributions. *Journal of Mathematical Economics, 7*, 271–285.

Fisk, P. R. (1961a). The graduation of income distributions. *Econometrica, 29*, 171–185.

Fisk, P. R. (1961b). Estimation of location and scale parameters in a truncated grouped sech-square distribution. *Journal of the American Statistical Association, 56*, 692–702.

Fraser, D. A. S. (1957). *Non-parametric methods in statistics*. New York: Wiley.

Gail, M. H., & Gastwirth, J. L. (1978). A scale-free goodness-of-fit test for the exponential distribution based on the Lorenz curve. *Journal of the American Statistical Association, 73*, 787–793.

Gastwirth, J. L. (1971). A general definition of the Lorenz curve. *Econometrica, 39*, 1037–1039.

Gastwirth, J. L. (1972). The estimation of the Lorenz curve and Gini index. *Review of Economics and Statistics, 54*, 306–316.

Gastwirth, J. L. (1974). Large sample theory of some measures of income inequality. *Econometrica, 42*, 191–196.

Gastwirth, J. L. (2016). Measures of economic inequality focusing on the status of the lower and middle income groups. *Statistics and Public Policy, 3*, 1–9.

Gini, C. (1914). Sulla misura della concentrazione e della variabilita dei caratteri. *Atti del R. Instituto Veneto di Scienze, Lettere ed Arti, 73*, 1203–1248.

Glasser, G. J. (1961). Relationship between the mean difference and other measures of variation. *Metron, 21*, 176–180.

Godwin, H. J. (1945). On the distribution of the estimate of mean deviation obtained from samples from a normal population. *Biometrika, 33*, 254–256.

Goldie, C. M. (1977). Convergence theorems for empirical Lorenz curves and their inverses. *Advances in Applied Probability, 9*, 765–791.

Good, I. J. (1968). Utility of a distribution. *Nature, 219*(5161), 1392.

Gómez-Déniz, E. (2016). A family of arctan Lorenz curves. *Empirical Economics, 51*, 1215–1233.

Greselin, F., & Pasquazzi, L. (2009). Asymptotic confidence intervals for a new inequality measure. *Communications in Statistics: Simulation and Computation, 38*, 1742–1756.

Gupta, M. R. (1984). Functional form for estimating the Lorenz curve. *Econometrica, 52*, 1313–1314.
Hardy, G. H., Littlewood, J. E., & Polya, G. (1929). Some simple inequalities satisfied by convex functions. *Messenger of Mathematics, 58*, 145–152.
Hardy, G. H., Littlewood, J. E., & Polya, G. (1959). *Inequalities*. London and New York: Cambridge University Press.
Hart, P. E. (1975). Moment distributions in economics: An exposition. *Journal of the Royal Statistical Society, Series A, 138*, 423–434.
Hill, W. G. (1976). Order statistics of correlated variables and implications in genetic selection programmes. *Biometrics, 32*, 889–902.
Hill, W. G. (1977). Order statistics of correlated variables and implications in genetic selection programmes II, Response to selection. *Biometrics, 33*, 703–712.
Hoeffding, W. (1948). A class of statistics with asymptotically normal distribution. *Annals of Mathematical Statistics, 19*, 293–325.
Hollander, M., Proschan, F., & Sethuraman, J. (1977). Functions decreasing in transposition and their applications in ranking problems. *Annals of Statistics, 5*, 722–733.
Holm, J. (1993). Maximum entropy Lorenz curves. *Journal of Econometrics, 59*, 377–389.
Horn, A. (1954). Doubly stochastic matrices and the diagonal of a rotation matrix. *American Journal of Mathematics, 76*, 620–630.
Huang, J. S., & Kotz, S. (1999). Modifications of the Farlie-Gumbel-Morgenstern distributions a tough hill to climb. *Metrika, 49*, 135–145.
Huffer, F. W., & Shepp, L. A. (1987). On the probability of covering the circle by random arcs. *Journal of Applied Probability, 24*, 422–429.
Jenkins, S. P. (2009). Distributionally sensitive inequality indices and The GB2 income distribution. *The Review of Income and Wealth, 55*, 392–398.
Joe, H. (1988). Majorization, entropy and paired comparisons. *Annals of Statistics, 16*, 915–925.
Joe, H., & Verducci, J. (1992). Multivariate majorization by positive combinations. In M. Shaked & Y. L. Tong (Eds.), *Stochastic inequalities. IMS lecture notes - Monograph series* (Vol. 22, pp. 159–181).
Jordá, V., Sarabia, J. M., & Prieto, F. (2014). On the estimation of the global income distribution using a parsimonious approach. In J. Bishop & J. G. Rodríguez (Eds.), *Economic well-being and inequality. Research on economic inequality* (Vol. 22, pp. 115–145). Emerald, Bingley, UK.
Kalbfleisch, J. D., & Prentice, R. L. (1980). *The statistical analysis of failure time data*. New York: Wiley.
Kakwani, N. C. (1977). Applications of Lorenz curves in economic analysis. *Econometrica, 45*, 719–728.
Kakwani, N. C. (1980a). *Income inequality and poverty, methods of estimation and policy applications*. New York: Oxford University Press.
Kakwani, N. C. (1980b). Functional forms for estimating the Lorenz curve: A reply. *Econometrica, 48*, 1063–1064.
Kakwani, N. C., & Podder, N. (1973). On the estimation of Lorenz curves from grouped observations. *International Economic Review, 14*, 278–292.
Kakwani, N. C., & Podder, N. (1976). Efficient estimation of the Lorenz curve and associated inequality measures from grouped observations. *Econometrica, 44*, 137–148.
Karamata, J. (1932). Sur une inegalite relative aux fonctions convexes. *Publ. Math. Univ. Belgrade, 1*, 145–148.
Karlin, S., & Rinott, Y. (1988). A generalized Cauchy-Binet formula and applications to total positivity and majorization. *Journal of Multivariate Analysis, 27*, 284–299.
Kemperman, J. H. B. (1973). Moment problems for sampling without replacement, I, II, III. *Nederl. Akad. Wetensch. Proc. Ser. A, 76*, 149–164; 165–180; 181–188.
Kleiber, C. (1996). Dagum vs. Singh-Maddala Income Distributions. *Economics Letters, 53*, 265–268.

Kleiber, C. (1999). On the Lorenz ordering within parametric of income distributions. *Sankhya B, 61*, 514–517.

Kleiber, C., & Kotz, S. (2002). A characterization of income distributions in terms of generalized Gini coefficients. *Social Choice and Welfare, 19*, 789–794.

Kleiber, C., & Kotz, S. (2003). *Statistical size distributions in economics and actuarial sciences.* Hoboken, NJ: Wiley.

Kolm, S. C. (1976). Unequal inequalities I. *Journal of Economic Theory, 12*, 416–442.

Kondor, Y. (1971). An old-new measure of income inequality. *Econometrica, 39*, 1041–1042.

Koshevoy, G. (1995). Multivariate Lorenz majorization. *Social Choice and Welfare, 12*, 93–102.

Koshevoy, G., & Mosler, K. (1996). The Lorenz zonoid of a multivariate distribution. *Journal of the American Statistical Association, 91*, 873–882.

Koshevoy, G., & Mosler, K. (1997). Multivariate Gini indices. *Journal of Multivariate Analysis, 60*, 252–276.

Krause, M. (2014). Parametric Lorenz curves and the modality of the income density function. *The Review of Income and Wealth, 60*, 905–929.

Lee, M.-L. T. (1996). Properties and applications of the Sarmanov family of bivariate distributions. *Communications in Statistics, Theory and Methods, 25*, 1207–1222.

Lefévre, C. (1994). Stochastic ordering of epidemics. In M. Shaked & G. J. Shanthikumar (Eds.), *Stochastic orders and their applications* (pp. 323–348). Boston: Academic.

Lerman, R. I., & Yitzhaki, S. (1984). A note on the calculation and interpretation of the Gini index. *Economics Letters, 15*, 363–368.

Lerman, R. I., & Yitzhaki, S. (1985). Income inequality effects by income source: A new approach and applications to the United States. *The Review of Economics and Statistics, 67*, 151–156.

Lombardo, E. (1979). Osservazioni sulla distribuzione campionaria di un indice di concentrazione dei fenomeni economici. *Note Economiche, 5*, 106–114.

Lomnicki, Z. A. (1952). The standard error of Gini's mean difference. *Annals of Mathematical Statistics, 23*, 635–637.

Lorenz, M. O. (1905). Methods of measuring the concentration of wealth. *Publication of the American Statistical Association, 9*, 209–219.

Lunetta, G. (1972). Di un indice di cocentrazione per variabili statistische doppie. *Annali della Facoltá di Economia e Commercio dell Universitá di Catania*, A 18.

Maasoumi, E., & Theil, H. (1979). The effect of the shape of the income distribution on two inequality measures. *Economics Letters, 4*, 289–291.

Mahfoud, M., & Patil, G. P. (1982). On weighted distributions. In G. Kallianpur, P. R. Krishnaiah, & J. K. Ghosh (Eds.), *Statistics and probability, Essays in honor of C. R. Rao* (pp. 479–492). Amsterdam: North Holland.

Marshall, A. W., & Olkin, I. (1974). Majorization in multivariate distributions. *Annals of Statistics, 2*, 1189–1200.

Marshall, A. W., & Olkin, I. (1979). *Inequalities: Theory of majorization and its applications*. New York: Academic.

Marshall, A. W., Olkin, I., & Arnold, B. C. (2011). *Inequalities: Theory of majorization and its applications* (2nd ed.). New York: Springer.

Marshall, A. W., Olkin, I., & Proschan, F. (1967). Monotonicity of ratios of means and other applications of majorization. In O. Shisha (Ed.), *Inequalities* (pp. 177–190). New York: Academic.

Marshall, A. W., Olkin, I., & Pukelsheim, F. (2002). A majorization comparison of apportionment methods in proportional representation. *Social Choice and Welfare, 19*, 885–900.

McDonald, J. B. (1981). Some issues associated with the measurement of income inequality. *Statistical Distributions in Scientific Work, 6*, 161–179.

McDonald, J. B. (1984). Some generalized functions for the size distribution of income. *Econometrica, 52*, 647–663.

McDonald, J. B., & Ransom, M. (2008). The generalized beta distribution as a model for the distribution of income. In D. Chotikapanich (Ed.), *Modeling income distributions and Lorenz curves* (pp. 147–166). New York: Springer.

Mehran, F. (1976). Linear measures of income inequality. *Econometrica, 44*, 805–809.
Meyer, P. A. (1966). *Probability and potentials*. Waltham, MA: Blaisdell.
Mirsky, L. (1958). Matrices with prescribed characteristic roots and diagonal elements. *Journal of the London Mathematical Society, 33*, 14–21.
Moothathu, T. S. K. (1983). Properties of Gastwirth's Lorenz curve bounds for general Gini index. *Journal of the Indian Statistical Association, 21*, 149–154.
Mosler, K. (2002). *Multivariate dispersion, central regions and depth: The lift Zonoid approach. Lecture notes in statistics* (Vol. 165). Berlin: Springer.
Moyes, P. (1987). A new concept of Lorenz domination. *Economics Letters, 23*, 203–207.
Muirhead, R. F. (1903). Some methods applicable to identities and inequalities of symmetric algebraic functions of *n* letters. *Proceedings of Edinburgh Mathematical Society, 21*, 144–157.
Muliere, P., & Scarsini, M. (1989). A note on stochastic dominance and inequality measures. *Journal of Economic Theory, 49*, 314–323.
Nair, U. S. (1936). The standard error of Gini's mean difference. *Biometrika, 28*, 428–436.
Nayak, T. K., Christman, M. C. (1992). Effect of unequal catchability on estimates of the number of classes in a population. *Scandinavian Journal of Statistics, 19*, 281–287.
Neuts, M. F. (1975). Computational uses of the method of phases in the theory of queues. *Computers & Mathematics with Applications, 1*, 151–166.
Nevius, S. E., Proschan, F., & Sethuraman, J. (1977). Schur functions in statistics, II. Stochastic majorization. *Annals of Statistics, 5*, 263–273.
Nygard, F., & Sandstrom, A. (1981). *Measuring income inequality*. Stockholm: Almqvist and Wiksell International.
O'Cinneide, C. A. (1991). Phase-type distributions and majorization. *Annals of Applied Probability, 1*, 219–227.
Ogwang, T., & Rao, U. L. G. (1996). A new functional form for approximating the Lorenz curve. *Economics Letters, 52*, 21–29.
Olkin, I., & Liu, R. (2003). A bivariate beta distribution. *Statistics and Probability Letters, 62*, 407–412.
Ord, J. K. (1975). Statistical models for personal income distributions. In G. P. Patil, S. Kotz, & J. K. Ord (Eds.), *Statistical distributions in scientific work* (Vol. 2, pp. 151–158). Dordrecht, Holland: Reidel.
Ord, J. K., Patil, G. P., & Taillie, C. (1978). *The choice of a distribution to describe personal incomes*. Unpublished manuscript.
Ord, J. K., Patil, G. P., & Taillie, C. (1981). Relationships between income distributions for individuals and for house-holds. In C. Taillie, G. P. Patil, & B. Baldessari (Eds.), *Statistical distributions in scientific work* (Vol. 6, pp. 203–210). Dordrecht: Reidel.
Ortega, P., Martín, A., Fernández, A., Ladoux, M., & García, A. (1991). A new functional form for estimating Lorenz curves. *The Review of Income and Wealth, 37*, 447–452.
Pakes, A. G. (1981). *On Income Distributions and Their Lorenz Curves*. Technical Report. Department of Mathematics, The University of Western Australia, Nedlands.
Palma, J. G. (2011). Homogenous middles vs. heterogenous tails, and the end of the "inverted-U": It's all about the share of the rich. *Development and Change, 42*, 87–153.
Patil, G. P., & Taillie, C. (1982). Diversity as a concept and its measurement. *Journal of the American Statistical Association, 77*, 548–561.
Perlman, M. D., & Rinott, Y. (1977). *On the unbiasedness of goodness-of-fit tests*. Unpublished manuscript.
Pham-Gia, T., & Turkkan, N. (1992). Determination of the beta distribution from its Lorenz curve. *Mathematical and Computer Modelling, 16*, 73–84.
Pietra, G. (1915). Delle relazioni tra gli indici di variabilita. Note I, II. *Atti del Reale Istituto Veneto di Scienze, Lettere ed Arti, 74*, 775–792; 793–804.
Pietra, G. (1948). *Studi di Statistica Metodologica*. Milano: Giuffré.
Pitt, L. D. (1977). A Gaussian correlation inequality for symmetric convex sets. *Annals of Probability, 5*, 470–474.

Pledger, G., & Proschan, F. (1971). Comparisons of order statistics and of spacings from heterogeneous distributions. In J. S. Rustagi (Ed.), *Optimizing methods in statistics* (pp. 89–113). New York: Academic.

Rao, C. R. (1965). On discrete distributions arising our of methods of ascertainment. In G. P. Patil (Ed.), *Classical and contagious discrete distributions* (pp. 320–332). Calcutta: Pergamon Press.

Rao, U. L. G., & Tam, A. Y. (1987). An empirical study of selection and estimation of alternative models of the Lorenz curve. *Journal of Applied Statistics, 14*, 275–280.

Rasche, R. H., Gaffney, J., Koo, A. Y. C., & Obst, N. (1980). Functional forms for estimating the Lorenz curve. *Econometrica, 48*, 1061–1062.

Rawlings, J. O. (1976). Order statistics for a special case of unequally correlated multinomial variates. *Biometrics, 32*, 875–887.

Rényi, A. (1961). On measures of entropy and information. In *Proceedings of 4th Berkeley Symposium on Mathematical Statistics and Probability* (Vol. I, pp. 547–561). Berkeley: University of California Press.

Rohde, N. (2009). An alternative functional form for estimating the Lorenz curve. *Economics Letters, 105*, 61–63.

Ross, S. M. (1981). A random graph. *Journal of Applied Probability, 18*, 309–315.

Ross, S. M. (1999). The mean waiting time for a pattern. *Probability in the Engineering and Informational Sciences, 13*, 1–9.

Runciman, W. G. (1966). *Relative deprivation and social justice*. London: Routledge and Kegan Paul.

Saari, D. G. (1995). *Basic geometry of voting*. Berlin: Springer.

Samuelson, P. A. (1965). A fallacy in the interpretation of the alleged constancy of income distribution. *Rivista Internazionale di Scienze Economiche e Commerciali, 12*, 246–253.

Sarabia, J. M. (1997). A hierarchy of Lorenz curves based on the generalized Tukey's lambda distribution. *Econometric Reviews, 16*, 305–320.

Sarabia, J. M. (2008a). Parametric Lorenz curves: Models and applications. In D. Chotikapanich (Ed.). *Modeling income distributions and Lorenz curves* (pp. 167–190). New York: Springer.

Sarabia, J. M. (2008b). A general definition of the Leimkuhler curve. *Journal of Informetrics, 2*, 156–163.

Sarabia, J. M. (2013). *Building new Lorenz curves from old ones*. Personal communication.

Sarabia, J. M., Castillo, E., Pascual, M., & Sarabia, M. (2005). Mixture Lorenz curves. *Economics Letters, 89*, 89–94.

Sarabia, J. M., Castillo, E., & Slottje, D. J. (1999). An ordered family of Lorenz curves. *Journal of Econometrics, 91*, 43–60.

Sarabia, J. M., Castillo, E., & Slottje, D. J. (2001). An exponential family of Lorenz curves. *Southern Economic Journal, 67*, 748–756.

Sarabia, J. M., Castillo, E., & Slottje, D. J. (2002). Lorenz ordering between McDonald's generalized functions of the income size distributions. *Economics Letters, 75*, 265–270.

Sarabia, J. M., Gómez-Déniz, E., Prieto, F., Jordá, V. (2016). Risk aggregation in multivariate dependent Pareto distributions. *Insurance: Mathematics and Economics, 71*, 154–163.

Sarabia, J. M., Gómez-Déniz, E., Sarabia, M., & Prieto, F. (2010). A general method for generating parametric Lorenz and Leimkuhler curves. *Journal of Informetrics, 4*, 524–539.

Sarabia, J. M., & Jordá, V. (2013). Modeling Bivariate Lorenz Curves with Applications to Multidimensional Inequality in Well-Being. *Fifth ECINEQ Meeting, Bari, Italy*, 201. Document available at: http://www.ecineq.org/ecineq_bari13/documents/booklet05.pdf

Sarabia, J. M., & Jordá, V. (2014a). Bivariate Lorenz curves based on the Sarmanov-Lee distribution. In V. B. Velas, S. Mignani, P. Monari, & L. Salmano (Eds.), *Topics in statistical simulation*. New York: Springer.

Sarabia, J. M., & Jordá, V. (2014b). Explicit expressions of the Pietra index for the generalized function for the size distribution of income. *Physica A: Statistical Mechanics and Its Applications, 416*, 582–595.

Sarabia, J. M., Jordá, V., & Remuzgo, L. (2017a). The Theil indices in parametric families of income distributions - A short review. *The Review of Income and Wealth, 63*, 867–880.

Sarabia, J. M., Jordá, V., & Trueba, C. (2017b). The Lamé class of Lorenz curves. *Communications in Statistics Theory and Methods, 46*, 5311–5326.
Sarabia, J. M., & Pascual, M., (2002). A class of Lorenz curves based on linear exponential loss functions. *Communications in Statistics Theory and Methods, 31*, 925–942.
Sarabia, J. M., Prieto, F., & Jordá, V. (2015). About the hyperbolic Lorenz curve. *Economics Letters, 136*, 42–45.
Sarabia, J. M., Prieto, F., & Sarabia, M. (2010). Revisiting a functional form for the Lorenz curve. *Economics Letters, 107*, 249–252.
Sarabia, J. M., & Sarabia, M. (2008). Explicit expressions for the Leimkuhler curve in parametric families. *Information Processing and Management, 44*, 1808–1818.
Scarsini, M. (1990). An ordering of dependence. In H. W. Block, A. R. Sampson, & T. H. Savits (Eds.), *Topics in statistical dependence. IMS lecture notes - Monograph series* (Vol. 16, pp. 403–414). Hayward: IMS.
Schur, I. (1923). Uber eine klasse von mittelbildungen mit anwendungen die determinaten. *Theorie Sitzungsber Berlin Math. Gesellschaft, 22*, 9–20.
Shaked, M. (1982). Dispersive ordering of distributions. *Journal of Applied Probability, 19*, 310–320.
Shaked, M., & Shanthikumar, J. G. (1998). Two variability orders. *Probability in the Engineering and Informational Sciences, 12*, 1–23.
Shannon, C. E. (1948). A mathematical theory of communication. *Bell Labs Technical Journal, 27*, 379–423.
Shorrocks, A. F. (1980). The class additively decomposable inequality measures. *Econometrica, 48*, 613–625.
Shorrocks, A. F. (1983). Ranking income distributions. *Economica, 50*, 3–17.
Shorrocks, A. F. (1984). Inequality decomposition by population subgroups. *Econometrica, 52*, 1369–1385.
Shorrocks, A. F., & Foster, J. E. (1987). Transfer sensitive inequality measures. *Review of Economic Studies, 54*, 485–497.
Singh, S. K., & Maddala, G. S. (1976). A function for the size distribution of incomes. *Econometrica, 44*, 963–970.
Sordo, M. A., Navarro, J., & Sarabia, J. M. (2013). Distorted Lorenz curves: Models and comparisons. *Social Choice and Welfare, 42*, 761–780.
Stevens, W. L. (1939). Solution to a geometrical problem in probability. *Annals of Eugenics, 9*, 315–320.
Stigler, S. M. (1974). Linear functions of order statistics with smooth weight functions. *Annals of Statistics, 2*, 676–693.
Strassen, V. (1965). The existence of probability measures with given marginals. *Annals of Mathematical Statistics, 36*, 423–439.
Taguchi, T. (1968). Concentration-curve methods and structures of skew populations. *Annals of Institute of Statistical Mathematics, 20*, 107–141.
Taguchi, T. (1972a). On the two-dimensional concentration surface and extensions of concentration coefficient and Pareto distribution to the two dimensional case-I. *Annals of Institute of Statistical Mathematics, 24*, 355–382.
Taguchi, T. (1972b). On the two-dimensional concentration surface and extensions of concentration coefficient and Pareto distribution to the two dimensional case-II. *Annals of Institute of Statistical Mathematics, 24*, 599–619.
Taillie, C. (1981). Lorenz ordering within the generalized gamma family of income distributions. In C. Taillie, G. P. Patil & B. Baldessari, (Eds.), *Statistical distributions in scientific work* (Vol. 6, pp. 181–192). Dordrecht, Holland: Reidel.
Thistle, P. D. (1990). Large sample properties of two inequality indices. *Econometrica, 58*, 725–728.
Tong, Y. L. (1982). Some applications of inequalities for extreme order statistics to a genetic selection problem. *Biometrics, 38*, 333–339.
Tong, Y. L. (1997). Some majorization orderings of heterogeneity in a class of epidemics. *Journal of Applied Probability, 34*, 84–93.

Tsui, K. Y. (1995). Multidimensional generalizations of the relative and absolute inequality indices: The Atkinson-Kolm-Sen approach. *Journal of Economic Theory, 67*, 251–265

Tsui, K. Y. (1999). Multidimensional inequality and multidimensional generalized entropy measures: An axiomatic derivation. *Social Choice and Welfare, 16*, 145–157.

Tzavelas, G., & Economou, P. (2012). Characterization properties of the log-normal distribution obtained with the help of divergence measures. *Statistics and Probability Letters, 82*, 1837–1840.

Villaseñor, J. A., & Arnold, B. C. (1984a). *The General Quadratic Lorenz Curve*. Technical Report, Colegio de Postgraduados, Chapingo, Mexico.

Villaseñor, J. A., & Arnold, B. C. (1984b). *Some Examples of Fitted General Quadratic Lorenz Curves*. Technical Report 130, Statistics Department, University of California, Riverside.

Villaseñor, J. A., & Arnold, B. C. (1989). Elliptical Lorenz curves. *Journal of Econometrics, 40*, 327–338.

Vitale, R. A. (1999). Majorization and Gaussian correlation. *Statistics & Probability Letter, 45*, 247–251.

Wang, Z., Ng, Y.-K. & Smyth, R. (2011). A general method for creating Lorenz curves. *The Review of Income and Wealth, 57*, 561–582.

Wang, Z., & Smyth, R. (2015). A hybrid method for creating Lorenz curves. *Economics Letters, 133*, 59–63.

Wang, Z., Smyth, R., & Ng, Y.-K. (2009). A new ordered family of Lorenz curves with an application to measuring income inequality and poverty in rural China. *China Economic Review, 20*, 218–235.

Weymark, J. A. (1979). *Generalized Gini Inequality Indices*. Discussion paper 79–112, Department of Economics, University of British Columbia, Vancouver, British Columbia.

Whitmore, G. A., & Findlay, M. C. (Eds.) (1978). *Stochastic dominance: An approach to decision-making under risk*. Lexington, MA: D. C. Heath.

Whitt, W. (1980). The effect of variability in the GI/G/s queue. *Journal of Applied Probability, 17*, 1062–1071.

Wilfling, B. (1996a). Lorenz ordering of generalized beta-II income distributions. *Journal of Econometrics, 71*, 381–388.

Wilfling, B. (1996b). Lorenz ordering of power-function order statistics. *Statistics and Probability Letters, 30*, 313–319.

Wilfling, B., & Krämer, W. (1993). Lorenz ordering of Singh-Maddala income distributions. *Economics Letters, 43*, 53–57.

Wilks, S. S. (1962). *Mathematical statistics*. New York: Wiley.

Wold, H. (1935). A study on the mean difference, concentration curves and concentration ratio. *Metron, 12*, 39–58.

Wyner, A. D., & Ziv, J. (1969). On communication of analog data from bounded source space. *Bell System Technical Journal, 48*, 3139–3172.

Yang, H.-J. (1982). On the variances of median and some other statistics. *Bulletin of the Institute of Mathematics, Academia Sinica, 10*, 197–204.

Yao, D. D. (1987). Majorization and arrangement orderings in open queueing networks. *Annals of Operations Research, 9*, 531–543.

Yitzhaki, S. (1979). Relative deprivation and the Gini coefficient. *The Quarterly Journal of Economics, 93*, 321–324.

Yitzhaki, S. (1983). On an extension of the Gini inequality index. *International Economic Review, 24*, 617–628.

Yntema, D. B. (1933). Measures of the inequality in the personal distribution of wealth or income. *Journal of the American Statistical Association, 28*, 423–433.

Zenga, M. (1984). Proposta per un indice di concentrazione basato sui rapporti tra quantili di popolazione e quantili reddito. *Giornale degli Economisti e Annali di Economia, 48*, 301–326.

Zenga, M. (2007). Inequality curve and inequality index based on the ratios between lower and upper arithmetic means. *Statistica e Applicazioni, 4*, 3–27.

Zografos, K., & Nadarajah, S. (2005). Expressions for Renyi and Shannon entropies for multivariate distributions. *Statistics & Probability Letter, 71*, 71–84.

Author Index

B. C. Arnold, J. M. Sarabia, *Majorization and the Lorenz Order with Applications in Applied Mathematics and Economics*, Statistics for Social and Behavioral Sciences,
https://doi.org/10.1007/978-3-319-93773-1

Subject Index

B. C. Arnold, J. M. Sarabia, *Majorization and the Lorenz Order with Applications in Applied Mathematics and Economics*, Statistics for Social and Behavioral Sciences,
https://doi.org/10.1007/978-3-319-93773-1

I

J

K

L

M

T

U

V

W

Z

Printed in the United States
By Bookmasters